Lecture Notes in Physics

The Editorial Policy for Proceedings

The series Lecture Notes in Physics reports new developments in physical research and teaching – quickly, informally, and at a high level. The proceedings to be considered for publication in this series should be limited to only a few areas of research, and these should be closely related to each other. The contributions should be of a high standard and should avoid lengthy redraftings of papers already published or about to be published elsewhere. As a whole, the proceedings should aim for a balanced presentation of the theme of the conference including a description of the techniques used and enough motivation for a broad readership. It should not be assumed that the published proceedings must reflect the conference in its entirety. (A listing or abstracts of papers presented at the meeting but not included in the proceedings could be added as an appendix.)

When applying for publication in the series Lecture Notes in Physics the volume's editor(s) should submit sufficient material to enable the series editors and their referees to make a fairly accurate evaluation (e.g. a complete list of speakers and titles of papers to be presented and abstracts). If, based on this information, the proceedings are (tentatively) accepted, the volume's editor(s), whose name(s) will appear on the title pages, should select the papers suitable for publication and have them refereed (as for a journal) when appropriate. As a rule discussions will not be accepted. The series editors and Springer-Verlag will normally not interfere with the detailed editing except in fairly obvious cases or on technical matters.

Final acceptance is expressed by the series editor in charge, in consultation with Springer-Verlag only after receiving the complete manuscript. It might help to send a copy of the authors' manuscripts in advance to the editor in charge to discuss possible revisions with him. As a general rule, the series editor will confirm his tentative acceptance if the final manuscript corresponds to the original concept discussed, if the quality of the contribution meets the requirements of the series, and if the final size of the manuscript does not greatly exceed the number of pages originally agreed upon. The manuscript should be forwarded to Springer-Verlag shortly after the meeting. In cases of extreme delay (more than six months after the conference) the series editors will check once more the timeliness of the papers. Therefore, the volume's editor(s) should establish strict deadlines, or collect the articles during the conference and have them revised on the spot. If a delay is unavoidable, one should encourage the authors to update their contributions if appropriate. The editors of proceedings are strongly advised to inform contributors about these points at an early stage.

The final manuscript should contain a table of contents and an informative introduction accessible also to readers not particularly familiar with the topic of the conference. The contributions should be in English. The volume's editor(s) should check the contributions for the correct use of language. At Springer-Verlag only the prefaces will be checked by a copy-editor for language and style. Grave linguistic or technical shortcomings may lead to the rejection of contributions by the series editors. A conference report should not exceed a total of 500 pages. Keeping the size within this bound should be achieved by a stricter selection of articles and not by imposing an upper limit to the length of the individual papers. Editors receive jointly 30 complimentary copies of their book. They are entitled to purchase further copies of their book at a reduced rate. As a rule no reprints of individual contributions can be supplied. No royalty is paid on Lecture Notes in Physics volumes. Commitment to publish is made by letter of interest rather than by signing a formal contract. Springer-Verlag secures the copyright for each volume.

The Production Process

The books are hardbound, and the publisher will select quality paper appropriate to the needs of the author(s). Publication time is about ten weeks. More than twenty years of experience guarantee authors the best possible service. To reach the goal of rapid publication at a low price the technique of photographic reproduction from a camera-ready manuscript was chosen. This process shifts the main responsibility for the technical quality considerably from the publisher to the authors. We therefore urge all authors and editors of proceedings to observe very carefully the essentials for the preparation of camera-ready manuscripts, which we will supply on request. This applies especially to the quality of figures and halftones submitted for publication. In addition, it might be useful to look at some of the volumes already published. As a special service, we offer free of charge LATEX and TEX macro packages to format the text according to Springer-Verlag's quality requirements. We strongly recommend that you make use of this offer, since the result will be a book of considerably improved technical quality. To avoid mistakes and time-consuming correspondence during the production period the conference editors should request special instructions from the publisher well before the beginning of the conference. Manuscripts not meeting the technical standard of the series will have to be returned for improvement.

For further information please contact Springer-Verlag, Physics Editorial Department V, Tiergartenstrasse 17, D-69121 Heidelberg, FRG

L. Päivärinta   E. Somersalo (Eds.)

Inverse Problems in Mathematical Physics

Proceedings of The Lapland Conference
on Inverse Problems
Held at Saariselkä, Finland, 14-20 June 1992

Springer-Verlag Berlin Heidelberg GmbH

Editors

Lassi Päivärinta
University of Oulu, Department of Mathematics
Linnanmaa, SF-90570 Oulu, Finland

Erkki Somersalo
Rolf Nevanlinna Institute, University of Helsinki
P. O. Box 26 (Teollisuuskatu 23)
SF-00014 Helsinki, Finland

ISBN 978-3-662-13928-8   ISBN 978-3-540-47947-5 (eBook)
DOI 10.1007/978-3-540-47947-5

© Springer-Verlag Berlin Heidelberg 1993
Originally published by Springer-Verlag Berlin Heidelberg New York in 1993
Softcover reprint of the hardcover 1st edition 1993

58/3140-543210 - Printed on acid-free paper

Contents

Introduction

Lassi Päivärinta and Erkki Somersalo

A concise way of describing inverse problems at large is that one attempts to retrieve information of unaccessible parameters by indirect measurements. In such general terms, the field of inverse problems seems like a widespread research area with little coherence, and a quick glance in the contents of this collection of works may give an impression of great diversity. In fact, the interpretation of every measurement can be seen as an inverse problem, a point of view that is in fact advocated by some researchers. Our belief, however, is that there is a strong feeling of uniformity and generality of ideas among the people working on various inverse problems, and this belief is strengthened especially on occasions like the *Lapland Conference of Inverse Problems* when these people are brought together to share their ideas. In the following introductory material to the collections of contributions, we try to indicate the ties, as we see them, between formally very different yet in the deeper sense related articles that are based on the presentations given in the meeting.

Let us start with a coarse classification of some widely studied inverse problems. The bulk of the material in the present volume fall mainly to three different but intimately related categories:

  I Inverse scattering problems,

 II Inverse boundary value problems,

III Inverse spectral problems.

During the past years, the active research carried out in the field of inverse problems has brought a lot of new insight into the deeper nature of these problems and especially to the interrelation between them. In the following, we describe and discuss briefly the main features of each of these classes, and indicate where in this collection the reader can find further information and references to current research articles.

I Inverse scattering problems form undoubtedly one of the most studied set of inverse problems. The setting is the following: Far away from the target having unknown physical properties, a wave field is sent in. It is assumed that the interaction mechanisms of the wave field with the target are qualitatively known. The scattered field is measured, and from this data one attempts to reconstruct the properties of the scatterer. A classical example of inverse scattering problems arise in quantum mechanics. Assume that we have scattering potential q in $\mathbb{R}^3$. The quantum mechanical scattering with fixed energy $E = k^2$, $k > 0$, is decribed by the Schrödinger equation

$$(-\Delta - k^2 + q(x))\psi(x) = 0. \tag{1}$$

The potential q should decay fast enough as x tends to infinity. The typical assumption about the field ψ is that it is a superposition of the incoming plane wave and the scattered radiating field satisfying Sommerfeld's radiation condition at infinity, i.e.,

$$\psi(x) = e^{ik\theta \cdot x} + \psi_{sc}(x), \tag{2}$$

where

$$\lim_{|x| \to \infty} |x|(\hat{x} \cdot \nabla - ik)\psi_{sc}(x) = 0, \quad \hat{x} = \frac{x}{|x|}. \tag{3}$$

An equivalent way of formulating the radiation condition (3) is to assume that

$$\psi_{sc}(x) = \frac{e^{ik|x|}}{|x|} A(\hat{x}, \theta, k) + o\left(\frac{1}{|x|}\right). \tag{4}$$

The function $A(\hat{x}, \theta, k)$ is called *the scattering amplitude*, and it is related to the scattering potential and the scattered field through

$$A(\hat{x}, \theta, k) = -\frac{1}{4\pi} \int_{\mathbb{R}^3} e^{-ik\hat{x} \cdot y} q(y)\psi(y)dy. \tag{5}$$

Depending on the type of measurements, one can now pose different inverse scattering problems, of which we list the following:

IP 1 Reconstruct the potential from the knowledge of

$$\{A(\hat{x}, \theta, k) \mid \hat{x}, \theta \in \mathbf{S}^2, \ k \in \mathbb{R}_+\}.$$

IP 2 Reconstruct the potential from the knowledge of

$$\{A(\hat{x}, \theta, k) \mid \hat{x}, \theta \in \mathbf{S}^2\}, \quad k > 0 \quad \text{fixed.}$$

IP 3 Reconstruct the potential from the knowledge of

$$\{A(-\theta, \theta, k) \mid \theta \in \mathbf{S}^2, \ k \in \mathbb{R}_+\},$$

The first one of the above inverse problems is the most classical one. The problem is formally overdetermined in the sense that the data set is indexed over a five dimensional space $\mathbf{S}^2 \times \mathbf{S}^2 \times \mathbb{R}_+$, while the unknown function q is over the three space. Based on this overdeterminacy, there is a rather simple way of seeing the uniqueness of the solution to this problem. Indeed, one can show that under relatively general assumption about the potential q, the scattering solution ψ to (1) behaves as

$$\psi(x) = e^{ik\theta \cdot x} + O\left(\frac{1}{k}\right), \quad \text{as } k \to \infty. \tag{6}$$

Therefore, if we choose $\xi \in \mathbb{R}^3$ and let k tend to infinity while keeping the vector $k(\theta - \hat{x}) = \xi$ fixed, we find that

$$\lim_{k \to \infty, k(\theta - \hat{x}) = \xi} A(\hat{x}, \theta, k) = \hat{q}(\xi), \tag{7}$$

i.e., the scattering amplitude tends towards the Fourier transform of the potential. Therefore, the data of IP 1 determines the potential q uniquely.

Despite this formally simple solution to IP 1, there are open problems that have obtained considerably attention in research literature. We mention the following two: 1. Given a function A definend on $\mathbf{S}^2 \times \mathbf{S}^2 \times \mathbb{R}_+$, how can one check whether the function is a scattering amplitude for an underlying potential q? This problem is usually referred to as *the characterization problem*. 2. By the reasoning above, the information of the potential q is contained in the high frequency behavior of the scattering amplitude. The question remains, how one can use effectively the overdetermined data to determine the high frequency behavior and thus the potential. One of the suggested methods of tackling these problems is the so called *generalized Machenko equation*. The roots of this method are in the classical one–dimensional Gelfand–Levitan–Marchenko method. We shall not review the generalized Marchenko method here. Instead, we refer to the article of **Newton** in this volume and references there.

Some of the technique used in studying the one dimensional problems, see **Kurasov**.

Next, we move to IP 2. Obviously, the overdeterminacy of the data is one dimension less and consequently the question of uniqueness of the solution is more difficult. Probably one of the most fruitful ideas in the study of inverse scattering problems in recent years has been the use of exponentially growing solutions and complex plane waves. More precisely, define a plane wave with a complex wave vector $\zeta \in \mathbb{C}^3$ as

$$\psi_0(x, \zeta) = e^{i\zeta \cdot x}, \quad \zeta \cdot \zeta = k^2 \in \mathbb{R}_+ \quad \text{fixed.} \tag{8}$$

The dot above denotes the inner product with no complex conjugation. Then, ψ_0 satisfies the free space Schrödinger equation. Furthermore, one can look for a solution to the full Schrödinger equation in the form

$$\psi(x, \zeta) = \psi_0(x, \zeta)(1 + v(x, \zeta)), \tag{9}$$

where v should satisfy the equation

$$(-\Delta - 2i\zeta \cdot \nabla + q(x))\psi(x, \zeta) = 0. \tag{10}$$

It turns out that equation 10 has a unique solution in certain weighted spaces under appropriate regularity conditions for the potential and what is more, one can prove a convergence result formally similar to (6),

$$v(x, \zeta) = O\left(\frac{1}{|\zeta|}\right), \quad \text{as } |\zeta| \to \infty, \zeta \cdot \zeta = k^2, \tag{11}$$

the convergence taking place in an appropriate function space. Assume now that we have two potentials, q_1 and q_2, that give rise to excactly the same scattering data $A(\hat{x}, \theta)$. Here, we dropped the k–dependence since the frequency is assumed to be fixed. An application of Green's formulas together with (5) give the orthogonality relation

$$\int_{\mathbb{R}^3} (q_1(x) - q_2(x))\psi_1(x, \theta_1)\psi_2(x, \theta_2)dx = 0, \tag{12}$$

where the functions $\psi_j(x, \theta_j)$, $j = 1, 2$ are scattering solutions to the equation (1) with the potentials q_j and incoming plane waves $\exp(ik\theta_j \cdot x)$, respectively. It turns out that the set of scattering solutions with different incoming directions θ is large enough for approximating the special solutions defined in (9). Therefore, it follows from (12) that for $\zeta_j \in \mathbb{C}^3$ satisfying $\zeta_j \cdot \zeta_j = k^2$, $j = 1, 2$, we eventually get

$$\int_{\mathbb{R}^3} (q_1(x) - q_2(x))\psi_1(x, \zeta_1)\psi_2(x, \zeta_2)dx = 0. \tag{13}$$

Next, one chooses the complex wave vectors so that $\zeta_1 + \zeta_2 = \xi \in \mathbb{R}^3$. This choice, together with the asymptotics (11) implies

$$\lim_{|\zeta_j| \to \infty} \int_{\mathbb{R}^3} (q_1(x) - q_2(x))\psi_1(x, \zeta_1)\psi_2(x, \zeta_2)dx = \hat{q}_1(\xi) - \hat{q}_2(\xi) = 0. \tag{14}$$

This holds for all $\xi \in \mathbb{R}^3$, so the potentials must coincide. The formal similarity of this uniqueness argument to that with varying frequency k is quite striking.

In the present volume, the article by **Ramm and Stefanov** gives a good idea of how large potential classes the method described above applies. The contribution of **Serov** reports certain extensions of the estimates of the type (11) previously obtained to more singular potential classes.

Note that when the frequency is fixed, the applicability ranges beyond quantum scattering because the equation formally coincides with the reduced wave equation of linear acoustics. This fact, of course, makes the inverse scattering problem more viable for engineering applications. There is a considerable literature of solving numerically the inverse scattering problem with fixed frequency. The paper by **Natterer and Wübbeling** gives one possible numerical approach to the problem as well as good reference for further reading.

Let us finally mention that the inverse scattering problems of the type IP 3 still have open problems with the global uniqueness and reconstructrion. In fact, IP 3 (called the inverse backscattering problem) is a formally determinend problem, i.e., the dimensionality of the data coincides with the dimensionality of the potential to be reconstructed. We want to emphasize that there are very few examples of completely solved multidimensional inverse problems with no overdetermination in data. We shall return to this remark when discussing the inverse boundary value problems.

So far, we have discussed only potential scattering, where the object of interest is described by a space dependent potential function. Another classical, very important and closely related type of inverse scattering problems deals with obstacle scattering. A typical inverse obstacle scattering can be formulated as follows: Assume that in space $\mathbb{R}^3$, there is an obstacle Ω, whose shape one tries to recover from far field measurements. Assume that the medium outside the obstacle is governed by the equations of linear acoustics, i.e., the pressure field u satisfies the Helmholtz equation

$$\Delta u(x) + k^2 u(x) = 0, \quad x \in \mathbb{R}^3 \setminus \overline{\Omega}. \tag{15}$$

The pressure field is assumed to satisfy a boundary condition at $\partial \Omega$, typically the Dirichlet ("sound soft"), Neumann ("sound hard") or a mixed ("impedance") condition. For penetrable obstacles, the appropriate boundary condition is a transmission condition. Again, one probes the target by sending in an initial field u_0, and the interacting total field is

$$u(x) = u_0(x) + u_{\mathrm{sc}}(x), \tag{16}$$

where the scattered field satisfies the outgoing radiation condition (3), or

$$u_{\mathrm{sc}}(x) = \frac{e^{ik|x|}}{|x|} u_\infty(\hat{x}) + O\left(\frac{1}{|x|^2}\right), \tag{17}$$

the function u_∞ being the *far field pattern* of the field which corresponds to the scattering amplitude here. The inverse scattering problem is now to reconstruct the shape of the object from far field patterns corresponding to a set of incoming initial fields.

Uniqueness proofs for large classes of inverse obstacle scattering problems are known. The present volume contains the article by **Kress** with an extensive discussion of different uniqueness proofs and their applicability. In this connenction, let us also mention the contribution of **Kirsch**, where scattering by periodic boundary structures are considered.

Besides the acoustic inverse obstacle scattering, one can look at the similar problem when the unknown target is illuminated with electromagnetic radiation. Since the electromagnetic fields satisfy the Helmholtz equation in vacuum, the transition from acoustics to electromagnetics does not seem so large. The boundary conditions, however, have vectorial nature and thus the variuos field components will be coupled. In the article by **Colton**, the electromagnetic inverse problem is discussed. The approach is closely related to the classical problem of target identification e.g. from radar measurements. The article of **Maponi, Misici and Zirilli** presents numerical work done for reconstructing the scattering obstacles from the far field data.

II Let us now move to the second large area of inverse problems, the inverse boundary value problems. The change in the setting when moving from inverse scattering problems to inverse boundary value problems is by no means sharp. Again, one has an object with unknown physical parameters and the objective is to find out these parameters in a non–invasive way.

Let us start with a concrete example of *impedance tomography*, extensively studied also in this volume: We ask if it is possible to make an image of the internal electromagnetic structure of a body (e.g. human body) by injecting electric currents into the body and measuring the voltages needed to maintain the current. In practice, one attaches a number of electrodes on the surface of the body and measures the voltages needed to maintain the current configuration. By the linearity of the governing equations, the dependence of the voltages of the currents is linear, so effectively the boundary data consists of a linear boundary map (or a matrix). Let us look at the mathematical idealization: The governing equation for the voltage distribution u in the body Ω is simply the equation of continuity,

$$\nabla \cdot \sigma \nabla u(x) = 0, \quad x \in \Omega, \tag{18}$$

where $\sigma(x) > 0$ is the conductivity distribution one is interested in. The current density through the boundary of the body is

$$j(x) = \sigma(x)\frac{\partial u}{\partial n}(x)\bigg|_{\partial\Omega}. \tag{19}$$

Assuming that the current density j is specified, one can solve the Neumann problem (18)–(19). In this way, one gets a complete collection of pairs

$$\{(j, u|_{\partial\Omega}) \mid u \text{ satisfies } (18)\text{ –}(19)\}, \tag{20}$$

or, equivalently, one knows the Neumann–to–Diriclet boundary map

$$R : \sigma\frac{\partial u}{\partial n}\bigg|_{\partial\Omega} \longmapsto u|_{\partial\Omega}. \tag{21}$$

The inverse problem is to reconstruct σ in Ω from the knowledge of R.

The close connection with the inverse scattering problems becomes more prominent if we modify the equation (18) slightly. Let us introduce the function ψ defined as

$$\psi(x) = \sigma(x)^{1/2}u(x). \tag{22}$$

A simple commutator relation

$$[\Delta, \sigma^{1/2}] = \sigma^{1/2}(L(\nabla) + q), \tag{23}$$

where

$$L(\nabla) \doteq \frac{\nabla\sigma}{\sigma} \cdot \nabla, \quad q = \frac{\Delta\sigma^{1/2}}{\sigma^{1/2}}, \tag{24}$$

shows that equation (18) can be rewritten for ψ as

$$(\Delta - q)\psi = 0. \tag{25}$$

Thus, we are back to the Schrödinger equation with zero energy.

Let us now apply Green's formula in Ω for the function ψ. Assuming that ψ_0 is harmonic in Ω, we obtain

$$\int_{\partial\Omega} \left(\psi_0\frac{\partial\psi}{\partial n} - \psi\frac{\partial\psi_0}{\partial n}\right) dS = \int_\Omega q\psi_0\psi dx. \tag{26}$$

Note that the left hand side depends completely on the boundary data. To calculate the unknown potential from the left hand integral, we use the same kind of idea as in the uniqueness proof of the fixed energy inverse scattering problem. Choose the harmonic function ψ_0 as

$$\psi_0(x, \zeta_1) = e^{i\zeta_1 \cdot x}, \quad \zeta_1 \cdot \zeta_1 = 0, \tag{27}$$

and the solution ψ as in equation (9), i.e.,

$$\psi(x, \zeta_2) = \psi(x, \zeta_2)(1 + v(x, \zeta_2)), \quad \zeta_2 \cdot \zeta_2 = 0, \tag{28}$$

and assume that the vectors $\zeta_j \in \mathbb{C}^3$ satisfy the extra condition $\zeta_1 + \zeta_2 = \xi \in \mathbb{R}^3$, ξ being arbitrary. By the asymptotic property (11) we find that

$$\lim_{|\zeta_j| \to \infty} \int_\Omega q\psi_0\psi \, dx = \hat{q}(\xi), \tag{29}$$

i.e., the right side of the equation (26) tends to the Fourier transform of the unknown q. Without going into details, we mention that the knowledge of the boundary map R ensures that the left side of the equation (26) can be calculated explicitly.

As the discussion here shows, there is a strong interrelation between the inverse scattering and inverse boundary value problems. In many cases they can be shown to be equivalent: The knowledge of the boundary map determines the far field data uniquely and vice versa.

This volume contains several presentations of the current status of various inverse boundary value problems and solution methods. The article by **Isaacson, Cheney and Newell** reviews the more practical aspects of the impedance tomography problem and gives a good idea of how close to the real world the more theoretical approaches can get. In the presentation by **Cheney, Isaacson and Somersalo**, a direct numerical layer stripping approach to the impedance imaging problem is decribed.

The works of **Alessandrini** and **Powell** deal with the same inverse boundary value problem except that instead of continuous conductivity distribution, one seeks to locate discontinuities in the medium. These problems have found applications especially in nondestructive material evaluation, where one tests material for cracks and impurities. The relation between sounding continuous and discontinuous media is similar to that between inverse potential scattering problems and inverse obstacle scattering problems.

In the article of **Sun**, the zero energy Scrödinger equation (25) is modified by adding a vector potential A to the Hamiltonian, the equation thus reading

$$((\nabla + A)^2 - q)\psi = 0. \tag{30}$$

This equation describes the scattering in presence of a magnetic field. The inverse problem is to reconstruct both q and curl$A$ from the boundary data. The article also discusses the inverse boundary value problem in two space dimensions. The two–dimensional problem can be viewed as a formally determined problem, and as the article indicates, thre is no complete solution in this case.

A different generalization of the Schrödinger operator is given in the contribution of **Isakov**. Instead of equation (25), one considers the semilinear differential equation

$$\Delta u + a(x, u) = 0 \quad \text{in } \Omega, \tag{31}$$

with the Dirichlet boundary condition at $\partial\Omega$. The problem is to identify the function a from the measurements of normal derivative of the solution u. The article discusses results with both single and several boundary measurements.

As in the inverse scattering, it is natural to look at the inverse boundary value problems for more general electromagnetic excitations than those obtained by current injection. The article by **Somersalo, Ola and Päivärinta** describes a general global uniqueness result for the inverse problem of the full set of Maxwell's equations. For Maxwell's equations, one replaces the scalar Neumann–to–Dirichlet map R by a boundary operator that could be called the *impedance map*,

$$\Lambda : n \wedge H|_{\partial\Omega} \mapsto n \wedge E|_{\partial\Omega}, \tag{32}$$

i.e., Λ maps the tangential component of the magnetic field to that of the electric field at the boundary. The inverse problem is to reconstruct the parameters ε (permittivity), σ (conductivity) and μ (permeability) in the medium.

Inverse boundary value problems get considerably more complicated if anisotropies in the medium are allowed. In fact, there are known obstructions of uniqueness for anisotropic inverse problems while the corresponding isotropic problems allow a unique solution. As an example, consider the anisotropic counterpart of equation (18),

$$\sum_{i,j=1}^{n} \frac{\partial}{\partial x_i} \gamma^{i,j} \frac{\partial}{\partial x_j} u = 0, \quad \text{in } \Omega \subset \mathbb{R}^n, \tag{33}$$

the potential u satisfying the boundary condition

$$\gamma^{i,j} \frac{\partial}{\partial x_j} u \bigg|_{\partial\Omega} = g^i. \tag{34}$$

These equations can be written in coordinate–free form using differntial forms as

$$\cdot\, d\gamma du = 0 \quad \text{in } \Omega, \tag{35}$$

$$\gamma du|_{\partial\Omega} = g. \tag{36}$$

From this formulation, it is evident that one can transform the map γ by a diffeomorphism that leaves the boundary $\partial\Omega$ untouched without affecting the boundary data. The question whether this is the only obstruction of uniqueness is still open. A thorough discussion of this topic and its extension to systems can be found in the contribution of **Sylvester**.

The article of **Grünbaum and Patch** also discusses the inverse boundary value problems but with a completely different starting point: Rather than describing the physical phenomena in terms of differential equations, the authors consider diffuse tomography, where the propagation of radiation is described in terms of transition probabilities between adjacent "cells".

Finally, the contribution of **Vainikko** deals with a related problem, the so called *ground water problem*, where the governing equation is similar to (18),

$$\nabla \cdot a\nabla u = f \quad \text{in } \Omega \qquad (37)$$

and the unknown parameter to be estimated is the filtration coefficient a. The treatment of the problem differs from the inverse conductivity problem because of the different type of data.

Before going to the discussion of inverse spectral problems, let us mention the four articles in this volume that deal with time dependent problems. It is found important, both from engineering and theoretical point of view, to understand transient phenomena and the related inverse problems. The contribution of **Kristensson and Rikte** treats the problem of reconstructing electromagnetic material parameters in a one–dimensionally layered structure from time-dependent boundary measurements of the fields. The constitutive equations have a rather general form in the sense that the electric and magnetic fields mix. This type of bi–isotropic media are increasingly important in engineering applications. It is interesting to notice that there is at least a formal similarity between the invariant embedding equation and the layer stripping method of impedance imaging mentioned earlier. The paper of **Hoffmann and Yamamoto** discusses inverse boundary value problems for certain parabolic differential equations. The model problem for parabolic inverse boundary value problem is the reconstruction of heat conductivity distribution of a body by surface measurements of the temperature. There is also a section in the work of Isakov discussing this type of problem. Finally, the article of **Choulli** considers the identification problem of inhomogenous terms in evolution equations using functional analytic methods.

III Let us move to the inverse spectral problems. The classical inverse spectral problem, as formulated by Mark Kac ("Can one hear the shape of the drum?"), gets an extensive review in the article by **Perry**, where some of the very recent results on this field are discussed. The recently discovered surprisingly simple method of constructing different isospectral domains (or drum heads with the same sound) is explained.

Some of the articles on inverse spectral problems in this collection come quite close to the inverse boundary value problems. The article of **Kurylev** deals with zero energy Schrödinger equation in a bounded domain having variable Riemannian metric. The data consists of the spectrum and the boundary values of the eigenfunctions at the boundary. The inverse problem is to reconstruct the potential as well as the metric in the domain. As it is pointed out in the article, the anisotropic conductivity γ in the system (33)– (34) plays a role similar to the Riemannian metric, and so some of the results have direct implications to the inverse boundary value problems described earlier.

Another article using spectral data for material parameter inversion is the one by **McLaughlin**. The inverse problem discussed is to reconstruct the variable sound speed in a domain from the interior transmission eigenvalues. The starting point of this article comes in fact from a numerical method of solving inverse scattering problems (see references in the cited article). Note that, as pointed out in the article of Colton, the transmission eigenvalues for a body are the values of γ for which the set $\{\gamma E_\infty + (1 - \gamma)H_\infty\}$ is incomplete in the space of square integrable tangential fields on the unit spere. Here E_∞ and H_∞ are the electric and magnetic far field patterns for the same body with a perfectly conducting (Dirichlet) boundary condition. This observation shows the strong interrelation between the inverse scattering and inverse spectral problems.

Besides those three categories of inverse problems, one can find articles on

IV Ill-posed problems,

 V The inverse scattering transform.

IV The history of inverse problems is intimately related to the question of ill-posedness. Hadamard defined the concept *ill–posed problem*, and as an example he used the problem of continuation of the Cauchy data of an elliptic equation from the boundary to the domain. It is interesting to notice that this original example of Hadamard shows up frequently in this book. For instance, the instability of the Riccati equation used in the layer stripping method in the article of Cheney *et al.* is nothing but a manifestation of this ill–posedness. As long as the history of ill–posed problems is the history of *regularization*, the procedure in which a trade–off of prior information and resolution is made. The articles of **Hofmann** and **Tautenhahn** discuss some of the more recent developments in treating regularization systematically and more generally.

V The applicability of inverse scattering methods to study the solutions of nonlinear partial differential equations has created a completely new branch of research in this field. The example now become a classic is the Korteweg–de Vries equation,

$$u_t(x,t) + u_{xxx}(x,t) - 6u_x(x,t)u(x,t) = 0, \qquad (38)$$

describing waves in shallow channels. It was found that if $x \mapsto u(x,t)$ is considered as the scattering potential of the stationary Schrödinger equation in one dimension, the time evolution of u corresponds to *linear* time evolution of the scattering data. One of the findings based on this observation is the soliton solutions, described by J. Scott Russell in 1844. Since the early days of inverse scattering transform, there has been a major hunt of nonlinear equations (and systems) that would allow an integration through the inverse scattering method. The article by **Sabatier** gives a unified an systematic view of how this hunt can proceed. (Incidentally, Sabatier's hunt takes the disguise of "killing parasites", as he describes it!).

Stability for the Crack Determination Problem

Giovanni Alessandrini*

Dipartimento di Scienze Matematiche, Università degli Studi di Trieste, Italy **, and Department of Mathematics, Northwestern University

1 Introduction

The inverse problem of crack determination was introduced by Friedman and Vogelius [F-V]. In its simplest form, it can be stated as follows:

Suppose that, in an electrically conducting planar body Ω, having constant conductivity, there is a crack, that is a curve σ where the conductivity is either infinite (perfectly conducting crack) or zero (perfectly insulating crack).

We are interested in recovering the shape and location of the crack from current and potential measurements at the boundary of the body Ω.

We may think of prescribing currents at the boundary and measuring the corresponding potentials or vice versa. Therefore it appears that we are dealing with four problems, but in fact, due to a duality argument (harmonic conjugation), it is possible to restrict the attention to the two cases when the crack σ is assumed to be either perfectly conducting or insulating and currents are assigned at the boundary.

Friedman and Vogelius have proved that in both cases if one assigns two appropriate currents at the boundary and measures the corresponding potentials then the crack is uniquely determined.

In [F-V] the importance of having stability estimates, that is, estimates on the continuous dependence of the crack from the measurements at the boundary, is also stressed. Some initial steps toward stability estimates are already present in [F-V]. In [A], a stability estimate for the case when the crack is perfectly conducting is proved. Our aim here is to outline the proof of that estimate and show how it can be improved (see Theorem 2.1 below).

The boundary value problem governing a potential u in Ω is the following

$$
\begin{cases}
\Delta u = 0 & \text{in } \Omega \setminus \sigma, \\
u = \text{constant} & \text{on } \sigma, \\
\dfrac{\partial u}{\partial \nu} = \psi & \text{on } \partial\Omega.
\end{cases}
$$

---

* Work performed within a 40% MURST research project.
** Permanent address

Here ν is the outer unit normal to $\partial\Omega$ and ψ is an assigned current profile.

The potential measurements will be taken on an open portion Γ on $\partial\Omega$:

$$g = u\big|_\Gamma.$$

Our choice of the assigned boundary conditions will be as follows:

We shall choose the boundary current ψ to be concentrated at two points (electrodes) P, Q on $\partial\Omega$

$$\psi = \delta_P - \delta_Q$$

and we shall measure g for two different choices of the points P, Q. In fact we shall leave P fixed and Q vary between two different positions.

The proof of Theorem 2.1 can be divided into two main steps: 1) estimates of the interior value of the potentials in terms of the boundary measurements; 2) estimates of the crack in terms of interior values of the potentials.

Step 1 (Theorems 2.2–2.3) consists of stability estimates on a Cauchy problem for Laplace's equation. Step 2 consists of a detailed study of the equipotential lines, Proposition 2.1.

The evidence of the fact that the continuous dependence of the crack from the data may be troublesome, comes from the observation that it seems necessary to pass through Step 1. In fact, it is well known that the Cauchy problem for Laplace's equation is severely ill-posed, as it was first shown by Hadamard [H]. Moreover, in our case, the region where u has to be continued is also unknown, since σ is unknown. These considerations must lead us to treat the crack determination problem as an ill posed one, which means estimates on continuous dependence are expected only when additional information is used. See [L-R-S] for a general theoretical setting of this issue. Our kind of *a priori* information will be concerned about the size and the smoothness of the crack, as already Friedman and Vogelius did in their discussion about stability, [F-V]. See Section 2 for a detailed description of the needed *a priori* information. Quite unfortunately, the rate of continuous dependence which was obtained in [A], was rather weak: a *log log* type modulus of continuity. Still our present rate is somewhat unsatisfactory, but it is improved to a *log* type modulus of continuity.

Let us conclude this introduction with some comments about prospective developments. It would be quite useful to have better stability estimates, possibly adding further *a priori* information. For instance, if we assume the analyticity of the crack then a Hölder estimate can be proved [A-V]. This result complements the numerical work towards the reconstruction of a linear crack which has been carried out by Santosa and Vogelius, [S-V], [L-S-V].

In [F-V] the uniqueness theorems are proven when the known background conductivity in the body varies analytically. It would be interesting to study uniqueness and stability when the known background conductivity varies, maybe smoothly, but not analytically.

Finally let us mention the challenging problem of treating the problem in the more realistic three dimensional setting.

The plan of the paper is as follows:

In Section 2 the *a priori* information is described, the main theorem is stated, Theorem 2.1, and an outline of the proof is given.

In Section 3 a stability estimate on the Cauchy problem for harmonic function is proven.

In Section 4 the central argument in the proof of Theorem 2.1 is given, which involves the geometrical study of equipotential lines.

2 The main theorem

We start by listing the *a priori* information we shall need. We will assume that constants L_1, L_2, $M > 0$, $0 < \alpha < 1$, are given such that the following facts hold. We shall refer to the set $\{L_1, L_2, M, \alpha\}$ as the *a priori* data.

Prior information on the body. Ω is a simply connected bounded open set in the plane $\mathbb{C}$, whose boundary $\partial\Omega$ is a $C^{2,\alpha}$ simple closed curve. Furthermore we shall assume

(2.1a) perimeter of $\Omega \le L_1$,

(2.1b) for every $z \in \partial\Omega$ there exists two circles of radius L_2, which are tangent in z, one is contained in $\overline{\Omega}$ and the other is contained in the complement.

(2.1c) if $z = z(s)$, $0 \le s \le \ell$ is the arclength parameterization of $\partial\Omega$ then we have

$$\|z\|_{C^{2,\alpha}[0,\ell]} \le M.$$

Prior information on the crack. The crack σ in a $C^{2,\alpha}$ simple curve in Ω satisfying the following hypotheses

(2.2a) distance of σ from $\partial\Omega \ge L_2$,

(2.2b) length of $\sigma \le L_1$,

(2.2c) if $z = z(s)$, $0 \le s \le m$, is the arclength parameterization of σ, then we have

$$|s_1 - s_2| \le M|z(s_1) - z(s_2)| \text{ for every } s_1, s_2 \in [0, m],$$

(2.2d) if $z = z(s)$ is as above, then we have

$$\|z\|_{C^{2,\alpha}[0,m]} \le M.$$

Prior information on the measurements. We will consider potential distributions u in Ω which are governed by the boundary value problem.

$$\begin{cases} \Delta u = 0 & \text{in } \Omega \setminus \sigma, & (2.3a) \\ u = \text{constant} & \text{on } \sigma, & (2.3b) \\ \dfrac{\partial u}{\partial \nu} = \delta_P - \delta_Q & \text{on } \partial\Omega. & (2.3c) \end{cases}$$

Here δ_P, δ_Q are Dirac masses on $\partial\Omega$ concentrated at distinct points $P, Q \in \partial\Omega$. We will measure the potential u on a fixed portion Γ of $\partial\Omega$, for two choices of the points P, Q as follows. Given three points $P_0, P_1, P_2 \in \partial\Omega$ we will consider the solutions u_1, u_2 to (2.3) when $(P, Q) = (P_0, P_1)$ and $(P, Q) = (P_0, P_2)$ respectively. Notice that $u_3 = u_1 - u_2$ is the solution u to (2.3) when $(P, Q) = (P_2, P_1)$. On Γ, P_0, P_1, P_2 the following assumptions will be used:

(2.4a) The distance between any two of the points P_0, P_1, P_2 is at least L_2,

(2.4b) Γ is a simple arc in $\partial\Omega$ whose length is at least L_2.

In the sequel we shall denote by d_H the Hausdorff distance between bounded closed sets in $\mathbb{C}$.

Theorem 2.1 *Let Ω satisfy (2.1). Let $\beta, 0 < \beta < 1/2$, be fixed. Let σ, σ' be two cracks both satisfying (2.2a)–(1.2d). Let u_1, u_2 be the solutions to (2.3) with the above mentioned choices of the points P, Q. Let u_1', u_2' be the corresponding solutions when σ is replaced by σ'. Given $\varepsilon > 0$, if the following error evaluation is known*

$$\max_{\Gamma} |u_1 - u_1'| + \max_{\Gamma} |u_2 - u_2'| \leq \varepsilon \tag{2.5}$$

then the two cracks satisfy

$$d_H(\sigma, \sigma') \leq \delta^\beta(\varepsilon) \tag{2.6}$$

and the function $\delta(\varepsilon)$ depends on the a priori *data only and satisfies*

$$\delta(\varepsilon) \leq A \left(|\log\varepsilon| \right)^{-1}, \tag{2.7}$$

for every ε, $0 < \varepsilon < 1/e$.

Here $A > 0$ depends on the a priori *data and on β only.*

Outline of the proof. First we shall need estimates on interior values of the potentials in terms of the Cauchy data.

Theorem 2.2 *Let the assumptions of Theorem 2.1 be satisfied, then we have*

$$\max_{i=1,2,3} \max_{\overline{\Omega}} |u_i - u_i'| \leq \eta(\varepsilon). \tag{2.8}$$

Here the function $\eta(\varepsilon)$ depends on the a priori *data only and we have*

$$\eta(\varepsilon) \leq A(\log|\log\varepsilon|)^{-\frac{1}{4}}, \quad \text{for every } \varepsilon, \ 0 < \varepsilon < 1/e. \tag{2.9}$$

In the theorem above no relation is imposed between the two cracks σ and σ'. This results in an estimate of a very weak type.

The proof of Theorem 2.2 can be found in [A], and we shall not repeat it here.

The next improved estimate assumes some form of correlation between the two cracks σ, σ' and yields a slightly better estimate. The proof is inspired to arguments previously used to Beretta and Vessella, in a related but different problem in inverse potential theory [B-V].

Theorem 2.3 *Let the assumptions of Theorem 2.1 be satisfied. Let γ, $0 < \gamma < \alpha$ be fixed. There exists $\delta_0 > 0$, depending on the a priori data and on β only, such that if σ, σ' have regular parameterizations $z = z(t)$, $z = z'(t)$, respectively, $0 \le t \le 1$, for which we have*

$$\|z - z'\|_{C^{2,\gamma}[0,1]} \le \delta_0 \tag{2.10}$$

then the following estimate holds

$$\max_{i=1,2,3} \max_{\overline{\Omega}} |u_i - u_i'| \le \delta^\beta(\varepsilon) \tag{2.11}$$

Here $\delta(\varepsilon)$ is a function depending only on the *a priori* data and satisfying (2.7). See Section 3 for a proof. The usefulness of the above Theorem will become clear after the following two propositions are stated.

Proposition 2.1 *Let all the assumptions of Theorem 2.1 be satisfied with the exception of (2.5). Let us suppose instead*

$$\max_{i=1,2,3} \max_{\overline{\Omega}} |u_i - u_i'| \le \delta \tag{2.12}$$

for some $\delta > 0$. Then we have

$$d_H(\sigma, \sigma') \le A\delta \tag{2.13}$$

where $A > 0$ depends on the a priori data only.

Proposition 2.2 *Let σ, σ' be two cracks satisfying (2.2a)–(2.2d). There exist regular parameterizations $z = z(t)$, $z = z'(t)$, $0 \le t \le 1$, of σ and σ' respectively, such that, for any γ, $0 < \gamma < \alpha$,*

$$\|z - z'\|_{C^{2,\gamma}[0,1]} \le B \left(d_H(\sigma, \sigma') \right)^{\frac{\alpha-\gamma}{2+\alpha}} . \tag{2.14}$$

Here $B > 0$ depends on the a priori data and on γ only.

While Proposition 2.2 consists of a more or less standard interpolation inequality (see [A, Corollary 1.1] for a proof), Proposition 2.1 contains the core argument for the final stability estimate. A sketch of a proof is given in Section 4.

By Theorem 2.2 and Proposition 2.1 we obtain (2.6) with $\delta^\beta(\varepsilon)$ replaced by a function $\eta(\varepsilon)$ satisfying (2.9). Using Proposition 2.2 we are in a position to apply Theorem 2.3 whenever ε is sufficiently small. By using Proposition 2.1 once more we get to (2.6). $\square$

3 Proof of Theorem 2.3

Our argument proceeds by fixing the index $i = 1, 2, 3$. Therefore it is convenient to drop the subscript i in the potentials u_i, u_i'.

It is easily checked that u, u' have single valued conjugate harmonic functions v, v' in $\Omega \setminus \sigma$, $\Omega \setminus \sigma'$ respectively. Moreover, in view of the *a priori* information, it is not difficult to obtain an $L^\infty(\Omega \setminus (\sigma \cup \sigma'))$ bound on $\phi = (u + iv) - (u' + iv')$ and a $C^{1/2}(\overline{\Omega})$ bound on $U = u - u'$ (for details, see [A, Lemma 3.1]).

The boundary condition (2.3c) for u and u' and the error estimate (2.5), imply

$$\max_{\Gamma} |\phi| \leq \varepsilon. \tag{3.1}$$

Let $\delta_0 > 0$ be a number which we shall choose later on, and let S_{δ_0}, be the δ_0-neighborhood of σ. By standard estimates on analytic continuation (see for instance [L-R-S]) we have

$$\max_{\overline{\Omega} \setminus S_{\delta_0}} |\phi| \leq C\varepsilon^\beta \tag{3.2}$$

where $C > 0$, $0 < \beta < 1$ depend on the *a priori* data only.

Due to (2.10) and to the *a priori* assumptions on σ and σ', for any α, $0 < \alpha < \pi$ we may choose a sufficiently small δ_0, in such a way that, for any $z_0 \in \sigma \cup \sigma'$, there exists a sector T with vertex at z_0, of radius $2\delta_0$ and aperture α such that $T \subset \Omega \setminus \{\sigma \cup \sigma'\}$ and the circular arc γ in ∂T is contained in $\Omega \setminus S_{\delta_0}$.

For any point $z \in T$, belonging to the line through z_0, which bisects T, we have

$$|\phi(z)| \leq \max_T |\phi| \left(\frac{\max_\gamma |\phi|}{\max_T |\phi|} \right)^{\left(\frac{|z-z_0|}{2\delta_0}\right)^{\pi/\alpha}}$$

see Carleman [C]. By the known bounds on U and ϕ and by (3.2), we obtain

$$\max_{\Omega \setminus (\sigma \cup \sigma')} |U| \leq C \left(r^{1/2} + \varepsilon^{Kr^{\pi/\alpha}} \right), \quad \text{for every } r, \ 0 < r < 2\delta_0$$

with $K, C > 0$ depending on the *a priori* data only. And (2.11) with $\delta(\varepsilon)$ satisfying (2.7) follows easily.   □

4 Proof of Proposition 2.1

Here we shall outline a proof of Proposition 2.1 trying to stress the geometrical aspects of the argument. We shall skip the quantitative details, about which we refer to [A, §2 and §4]. We start by considering the solution u to (2.3) for a given crack σ.

We are interested in describing the level line of u in Ω which passes through σ. Since the solution u of (2.3) is unique up to an additive constant, we may suppose at this stage that $u|_\sigma = 0$. Notice that this normalization will not be permitted when potential measurements will be taken at the boundary: in such a case the constant $u|_\sigma$ must be considered as an unknown quantity.

It is convenient to introduce a conformal mapping f which maps the doubly connected domain $\overline{\Omega} \setminus \sigma$ into an annulus $\overline{B_R(0)} \setminus B_1(0)$. Of course $R > 1$ will have to depend on Ω and σ, but its size can be controlled in terms of the *a priori* data.

Let $\tilde{P} = f(P)$, $\tilde{Q} = f(Q)$ and $v = u(f^{-1})$. The function v is the solution of the following problem

$$
\begin{cases}
\Delta v = 0 & \text{in } B_R(0) \setminus \overline{B_1(0)}, \\
v = 0 & \text{on } \partial B_1(0), \\
\dfrac{\partial v}{\partial \nu} = \delta_{\tilde{P}} - \delta_{\tilde{Q}} & \text{on } \partial B_R(0),
\end{cases}
$$

and it is readily seen to be odd symmetric with respect to the line λ which passes through the origin and which is the axis of symmetry between $\tilde{P}$ and $\tilde{Q}$. Notice that if we keep $\tilde{P}$ fixed and move $\tilde{Q}$ counterclockwise on $\partial B_R(0)$ then the line λ rotates in the same direction. Therefore, pulling back this information to the original coordinates we obtain that the level line $\{z \in \Omega \mid u(z) = u|_\sigma\}$ is composed by three simple curves are in $\Omega \setminus \sigma$ and σ, τ_1, τ_2. The curves τ_1, τ_2 do not intersect. τ_j, $j = 1, 2$, has one endpoint B_j on σ, and the other one R_j on $\partial \Omega$. If we move Q counterclockwise around $\partial \Omega$ while P is kept fixed, then B_1 and B_2 rotate around σ. It is convenient here to imagine σ as a degenerate closed curve, thus distinguishing the one sided limits as $z \to \sigma$, $z \in \Omega \setminus \sigma$. In particular, the rotating feature of the points B_1, B_2, which we shall call branching points, implies that there exist one end point V of σ and at least two choices among the pairs (P_0, P_1), (P_0, P_2), (P_2, P_1) for which the branching points of the corresponding potentials are far from V. How far? This can be evaluated in terms of the *a priori* data. Now we make use of hypothesis (2.12). Let us call

$$
c_i = u_i|_\sigma, \quad c_i' = u_i'|_{\sigma'} \quad i = 1, 2, 3
$$

By (2.12) we obtain

$$
\sigma' \subset \bigcap_{i=1}^{3} \{ z \in \Omega \mid |u_i - c_i'| < \delta \}
$$

Suppose that the pairs (P_0, P_1), (P_0, P_2) are those for which the branching points of u_1, u_2 are far from the endpoint V of σ. It is easily seen that the gradients of u_1, u_2 have a singularity at V of the type $1/\sqrt{z - V}$. Therefore V cannot be too far from σ', otherwise u_1', u_2' would be harmonic in a neighborhood of V thus contradicting (2.12). The distance of V from σ' can be in fact evaluated to be of the same order of δ^2. Using the $C^{1/2}$ regularity of the potentials near the crack we deduce

$$
|c_1 - c_1'| + |c_2 - c_2'| \le C\delta.
$$

Therefore

$$
\sigma' \subset \bigcap_{i=1}^{2} \{ z \in \Omega \mid |u_i - c_i| \le C\delta \}.
$$

Using once more the conformal mapping into the annulus, we easily see that the intersection of the right hand side is contained into an $O(\delta)$-neighborhood of σ and therefore

$$\sigma' \subset \{z \in \Omega \mid \operatorname{dist}(z, \sigma) < C\delta\}.$$

We may revert the roles of σ and σ' and (2.13) follows.   □

References

[A]   Alessandrini, G.: Stable determination of a crack from boundary measurements. To appear in Proc. Royal Soc. Edinb. Ser. A.

[A-V]   Alessandrini, G., Vessella, S.: In preparation.

[B-V]   Beretta, E., Vessella, S.: Stability results for an inverse problem in potential theory. Ann. Mat. Pura Appl. IV 156 (1990) 381–404.

[C]   Carleman, T.: Les Fonctions Quasi-Analytiques, Gauthier-Villers, Paris 1926.

[F-V]   Friedman, A., Vogelius, M.: Determining cracks by boundary measurements. Indiana Univ. Math. J. 38 (1989) 527–556.

[H]   Hadamard, J.: Lectures on Cauchy's Problem, Dover, New York 1952.

[L-S-V]   Liepa, V., Santosa F., Vogelius, M.: Crack determination from boundary measurements—Reconstruction using experimental data. Preprint.

[L-R-S]   Lavrentiev, M. M., Romanov V. G., Sisatskii S. P.: Problemi Non Ben Posti in Fisica Matematica e Analisi, Pubblicazioni dell'Istituto di Analisi Globale e Applicazioni, Serie Problemi non ben posti ed inversi, Firenze, 1983. Italian translation.

[S-V]   Santosa, F., Vogelius, M.: A computational algorithm to determine cracks from electrostatic boundary measurements. Int. J. Engng. Sci. 29 (1991) 917–937.

Layer-stripping Reconstruction Algorithms in Impedance Imaging

Margaret Cheney[1], David Isaacson[1] and Erkki Somersalo[2]

[1] Department of Mathematical Sciences, Rensselaer Polytechnic Institute, Troy, NY 12180, USA
[2] Rolf Nevanlinna Institute, P.O. Box 26 (Teollisuuskatu 23), 00014 UNIVERSITY OF HELSINKI, Finland

Impedance imaging systems apply currents to electrodes placed on the surface of a body and measure the voltages induced on these electrodes. These data are then used to reconstruct an approximation to the electric conductivity and permittivity in the interior of the body. This reconstruction problem is nonlinear and ill-posed.

Impedance imaging systems have many applications. The first is in medical imaging: because different tissues in the body have different electrical properties, an image of the conductivity and permittivity would provide an image of the different organs in the body. Another application is in nondestructive evaluation; at the General Electric Research and Development Center, for example, impedance imaging systems have made images of cracks in metals and have shown regions of non-uniformity in nuclear fuel rods. Impedance imaging systems are also being used for imaging multiphase fluid flow. For example, such systems are being used to ascertain the amount of water, gas, and sludge in pipes leading from offshore oil rigs. Finally, impedance imaging systems can be used for geophysical prospecting, especially in looking for mineral deposits.

Mathematical work, such as the design of optimal systems and development of reconstruction algorithms, requires a mathematical model. We obtain a mathematical model from Maxwell's equations by making the assumption that the frequency multiplied by the magnetic permeability is small. The leading order term gives us the equation

$$\nabla \cdot (\sigma + i\omega\epsilon)\nabla u = 0 \quad \text{inside the body } \Omega. \tag{1}$$

Here u is the electric potential, σ is the electric conductivity, ϵ is the electric permittivity, and ω is the temporal frequency of the applied currents. The naive boundary condition corresponding to the application of currents to the boundary is

$$(\sigma + i\omega\epsilon)\frac{\partial u}{\partial \nu} = j \quad \text{on the surface } \partial\Omega. \tag{2}$$

Here ν denotes the outward unit normal to Ω and j denotes the applied current density. The current density must satisfy the conservation of charge condition

$$\int_{\partial\Omega} j = 0. \tag{3}$$

Measuring the voltages corresponds to measuring the electric potential u on the boundary; the restriction of u to the boundary we denote by v. A solution of (1) and (2) is not uniquely determined until we specify a choice of ground or reference potential; we do this by requiring

$$\int_{\partial \Omega} v = 0. \tag{4}$$

In the rest of this paper, we will take $\omega = 0$ for simplicity.

The inverse problem we consider is the following: given all pairs (j, v), reconstruct an approximation to σ in the interior of Ω. Knowledge of all pairs (j, v) is equivalent to knowing the map

$$R : j \longmapsto v. \tag{5}$$

Sylvester and Uhlmann [SU] have shown that the map R uniquely determines σ, at least in the three-dimensional case.

The kernel of the map R is the restriction to the boundary of the Green's function G satisfying

$$\nabla \cdot \sigma \nabla G = \delta \text{ in } \Omega, \tag{6}$$

$$\sigma \frac{\partial G}{\partial \nu} = -\frac{1}{|\partial \Omega|} \text{ on } \partial \Omega, \tag{7}$$

$$\int_{\partial \Omega} G = 0, \tag{8}$$

where $|\partial \Omega|$ denotes the length of the boundary $\partial \Omega$ and δ denotes the Dirac delta function. That G restricted to the boundary is R can be seen from the (Green's) identity

$$u(x) = \int_{\Omega} (G \nabla \cdot \sigma \nabla u - u \nabla \cdot \sigma \nabla G) = \int_{\partial \Omega} (G \sigma \frac{\partial u}{\partial \nu} - u \sigma \frac{\partial G}{\partial \nu}). \tag{9}$$

Our choice of ground, together with the boundary conditions for G, cause the last term on the right side to vanish; on the other hand, $\sigma \frac{\partial u}{\partial \nu}$ is j. This shows that G restricted to the boundary is the kernel of R.

For a different problem, Kristensson and Cheney [CK] had derived a Green's function invariant imbedding equation that suggested a layer-stripping reconstruction algorithm. Because the Green's function here is so closely related to the map R, we decided to try a similar approach to this problem.

The layer-stripping algorithm comprises the following steps.

1) Make measurements j and v on the surface of the body, and assemble them into the map R.

2) Find σ on the surface.

3) Synthesize measurements (i.e., R) on a subsurface.

4) Repeat, starting with step 2.

To accomplish step 3, we think of the body as the union of surfaces, each surface being labelled by the parameter a. We calculate $\partial R/\partial a$ (see below), and then approximate R on the subsurface $a + \Delta$ by

$$R(a + \Delta) = R(a) + \Delta \frac{\partial R}{\partial r}. \tag{10}$$

For simplicity, we will take the body Ω to be a disk, so that the parameter a can be taken to be the radial variable r. Computing $\partial R/\partial a$ then reduces to computing $\partial R/\partial r$.

To understand how R_r changes with r, we simply differentiate (5) with respect to r. This gives us

$$\partial_r u = (\partial_r R)\, j + R\, (\partial_r j)\,. \tag{11}$$

The last term of (11) contains the expression

$$\partial_r j = \partial_r \sigma \partial_r u,$$

which arises in the polar form of the differential equation (1). In the two-dimensional case this polar form is

$$\partial_r \sigma \partial_r u + \frac{\sigma}{r} \partial_r u + \frac{1}{r^2} \partial_\theta \sigma \partial_\theta u = 0, \tag{12}$$

where ∂_θ is the angular derivative. We therefore use (12) in (11), obtaining

$$\partial_r u = (\partial_r R)\, j - R\left(\frac{\sigma}{r}\partial_r u\right) - R\left(\frac{1}{r^2}\partial_\theta \sigma \partial_\theta u\right). \tag{13}$$

Finally, we use the boundary condition (2) in (13), which gives

$$(\partial_r R) = \frac{I}{\sigma} + \frac{1}{r}R + \frac{1}{r^2}R\,(\partial_\theta \sigma \partial_\theta R)\,, \tag{14}$$

where I denotes the identity operator. This is an operator Riccati equation.

Next we would like to find σ on the surface. This can be done by using the ideas of [KV] and [SU] of applying highly oscillatory boundary data. When the boundary data is highly oscillatory, most of the current flow remains near the boundary, so one expects to be able to extract information about the conductivity near the boundary.

We obtain formulas for σ on the surface by first recalling that the kernel of R is the Green's function restricted to the boundary. The equation satisfied by the Green's function, however, can be written

$$\sigma \nabla^2 G + \nabla \cdot \nabla G = \delta. \tag{15}$$

The highest order terms in this equation determine the singularities of G; thus G looks like

$$G = \frac{1}{\sigma}G_0 + \text{smoother stuff}, \tag{16}$$

where G_0 corresponds to the case $\sigma = 1$. We can therefore recover σ on the boundary by

$$\frac{1}{\sigma(\theta)} = \lim_{\phi \to \theta} \frac{G(\theta, \phi)}{G_0(\theta, \phi)}. \tag{17}$$

Another formula can be obtained by recalling that since G is the kernel of R, we can apply R to the current density $\exp(in\phi)$ to obtain the boundary potential v^n:

$$v^n(\theta) = \int_{\partial\Omega} G(\theta, \phi) \exp(in\phi) d\phi \tag{18}$$

$$= \frac{1}{\sigma(\theta)} \int_{\partial\Omega} G_0(\theta, \phi) \exp(in\phi) d\phi + (\text{ terms that vanish as } n \to \infty).$$

The last integral, however, is just the boundary potential $v_0^n(\theta)$ that would result from applying $\exp(in\phi)$ if σ were identically one. Thus we have the alternate formula

$$\frac{1}{\sigma(\theta)} = \lim_{n \to \infty} \frac{v^n(\theta)}{v_0^n(\theta)}. \tag{19}$$

We found that this latter formula gave slightly better numerical results, at least when tested on bodies that were invariant under rotations.

We now have the following algorithm.

1) Assemble the measurements into the operator $R(r_0)$.

2) Find σ on the surface r_0 by the formula

$$\sigma(r_0, \theta) \approx \frac{v_0^N(\theta)}{v^N(\theta)}. \tag{20}$$

3) Find $R(r_0 - \Delta)$ from

$$R(r_0 - \Delta) = R(r_0) - \Delta \frac{\partial R}{\partial r}, \tag{21}$$

where $\partial R/\partial r$ is given by (14).

4) Replace r_0 by $r_0 - \Delta$ and repeat, starting with step 2.

The layer-stripping algorithm is a noniterative method for getting an approximate solution to the full nonlinear inverse problem. It has the following possible advantages. First, it has no difficulty with extraneous local minima as do most optimization-type methods. Second, in principle at least, it solves the full nonlinear problem. Third, it requires fewer computations and less storage than naive optimization-type methods. Fourth, with only minor modifications, it can be used for different geometries and for three-dimensional problems.

But will it work?

In our numerical tests, we used the operator $W = R/r$, which satisfies the slightly simpler equation

$$r\frac{\partial W}{\partial r} = \frac{1}{\sigma} + W\frac{\partial}{\partial \theta}\sigma\frac{\partial W}{\partial \theta}. \tag{22}$$

For our computations, we expanded all quantities in Fourier series.

We soon found that the naive algorithm is unstable. This can be seen easily in the case $\sigma = \sigma_0$, a constant. In this case, W is diagonal in the Fourier basis, and its diagonal entries are $w_n = 1/(\sigma_0|n|)$. The Riccati equation reduces to the set of uncoupled scalar equations

$$r\frac{dw_n}{dr} = \frac{1}{\sigma_0} - n^2\sigma_0^2 w_n^2. \tag{23}$$

These equations can be solved exactly; if the initial data contains a small error, then the solutions either blow up at a finite radius, or become negative, which is unphysical. The radius at which these unpleasant things happen moves outward as n increases.

This instability is only to be expected; after all, the problem is ill-posed. To overcome the ill-posedness, we must regularize. One way to regularize is to reduce the number of degrees of freedom sought. We can do this by updating σ less frequently. This, as it turns out, makes a small improvement in the stability.

Another way to regularize is to do smoothing. This we can do by dropping Fourier modes as we step in. But where, exactly, should which Fourier mode be dropped? A naive guess is to drop modes, in descending order, at equally spaced intervals, a strategy we refer to as the "arithmetic" one. A more informed strategy drops modes on the basis of the information content of each mode. When measurements are made with limited precision, each Fourier mode essentially probes only a certain region near the electrodes. After the layer-stripping algorithm has passed this region, the mode in question should be dropped. The following calculation suggests a formula to use.

Consider two cylindrical bodies of radius r_0. Assume that the conductivity distribution of the first is σ_0, a constant, and the conductivity distribution of the second body is σ, where in cross section σ is composed of two concentric disks, the inner disk (of radius r_1) being of constant conductivity σ_1, and the outer ring being of conductivity σ_0. To the outer boundary of both, we apply the current density $j = \exp(in\theta)$. We can use the method of separation of variables to compute the corresponding voltages v_0 and v on the boundary. We find that

$$v(\theta) - v_0(\theta) = \frac{r_0}{\sigma_0|n|}\frac{2\mu\Gamma^{2|n|}}{1 + \mu\Gamma^{2|n|}}\exp(in\theta), \tag{24}$$

where

$$\mu = \frac{\sigma_1 - \sigma_0}{\sigma_1 + \sigma_0} \tag{25}$$

and $\Gamma = r_1/r_0$. The two conductivity distributions cannot be distinguished from one another if the magnitude of this voltage difference in less than ϵ, the measurement precision. The condition that we be able to distinguish the two distributions gives us an inequality for Γ in terms of ϵ:

$$\frac{r_1}{r_0} = \Gamma \geq \left(\frac{\dfrac{|n|\sigma_0\epsilon}{2r_0}}{1 + \dfrac{|n|\sigma_0\epsilon}{2r_0}}\right)^{1/2|n|} \tag{26}$$

If we assume that $|n|\epsilon$ is constant (i.e., higher frequency measurements can be made more accurately), then this formula suggests that we should drop the nth Fourier mode at radius

$$r \approx \alpha^{1/2|n|} r_0, \tag{27}$$

where α is a regularization parameter representing the size of the smallest detectable object. We refer to this mode-dropping strategy as the "geometric" strategy.

If, on the other hand, we assume that the measurement precision ϵ is constant, then we are led to drop the nth Fourier mode at radius

$$r \approx |n|^{1/(2|n|)} \alpha^{1/(2|n|)} r_0. \tag{28}$$

This mode-dropping strategy we refer to as the "corrected" one.

We found that the "geometric" mode-dropping strategy yielded a dramatic improvement in stability over the "arithmetic" one, and the "corrected" method showed an equally dramatic improvement over the "geometric" one.

We also compared computations done by solving the Riccati equation to different accuracies, computations using different formulas for σ at the boundary, and different numbers of Fourier modes. We found that none of these modifications had as dramatic an effect on the stability as the positioning of the radii where Fourier modes were dropped.

We have also tried the method on experimental data. There we have the difficulty that the naive boundary conditions (2) do not adequately account for the presence of the electrodes on the boundary [CING, SCI, IC]. Consequently, there is a large error in the reconstruction at the boundary, and this error pollutes the reconstruction in the interior. The resulting reconstructions, especially the difference images, are recognizable, but they are not as good as our reconstructions from NOSER, our one-step Newton optimization algorithm [CINGS, IC].

Many questions about layer-stripping remain to be answered. How can we incorporate the more accurate boundary conditions [CING, SCI]? Can we find a better formula for the conductivity on the boundary? Is the Fourier basis the best basis to use, or would, say, a wavelet basis be better? Is there a better way to regularize? How can we ensure that our reconstruction is consistent with the data? John Sylvester has used layer-stripping ideas to prove existence and uniqueness results [S] in the case of rotational symmetry; can layer-stripping ideas be used to prove theorems in the general case?

Acknowledgments We would like to thank John Sylvester for sharing with us his ideas about a similar approach. In particular, our derivation for $\partial R/\partial r$ is essentially the same as one he showed us. We are also grateful to Jon Newell, Dave Gisser, Gary Saulnier, and the rest of the Rensselaer group for their ongoing help in building and testing impedance imaging systems.

References

[CK]   Cheney, M., Kristensson, G.: Three-dimensional inverse scattering: layer-stripping formulae and ill-posedness results. Inverse Problems 4 (1988) 625–642.

[CI]   Cheney, M., Isaacson, D.: Invariant imbedding, layer-stripping and impedance imaging. In: Corones, J., Kristensson, G., Nelson, P., Seth, D. (eds.): em Inverse Problems and Invariant Imbedding, SIAM, Philadelphia 1992.

[CING]   Cheng,K.-S., Isaacson, D., Newell, J.C., Gisser, D. G.: Electrode models for electric current computed tomography, IEEE Trans. Biomed. Engr. 36 (1989) 918–924.

[CINGS]   Cheney, M., Isaacson, D., Newell, J., Goble, J., Simske, S: NOSER: An algorithm for solving the inverse conductivity problem. Internat. J. Imaging Systems and Technology 2 (1990) 66–75.

[CISI]   Cheney, M., Isaacson, D., Somersalo, E., Isaacson, E. L.,: A layer-stripping approach to impedance imaging. In: *7th Annual Review of Progress in Applied Computational Electromagnetics*, Naval Postgraduate School, Monterey 1991.

[CISIC]   Cheney, M., Isaacson, D., Somersalo, E., Isaacson, E. L., Coffey, E. J.: A layer-stripping reconstruction algorithm for impedance imaging. IEEE-EMBS Proceedings 13 (1991).

[IC]   Isaacson, D., Cheney, M.: Problems in impedance imaging. This volume.

[KV]   Kohn, R., Vogelius, M.: Determining conductivity by boundary measurements. Comm. Pure Appl. Math. 37 (1984) 113–123.

[S]   Sylvester, J.: A convergent layer stripping algorithm for the radially symmetric impedance tomography problem. To appear in Comm. Partial Diff. Eqs.

[SCI]   Somersalo, E., Cheney, M., Isaacson, D.: Existence and Uniqueness for Electrode Models for Electric Current Computed Tomography, SIAM J. Appl. Math. 52 (1992) 1023–1040.

[SCII]   Somersalo, E., Cheney, M., Isaacson, D., Isaacson, E. L.: Layer Stripping: A Direct Numerical Method for Impedance Imaging, Inverse Problems 7 (1991) 899–926.

[SU]   Sylvester, J., Uhlmann, G: A uniqueness theorem for an inverse boundary value problem in electrical prospection", Comm. Pure Appl. Math. 39 (1986) 91–112; A global uniqueness theorem for an inverse boundary value problem. Ann. of Math. 125 (1987) 153–169; Inverse boundary value problems at the boundary — continuous dependence. Comm. Pure and Appl. Math. 41 (1988) 197–221.

Determination of the Inhomogeneous Term in Evolution Equations

M. Choulli

Laboratoire de mathématiques, Université de Franche–Comté, 25030 Besançon, France

Abstract *We consider the determination of the inhomogeneous term in evolution equations from an overspecified data. We present some existence and uniqueness results established by the author in [1,2]. We also show how this results can be applied to some inverse problems associated to the classical partial differential equations.*

1 Problems for which the final data is overspecified

Let X be a Banach space, A the generator of a linear strongly continuous semi-group $(S(t))_{t\geq 0}$, and $T > 0$. Let g be continuously differentiable real valued function defined on IR_{+}. In this section we are concerned by the study of the range and the null space of the operator

$$\Phi : y \in X \rightarrow u(T),$$

where u is the solution of the following Cauchy problem:

$$\begin{cases} u'(t) = Au(t) + g(t)y, & 0 \leq t \leq T, \\ u(0) = 0. \end{cases} \tag{1}$$

We will denote respectively by $R(\Phi)$ and $N(\Phi)$ the range and the null space of Φ. Since the classical solution of the initial value problem (1) is given by:

$$u(t) = \int_0^t g(t-s)S(s)yds, \ 0 \leq t \leq T,$$

it follows that $\Phi = \int_0^T g(T-t)S(t)dt$ and $R(\Phi) \subset D(A)$.

In what follows, we assume that $g = 1$ (see [2] for the case in which g is not equal to a constant).

Proposition 1. *Suppose that there exists a subset $X_0 \subset X$, $X_0 \cap D(A) \neq \emptyset$ and $X_0 \cap A(X_0 \cap D(A)) \neq \emptyset$ such that*

$$\sum_{n \geq 0} S(nt)x \quad \text{converges strongly for each } x \in X_0, \text{ and } t > 0.$$

Then $R(\Phi) \supset \{z \in D(A) \cap X_0, \ Az \in X_0\}$.

Corollary 2. *If A is a closed linear operator with the following property:*

there exists $\theta \in]\frac{\pi}{2}, \pi[$, and $M > 0$ such that

$S_\theta = \{z \in \mathbb{C}, \ z \neq 0, \ |\arg z| \leq \theta\} \subset \rho(A)$ *and* $\| \lambda(\lambda - A)^{-1} \| \leq M, \ \lambda \in S_\theta$,

then $R(\Phi) \supset D(A) \cap R(A^2)$.

Corollary 3. *If there exists $M \geq 1$, and $\lambda > 0$ such that*

$$\|S(t)\| \leq M e^{-\lambda t}, t \geq 0,$$

Then $R(\Phi) = D(A)$.

Example 1. Let $X = BUC(\mathbb{R})$ or $X = C_0(\mathbb{R})$. $BUC(\mathbb{R})$ is the space of bounded uniformly continuous functions on $\mathbb{R}$ and $C_0(\mathbb{R})$ consits of those functions in $BUC(\mathbb{R})$ which vanish at $\pm\infty$.

From [4] we know that the operator $A = \dfrac{d}{dx}$ having as domain

$$D(A) = \{f \in X, \ f \text{ absolutely continuous}, \ f' \in X\}$$

generates a strongly continuous group of isometries on X:

$$(S(t)f)(x) = f(x + t), \ t \in \mathbb{R}, \ x \in \mathbb{R}, \text{ and } f \in X.$$

For this example, it is not hard to see that

$$N(\Phi) = \begin{cases} \{0\} \text{ if } X = C_0(\mathbb{R}) \\ \\ \{f \in X, \ f \text{ is T-periodic}\} \text{ if } X = BUC(\mathbb{R}) \end{cases}$$

and from the previous proposition that

$$R(\Phi) \supset \{f \in D(A), \ x \to \sum_{n \geq 0} f(x + nT), \ \sum_{n \geq 0} f'(x + nT) \in X\} \supset S.$$

Here, S is the Schwartz class.

Example 2. Let Ω be a bounded open domain of $\mathbb{R}^d$, and let V be a bounded closed domain of $\mathbb{R}^d$ such that

$$\|v\| \geq \alpha, \text{ for each } v \in V,$$

for some $\alpha > 0$.

Let $\Gamma_- = \{(x, v) \in \partial\Omega \times V, vn(x) < 0\}$, where n denotes the outward normal to $\partial\Omega$.

For $\sigma \in L^\infty(V)$, $\sigma \geq 0$, define the operator $A = -v\nabla - \sigma.v$ with

$$D(A) = \{u \in L^p(\Omega \times V), v\nabla u \in L^p(\Omega \times V), u_{|\Gamma_-} = 0\}$$

It is well known (see for instance [3]) that A generates a strongly continuous semigroup of contractions $(S(t))_{t\geq 0}$ in $L^p(\Omega \times V)$ given by:

$$(S(t)u)(x, v) = \begin{cases} u(x - vt, v)e^{-\sigma(v)t} & t \leq t(x, v), \\ 0 & \text{otherwise} \end{cases} \quad u \in L^p(\Omega \times V),$$

where, $t(x, v) = \sup\{t, x - vs \in X, 0 \leq s < t\}$. The assumption on V implies that $t(.,.)$ is bounded and then $S(t) = 0$, $t \geq T_0$, for some $T_0 > 0$. Hence, $R(\Phi) = D(A)$ according to the proposition 1.

Example 3. Let Ω be a bounded open domain of $\mathbb{R}^d$ whose boundary is smooth.

Let A be the operator $A = \Delta$ with domain $D(A) = H^2(\Omega) \cap H_0^1(\Omega)$. It is a classical result (see for instance [3]) that $0 \in \rho(A)$ and the spectrum of A

$$\sigma(A) = \{-\lambda_n, \ n \geq 1\},$$

where

$$-\infty < -\lambda_n \leq \ldots \leq -\lambda_1 < 0, \ \lambda_n \to +\infty.$$

Let x_n be the normalized eigenfunction associated to $-\lambda_n$. Since (x_n) is a basis of $L^2(\Omega)$, a simple computation show that the semigroup generated by A is given by:

$$S(t) = \sum_{n\geq 1} e^{-\lambda_n t}(., x_n)x_n.$$

Here, $(.,.)$ is the scalar product in $L^2(\Omega)$.

Hence $\|S(t)\| \leq e^{-\lambda_1 t}$, $t \geq 0$, and

$$\int_0^T S(t)dt = \sum_{n\geq 1} c_n(., x_n)x_n, \tag{2}$$

where $c_n = \dfrac{1 - e^{-\lambda_n T}}{\lambda_n} > 0$. Now, corollary 2 and (2) show that $R(\Phi) = D(A)$, and $N(\Phi) = \{0\}$.

Proposition 4. *If A is injective and $S(T)$ is compact then*

$$R(\Phi) = \{z \in D(A), \ -Az \in N(I - S(T)^*)^\perp\}.$$

Proposition 5. *Assume that X is a reflexive Banach space and that $0 \in \rho(A)$. If Φ is injective then for each $z \in D(A)$, there exists a sequence in $(z_n) \in R(\Phi)$ such that z_n converges weakly to z.*

Application Second order evolution equation in a real Hilbert space.

Let H and V be two real Hilbert spaces such that $V \hookrightarrow H$ continuously and densely. Let B be a linear operator defined by a symmetric bilinear form which is continuous in $V \times V$ and V-coercive.

We define the operator

$$\Psi : y \in H \to (u(T), u'(T))$$

where u is the solution of the Cauchy problem:

$$\begin{cases} u''(t) + Bu(t) = y, & 0 \le t \le T, \\ u(0) = 0, \\ u'(0) = 0, \end{cases} \tag{3}$$

Since the initial value problem (3) can be converted into (1), the next statement follows from the proposition 3.

Corollary 6. *For each $z \in V \times H$, there exists a sequence $(z_n) \in R(\Psi)$ such that z_n converges weakly to z.*

2 Problems for which the overspecified data is time-dependent.

Let X and A be as in the previous section. Let Y be a Banach space, $K : X \to Y$ and $B : Y \to D(A)$ be a linear operators. We consider the following abstract Cauchy problem:

$$\begin{cases} u'(t) = Au(t) + Bf(t), & 0 \le t \le T, \\ u(0) = x. \end{cases} \tag{4}$$

This section is devoted to the problem of finding $f : [0, T] \to Y$ such that

$$Ku(t) = g(t), \quad 0 \le t \le T, \tag{5}$$

where $u = u(t)$ is the solution of the previous Cauchy problem corresponding to f and $g : [0, T] \to Y$ is a given function.

Proposition 7. *Let $x \in D(A)$, $g \in C^1([0, T] : Y)$ and assume that K and B are bounded operators. If 0 belongs to the resolvent set of KB and the compatibility condition $g(0) = Kx$ is satisfied, then (5) has a unique solution $f \in C([0, T] : Y)$. In addition, f depend continuously on the data g and x.*

Remark. The conclusion of the proposition 7 is valid for second order evolution equation in the case that A is the generator of strongly continuous cosine family (see [2]).

Application Let $(z_1, \ldots, z_n) \in D(A)^n$, and define the operators B and K as follows.

$$B : (\xi_1, \ldots, \xi_n) \in \mathbb{C}^n \rightarrow \xi_1 z_1 + \ldots + \xi_n z_n \in D(A),$$

$$K : x \in X \rightarrow (K_1 x, \ldots, K_n x) \in \mathbb{C}^n,$$

where K_p belongs to $\mathcal{L}(X, \mathbb{C})$, $1 \leq p \leq n$. Clearly, KB maps $\mathbb{C}^n$ into itself and

$$KB(\xi_1, \ldots, \xi_n) = (\sum_{i=1}^{n} \xi_i K_1 z_i, \ldots, \sum_{i=1}^{n} \xi_i K_n z_i), \quad (\xi_1, \ldots, \xi_n) \in \mathbb{C}^n.$$

Therefore, $0 \in \rho(KB)$ if and only if the matrix $(K_j z_i)_{1 \leq i,j \leq n}$ is invertible.

Example 4. Let Ω be an open and bounded set ot $\mathbb{R}^n$ with a smooth boundary $\partial\Omega$. Let $a_{ij}, b_j, c : \Omega \rightarrow \mathbb{R}, i, j = 1, \ldots, n$, be a continuous functions such that

$$\exists\, \alpha > 0, \ \sum_{i,j=1}^{n} a_{ij}(x) y_i y_j > \alpha \sum_{i=1}^{n} y_i^2, \ \text{for each } (y_1, \ldots, y_n) \in \mathbb{R}^n, \ \text{and } x \in \overline{\Omega}.$$

We define the elliptic operator A on $C(\overline{\Omega})$ by

$$(Au)(x) = \sum_{i,j=1}^{n} a_{ij}(x) \frac{\partial^2}{\partial x_i \partial x_j} u(x) + \sum_{i=1}^{n} b_i(x) \frac{\partial}{\partial x_i} u(x) + c(x)u(x), \quad x \in \overline{\Omega},$$

and

$$D(A) = \{u \in W^{2,p}(\Omega), \ Au \in C(\overline{\Omega}), \ u = 0 \ \text{in } \partial\Omega\} \quad (p > n).$$

From [6], we have that A generates an holomorphic semigroup. This semigroup is not strongly continuous at the origin because A is not densely defined on $C(\overline{\Omega})$ $(\overline{D(A)} = C_0(\overline{\Omega}))$. But, the conclusion of the proposition 4 remains valid for this case (see [1] for more details).

Let us consider the problem of finding $u : [0, T] \times \overline{\Omega} \rightarrow \mathbb{R}$ and $f_i : [0, T] \rightarrow \mathbb{R}$, $1 \leq i \leq n$, such that

$$\begin{cases} \frac{\partial}{\partial t}u(t, x) = Au(t, x) + \sum_{i=1}^{n} f_i(t)y_i(x), & (t, x) \in [0, T] \times \overline{\Omega}, \\ u(0, x) = u_0(x), & x \in \overline{\Omega}, \\ u(t, x) = 0, & (t, x) \in [0, T] \times \partial\Omega, \end{cases} \tag{6}$$

and

$$u(t, x_i) = g_i(t), \quad t \in [0, T], \quad 1 \leq i \leq n. \tag{7}$$

where $x_i \in \overline{\Omega}$, $1 \leq i \leq n$. if we assume that $u_0 \in D(A)$, $g_i \in C^1[0, T]$, $g_i(0) = u_0(x_i)$, $y_i \in D(A)$, $1 \leq i \leq n$, and that the matrix $(y_i(x_j))_{1 \leq i,j \leq n}$ is invertible, then it follows from the proposition 4 that there is a unique $(f_1, \ldots, f_n) \in C[0, T]^n$ and $u \in C([0, T] : D(A)) \cap C^1([0, T] : C(\overline{\Omega}))$ such that (6) and (7) are satisfied.

Example 5. We consider the wave equation:

$$\begin{cases} \dfrac{\partial^2}{\partial t^2} u(t, x) = \dfrac{\partial^2}{\partial x^2} u(t, x) + f(t)q(x), & x \in \mathbb{R}, \ 0 \le t \le T \\[2mm] u(0, x) = \dfrac{\partial}{\partial t} u(0, x) = 0, & x \in \mathbb{R}. \end{cases} \tag{8}$$

Let $x_0 \in \mathbb{R}$, q and g be given. We seek $f = f(t)$ such that

$$u(t, x_0) = g(t), \quad 0 \le t \le T. \tag{9}$$

where u is the solution of (8) corresponding to f.

Let $X = C_b(\mathbb{R})$ the space of bounded continuous functions defined on $\mathbb{R}$
The cosine family $(C(t))_{t \in \mathbb{R}}$ given by

$$(C(t)f)(x) = \frac{1}{2}(f(x + t) + f(x - t)), \quad x, t \in \mathbb{R}.$$

is strongly continuous. Its generator is the operator $A = \dfrac{d^2}{dx^2}$ with the domain

$$D(A) = \{f \in X, t \to C(t)f \text{ is twice continuously differentiable}\}$$

If $q \in C^2(\mathbb{R})$, $q(x_0) \ne 0$, $g \in C^2[0, T]$ and $g(0) = g'(0) = 0$ then the proposition
4 and the remark following it, (9) has a unique solution $f \in C[0, T]$. The inverse

problem (9) can be solved without using the proposition 4. We can show, under
the previous assumptions, that f is a solution of (9) if and only if f solve the
Volterra integral equation:

$$f(t) + \int_0^t \frac{q'(x_0 + (t - s)) - q'(x_0 - (t - s))}{2q(x_0)} f(s)ds = \frac{g''(t)}{2q(x_0)}, \quad 0 \le t \le T.$$

References

1. Choulli, M: An abstract inverse problem and application. J. Math. Anal. Appl. **160** (1991) 190–202.
2. Choulli, M.: Inverse problems for inhomogeneous term in partial differential equations. To appear.
3. Dautray, R., Lions J. L. (ed):*Analyse mathématiques et calcul numérique pour les sciences et les techniques.* Masson, 1984.
4. Goldstein, J. A.: *Semigroups of linear operators and applications.* Oxford Mathematical Monographs, 1985.
5. Nagel, R.: *One parameter semigroup of positive operators.* Springer-Verlag, LN 1184, Berlin, 1986.
6. Stewart, H. B.: Generation of analytics semigroups by strongly elliptic operators. Trans. Amer. Math. Soc. **199** (1974) 141–162.
7. Travis, C. C., Webb, G. F.: Cosine families and abstract nonlinear second order differential equations. Acta Mathematica, Academiae Scientiarum Hungaricae **32** (1978) 75–79.
8. Yosida, K.: *Functional analysis*, (Fifth edition). Springer-Verlag, Berlin 1978.

Target Signatures for Maxwell's Equations*

David Colton

Department of Mathematical Sciences, University of Delaware, Newark, Delaware 19716, USA

1 Introduction

Inverse scattering problems can be broadly divided into two distinct classes. The first of these assumes minimal a priori knowledge of the scattering object and attempts to reconstruct the shape of an obstacle or the function values of constitutive parameters from an inexact knowledge of the far field pattern. The mathematical basis for this class of problems is extensively discussed in my recent monograph with Rainer Kress ([6]). The second class of inverse scattering problems is concerned with the determination of 'target signatures,' i.e. eigenvalues that can be determined from the scattering data that are characteristic of the scattering object but not the incident field. In this case, the purpose is not to reconstruct the scattering obstacle or constitutive parameters but rather to either distinguish a specific object from a known set of objects or to detect the existence of an anomaly in a given background configuration. Typical of this second class of inverse scattering problems is the singularity expansion method as discussed in [2] and [8]. The purpose of this short survey talk is to acquaint the reader with some recent (and not so recent) results on target signatures associated with the scattering of electromagnetic waves by a bounded obstacle.

I shall begin my talk by recalling the salient features of the singularity expansion method. Although this method is well known to electrical engineers in connection with radar applications ([2]) and to mathematicians in connection with the mathematical theory of scattering (c.f. the epilogue to [10]), there seems to be very little communication between the two groups (an exception is [8]). The basic idea of the singularity expansion method is to determine the rate of decay of the scattered electric field with respect to time. This rate of decay is determined by the scattering frequencies associated with Maxwell's equations defined in the exterior of a scattering obstacle, i.e. the target signatures are the scattering frequencies. Due to the difficulty in actually measuring the rate of decay of the scattered field, the only scattering frequencies of much practical interest are those nearest the origin.

* This research was supported in part by the Air Force Office of Scientific Research

The singularity expansion method requires a broad range of frequencies (corresponding, for example, to a broad band pulse as an incident field) for its practical application. By contrast, Colton and Monk ([7]) and Colton and Kirsch ([4],[5]) have recently introduced a new set of target signatures which is applicable for a single fixed frequency. This set is determined by the fact that a convex combination of the electric and magnetic far field patterns for time harmonic fields with arbitrary orthogonal polarizations and arbitrary directions is incomplete for a discrete set of the convexity parameter γ. This set of parameters $\{\gamma_j\}$ can be used as a set of target signatures. Since the accuracy of the determination of the set $\{\gamma_j\}$ is enhanced by increasing the number of incident fields and the number of points at which the far field pattern is measured, the set of target signatures $\{\gamma_j\}$ is probably more suitable for an area such as nondestructive testing rather than the usual radar applications. On the other hand, it is undoubtedly premature to make any firm statements on applicability at this stage since so far only a few preliminary experiments concerning the calculation of the set $\{\gamma_j\}$ from synthetic data have been made ([7]). In addition, the mathematical investigation of the set $\{\gamma_j\}$ has just begun. A striking result which has been discovered is that for an imperfect conductor the set $\{\gamma_j\}$ is contained in a region in the complex γ–plane whose geometry depends only on the surface impedance of the scattering obstacle ([5]).

2 The Singularity Expansion Method

Let $\mathcal{E}, \mathcal{H}$ be the electric and magnetic fields in the exterior of a perfectly conducting scattering obstacle $D \subset I\!\!R^3$, i.e. after suitable normalization $\mathcal{E}$ and $\mathcal{H}$ satisfy Maxwell's equations

$$\operatorname{curl} \mathcal{E} + \frac{\partial \mathcal{H}}{\partial t} = 0 \qquad \operatorname{curl} \mathcal{H} - \frac{\partial \mathcal{E}}{\partial t} = 0$$

$$\operatorname{div} \mathcal{E} = 0 \qquad \operatorname{div} \mathcal{H} = 0$$

$$(2.1)$$

in the exterior of D and

$$\nu \times \mathcal{E} = 0 \tag{2.2}$$

on the boundary ∂D of D where ν is the unit outward normal to ∂D. The singularity expansion method is based on the observation that under certain assumptions on D (e.g. D is 'nontrapping') $\mathcal{E}$ has the local asymptotic expansion

$$\mathcal{E}(x, t) \sim \sum_{j=1}^{\infty} a_j e^{-ik_j t} E_j(x) \tag{2.3}$$

for $x \in I\!\!R^3 \backslash \bar{D}$ where the a_j are constants (or possibly polynomials in t) and the k_j and E_j are the eigenvalues and eigenfunctions of the non–self adjoint eigenvalue problem

$$\operatorname{curl} E_j - ik_j H_j = 0 \qquad \operatorname{curl} H_j + ik_j E_j = 0 \quad \text{in } I\!\!R^3 \backslash \bar{D} \tag{2.4}$$

$$\nu \times E_j = 0 \quad \text{on } \partial D$$

where E_j satisfies an 'outgoing' radiation condition. By Rellich's lemma, this implies that $\operatorname{Im} k_j < 0$. It is further known that the set $\{k_j\}$ is discrete and, under appropriate assumptions, $\operatorname{Im} k_j \to -\infty$ ([3]).

The eigenvalues k_j defined above are known as the *scattering frequencies* of the perfectly conducting obstacle D and have been extensively investigated by numerous investigators, particularly in the case of the scalar analogue of (2.4) (c..f. the epilogue of [10]). The idea of the singularity expansion method is to use (2.3) to measure the k_j and from this knowledge identify the target D. Since the terms in (2.3) are exponentially decaying, this means in practice that only the first few scattering frequencies can be measured from experimental data. An obvious question to ask is what do the scattering frequencies say about the geometry of D? In the scalar case, this question was first addressed by Lax and Phillips ([9]) who showed that, roughly speaking, the purely imaginary scattering frequencies are influenced by the bulk properties of the obstacle whereas those with a real part are sensitive to surface details. The ideas of Lax and Phillips were extended to the case of Maxwell's equations by Beale ([3]) who proved the following striking result on the existence and monotonicity of the purely imaginary scattering frequencies. Since the scattering frequencies for a sphere can be explicitly computed, Beale's result shows that the smallest purely imaginary scattering frequency gives an indication of the volume of the scattering obstacle.

Theorem 1 ([3]). *There exist scattering frequencies that are purely imaginary. Suppose D_1 and D_2 are obstacles such that $\bar{D}_1 \subset D_2$ and D_1 is star–shaped. If the purely imaginary scattering frequencies of D_ℓ are written as $k_j^{(\ell)} = -i\sigma_j^{(\ell)}, j = 1, 2, \ldots$, so that $\sigma_j^{(\ell)} \le \sigma_{j+1}^{(\ell)}$ and repeated according to multiplicities, then $\sigma_j^{(1)} \ge \sigma_j^{(2)}$ for each j.*

Beale was able to extend his theorem to the case when the perfectly conducting boundary condition (2.2) is replaced by $(\mathcal{E} - \lambda \mathcal{H}) \times \nu = 0$ where λ is a real constant. However, this does not include the case of the impedance boundary condition $\nu \times (\nu \times \mathcal{H}) - \lambda(\nu \times \mathcal{E}) = 0$. Hence, the analogue of Beale's theorem for an imperfect conductor has not been established although some results are available for the scalar case ([3]).

3 Target Signatures for a Fixed Frequency

Are there target signatures analogous to the scattering frequencies for time–harmonic scattering problems, i.e. the scattering of electromagnetic waves at fixed frequency? As the reader may suspect from the title of this section, the answer is yes. However, since the frequency is fixed, it turns out that the eigenvalues no longer appear in the differential equation as in (2.4) but rather in the boundary condition of an associated scattering problem. To understand how this associated problem arises, consider the scattering of a time harmonic electromagnetic wave by a perfectly conducting obstacle $D \subset \mathbb{R}^3$. Factoring out the time–harmonic part and letting $k > 0$ denote the wave number, it is well known ([6]) that after appropriate normalization the electric field E and magnetic field H satisfy Maxwell's equations

$$\text{curl } E - ikH = 0 \qquad \text{curl } H + ikE = 0 \quad \text{in} \quad \mathbb{R}^3 \backslash \bar{D} \tag{3.1}$$

where $E = E^i + E^s$, $H = H^i + H^s$ with the incident electromagnetic field E^i, H^i being defined by Maxwell's equations and

$$E^i(x) = pe^{ikd \cdot x}, \quad p \cdot d = 0 \tag{3.2}$$

for $x \in \mathbb{R}^3$ and constant vectors $p \in \mathbb{C}^3$, $d \in \mathbb{R}^3$, $|d| = 1$, and the scattered electromagnetic field E^s, H^s satisfies the Silver-Müller radiation condition

$$H^s \times \hat{x} - E^s = 0 \left(\frac{1}{|x|^2} \right), \quad |x| \to \infty \tag{3.3}$$

with $\hat{x} = x/|x|$. On the boundary of the perfect conductor E must satisfy the boundary condition

$$\nu \times E = 0. \tag{3.4}$$

Under these conditions, E^s and H^s exist and have the asymptotic behavior

$$E^s(x) = \frac{e^{ik|x|}}{|x|} E_\infty(\hat{x}; d, p) + 0 \left(\frac{1}{|x|^2} \right)$$
$$H^s(x) = \frac{e^{ik|x|}}{|x|} H_\infty(\hat{x}; d, p) + 0 \left(\frac{1}{|x|^2} \right) \tag{3.5}$$

where E_∞ is the *electric far field pattern* and H_∞ is the *magnetic far field pattern*.
Suppose it is now asked for what values of γ is the set

$$\gamma E_\infty(\hat{x}; d, d \times p) + (1 - \gamma) H_\infty(\hat{z}; d, d \times (p \times d)) \tag{3.6}$$

for $d \in \Omega = \{x : |x| = 1\}$ an arbitrary direction and $p \in \mathbb{C}^3$ an arbitrary polarization incomplete in the space $T^2(\Omega)$ of square integrable tangential fields on Ω? Before trying to answer this question, it is perhaps worthwhile to pause and ask how can it be decided in practice whether or not (3.6) is incomplete? As is well known, the set (3.6) is incomplete if and only if there exists $g \in T^2(\Omega)$ such that

$$\int_\Omega g(\hat{x}) \cdot [\gamma E_\infty(\hat{x}; d, d \times p) + (1 - \gamma) H_\infty(\hat{x}; d, d \times (p \times d))] ds(\hat{x}) = 0 \tag{3.7}$$

for all $d \in \Omega, p \in \mathbb{C}^3$. Now choose N linearly independent directions and three linearly independent polarizations and for each fixed direction and polarization approximate each component of E_∞ and H_∞ by a finite Fourier series of N spherical harmonics. If E_∞^N and H_∞^N are the matrices whose columns are these approximations, then the discrete analogue of the eigenvalue problem (3.7) is

$$\det[\gamma E_\infty^N + (1 - \gamma) H_\infty^N] = 0. \tag{3.8}$$

Note that (3.8) is a finite dimensional *linear* eigenvalue problem and hence is relatively easy to solve. How accurately the eigenvalues of (3.8) approximate those of (3.7) (if they exist at all!) remains to be seen, although preliminary computations hold out some hope that for γ bounded away from $\gamma = 1/2$ the error may not be too large ([7]). The significance of $\gamma = 1/2$ will become apparent in the sequel.

Returning now to the completeness properties of the set (3.6), a few preliminary definitions are needed in order to proceed. An *electromagnetic Herglotz pair* ([6]) is a solution to Maxwell's equations in $I\!R^3$ such that the electric field E can be represented in the form

$$E(x) = \int_\Omega g(\hat{y}) e^{ikx \cdot \hat{y}} ds(\hat{y}) \tag{3.9}$$

where $g \in T^2(\Omega)$. Values of γ such that there exists a nontrivial classical solution of the homogeneous transmission problem

$$\operatorname{curl} E^+ - ikH^+ = 0, \qquad \operatorname{curl} H^+ + ikE^+ = 0 \quad \text{in} \quad I\!R^3 \backslash \bar{D},$$

$$\operatorname{curl} E^- - ikH^- = 0, \qquad \operatorname{curl} H^- + ikE^- = 0 \quad \text{in} \quad D,$$

$$\nu \times E^+ - \nu \times E^- = 0$$

$$\gamma(\nu \times \operatorname{curl} E^+) + (1 - \gamma)(\nu \times \operatorname{curl} E^-) = 0 \qquad \text{on } \partial D, \tag{3.10}$$

$$H^+ \times \hat{x} = E^+ = 0\left(\frac{1}{|x|^2}\right), \qquad\qquad |x| \to \infty$$

are called *transmission eigenvalues* and the nontrivial solution E^-, H^- is called a *transmission eigenfunction*. The following theorem shows that the non–selfadjoint eigenvalue problem (3.10) is the fixed frequency analogue of the eigenvalue problem (2.4) for the scattering frequencies.

Theorem 2 ([4], [7]). *Let E_∞ and H_∞ be the electric and magnetic far field patterns corresponding to a perfectly conducting obstacle D. Then the set (3.6) is incomplete in $T^2(\Omega)$ if and only if γ is a transmission eigenvalue and at least one of the corresponding eigenfunctions is an electromagnetic Herlgotz pair.*

The completeness properties of the set (3.6) are now obtained by investigating the eigenvalue problem (3.10). In particular, a simple application of Green's formula gives the following result:

Theorem 3 ([4], [7]). *If γ is a transmission eigenvalue and $\gamma \neq 0, 1$ then $Im\,\gamma \neq 0$.*

In the two–dimensional scalar case, i.e. when D is an infinite cylinder, it was shown by Colton and Monk ([7]) that the set of transmission eigenvalues is discrete and the only possible accumulation point is $\gamma = 1/2$, thus explaining the difficulties in computing the eigenvalues of (3.8) near $\gamma = 1/2$. It is conjectured that the same result is true in the three dimensional vector case. It is also conjectured that the transmission eigenvalues lie in a disk in the complex plane whose radius depends only on the volume of D and the surface area and curvature of ∂D (I have shown that this is true in the scalar case).

The above results for a perfect conductor have been extended by Colton and Kirsch ([5]) to the case of an imperfect conductor, i.e. the perfectly conducting boundary condition (3.4) is now replaced by the impedance boundary condition

$$\nu \times (\nu \times \operatorname{curl} E) - ik\lambda(\nu \times E) = 0 \qquad (3.11)$$

where the surface impedance λ is positive. As already mentioned in the Introduction, the behavior of the eigenvalues associated with the imperfect conductor is quite different from that of the perfect conductor. To see this, let E_∞ and H_∞ be the electromagnetic far field patterns associated with the scattering of the incident field (3.2) by an imperfect conductor and again consider the set (3.6) in $T^2(\Omega)$. Instead of the transmission problem (3.10), it is now necessary to consider the following conductive boundary value problem (c.f. [1]): Find a nontrivial classical solution of

$$\operatorname{curl} E^+ - ikH^+ = 0, \qquad \operatorname{curl} H^+ + ikE^+ = 0 \qquad \text{in} \quad \mathbb{R}^3 \backslash \bar{D},$$

$$\operatorname{curl} E^- - ikH^- = 0, \qquad \operatorname{curl} H^- + ikE^- = 0 \qquad \text{in} \quad D,$$

$$\nu \times (\nu \times (\operatorname{curl} E^- + \operatorname{curl} E^+)) - ik\lambda\nu \times (E^- + E^+) = 0,$$

$$\nu \times (\nu \times (\operatorname{curl} E^- + \eta\operatorname{curl} E^+)) - \frac{ik}{\lambda}\nu \times (E^- + \eta E^+) = 0, \qquad \text{on} \quad \partial D, \qquad (3.12)$$

$$H^+ \times \hat{x} - E^+ = 0\left(\frac{1}{|x|^2}\right), \qquad |x| \to \infty$$

where $\eta = \gamma/\gamma - 1$ and it is assumed that $\gamma \neq 0, 1$. Values of η such that there exists a nontrivial solution of (3.12) are called *conductive eigenvalues* and the nontrivial solution E^-, H^- is called a *conductive eigenfunction*. The non–selfadjoint eigenvalue problem (3.12) is the analogue for an imperfect conductor of the eigenvalue problem (3.10) for a perfect conductor. In particular, the following theorems are true:

Theorem 4 ([5]). *Let $\gamma \neq 0, 1$ and let E_∞ and H_∞ be the electric and magnetic far field patterns corresponding to an imperfectly conducting obstacle D. Then the set (3.6) is incomplete in $T^2(\Omega)$ if and only if η is a conductive eigenvalue and at least one of the corresponding eigenfunctions is an electromagnetic Herglotz pair.*

Theorem 5 ([5]). *Suppose $\gamma \neq 0, 1$ and $\lambda \neq 1$. Then if η is a conductive eigenvalue η must be in the interior of the curve $\eta = \eta(\varphi)$ where*

$$\eta(\varphi) = 1 + \left\{\frac{1}{2}|\lambda^2 - \frac{1}{\lambda^2}| + \frac{1}{2}\left(\lambda^2 + \frac{1}{\lambda^2} - 2\right)\cos\varphi\right\}e^{i\varphi}$$

and $-\pi \leq \varphi \leq \pi$.

Note that for λ near one the curve $\eta = \eta(\varphi)$ is a simple closed curve contained in a small neighborhood of $\eta = 1$ whereas as $\lambda \to \infty$ only points on the negative real axis are excluded from being in the interior of $\eta = \eta(\varphi)$. Note also that the curve $\eta = \eta(\varphi)$ is independent of the domain D. Graphs of the curve $\eta = \eta(\varphi)$ for $\lambda = 1.2$ and $\lambda = 100$ can be found in [5]. In the two dimensional scalar case it was shown in [5] that the set of conductive eigenvalues is discrete and the only possible accumulation point is $\eta = 1$ (corresponding to $\gamma = \infty$). It is conjectured that this is also true in the three dimensional vector case.

References

1. Angell, T. S., Kirsch, A.: The conductive boundary condition for Maxwell's equations. To appear in SIAM J. Appl. Math.
2. Baum, C. E., Rothwell, E. J., Chen, K. M., Nyquist, D. P.: The singularity expansion method and its application to target identification. Proc. IEEE **70** (1991) 1481–1492.
3. Beale, J. T.: Purely imaginary scattering frequencies for exterior domains. Duke Math. J. **41** (1974) 607–637.
4. Colton, D., Kirsch, A.: The use of polarization effects in electromagnetic inverse scattering problems. Math. Methods Applied Science **15** (1992) 1–10.
5. Colton, D., Kirsch, A.: Target signatures for imperfectly conducting obstacles at fixed frequency. To appear.
6. Colton, D., Kress, R.: *Inverse Acoustic and Electromagnetic Scattering Theory*, Springer–Verlag, Berlin, 1992.
7. Colton, D., Monk, P.: The scattering of electromagnetic waves by a perfectly conducting infinite cylinder. Math. Methods Applied Science **12** (1990) 503–518.
8. Dolph C. L., Cho, S. K.: On the relationship between the singularity expansion method and the mathematical theory of scattering. IEEE Trans. Antennas Prop. **AP–28** (1980) 888–897.
9. Lax, P. D., Phillips, R. S.: Decaying modes for the wave equation in the exterior of an obstacle. Comm. Pure Applied Math **22** (1969) 737–787.
10. P.D. Lax, P. D., Phillips, R. S.: *Scattering Theory*, Academic Press, New York, 1989 (revised edition).

The Use of Graßmann Identities for Inversion of a General Model in Diffuse Tomography[*]

F. Alberto Grünbaum and S. K. Patch

Mathematics Department, University of California, Berkeley, California 94720, USA

1 Introduction

The idea of diffuse tomography has been recently introduced as a way of modeling an imaging problem using photons with very low energy ([1], [2]). It can be seen as a far reaching extension of the standard tomographic problem where photons are assumed to travel in a straight line. Although any real-life application will require the solution to the three-dimensional problem, we start with the two-dimensional problem.

The immediate mathematical consequence of this richer model is that in handling either the direct or (the more important) inverse problem one must solve NONLINEAR equations. In some instances these equations have been solved explicitly for relatively simple models ([1], [2], [3]).

Here we undertake a more ambitious program. We consider a VERY GENERAL (but very small) model. We observe that under minimal restrictions we are faced with a system of 48 nonlinear equations in 48 variables. Fortunately, the equations are linear in each variable separately. Given a permissible data set, (64 rational numbers), we are able solve the system for its unique physical solution. (There are spurious solutions which we reject.)

In this paper, we show how to use "Plücker", or more properly, Graßmann, relations to simplify the equations for a general data set. By a general data set we mean that the data is an 8×8 matrix Q, where for $i, j = 1, 2, ..8, Q[i, j]$ are variables, not numbers. Using Graßmann relations we are able to reduce the original system of 48 equations in 48 unknowns to a system of two equations in two unknowns. The last two equations are very cumbersome. So cumbersome that we run out of computing power when trying to solve them.

2 Description of the model

We begin our investigation with the smallest interesting case, namely a 2×2 object consisting of four pixels (Figure 1). On each of the eight outer edges there

---

[*] Research partially supported by AFOSR under Contract FDF-49620-92-J-0067-11792 and by the NSF under Grant DMS91-01224.

are two devices. One device can be used to shoot photons across the outside edge into the neighboring pixel; the other device detects photons as they leave the system. Photons change direction by turning an integral multiple of $\pi/2$. We assume that photons do not interact. Another property of this model is that a photon may die within a pixel.

Photons travelling according to the above rules are simply moving according to a 2-step Markov process. When a photon enters pixel i, j from a particular direction it either dies or continues its journey. The probabilities with which the photon moves forward, backwards, left or right are functions of its previous as well as present location. For our analysis we recast the problem as a one-step Markov process. In our new state space, the previous as well as present location of a photon define its state. There are 24 "living" states and one "dead" (i.e. absorbed) state. There are three classes of "living" states: incoming, outgoing, and hidden. These are indicated in Figure 2 by three sets of labels running from 1 to 8. There are $4 \times 4 = 16$ probabilities for each of the four pixels, yielding a total of 64 unknown probabilities.

Notation The probability that the photon will travel east into pixel $1, 1$ and continue east into pixel $1, 2$ is written as $e11e$. The probability that it will turn right and travel into pixel $2, 1$ is written as $e11s$.

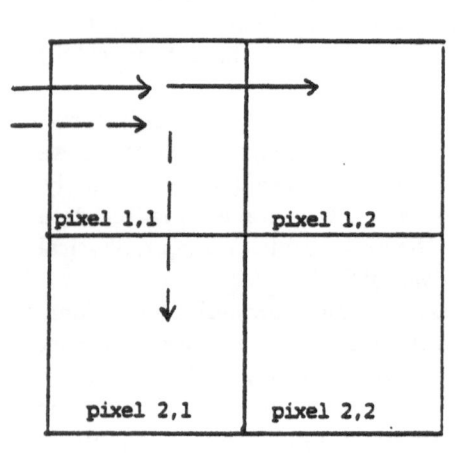

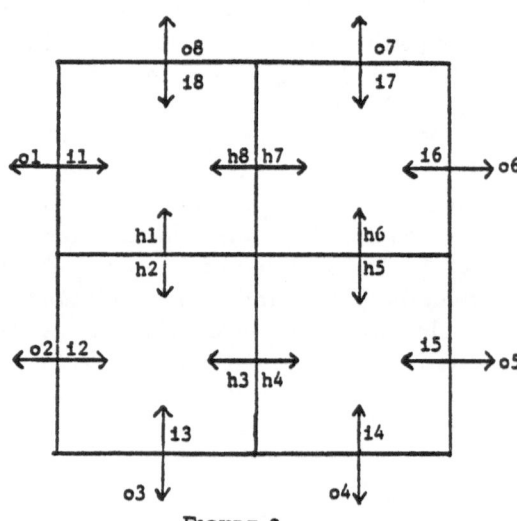

FIGURE 1

$e11e$ represented by solid arrows
$e11s$ represented by dashed arrows

FIGURE 2

Incoming, hidden, and outgoing states are represented by arrows labeled with i's, h's, and o's, respectively.

Now that we understand the objects we seek, let us turn our attention to the tools we have at our disposal. Recall the eight detectors positioned around the outer edges of the system. When a photon is shot into the system through an outer edge, the photon either dies somewhere inside the system or else it is detected as it leaves the system. By collecting data on many photons which enter through the same edge, we can calculate the probability that a photon entering the system through edge s will exit through edge t (here $s, t = 1, 2, \ldots, 8$). Hence we have $8 \times 8 = 64$ such pieces of data and may hope that there is an analytic solution to the problem.

Our transition probability matrix for the Markov chain breaks up naturally into several blocks: P_{IO}, P_{HH}, P_{HO} and P_{IH}. For example, $P_{IO}[s, t] =$ the probability of a photon moving from incoming state s directly to outgoing state t; $P_{IH}[s, t] =$ the probability of a photon moving from incoming state s directly to hidden state t. Note their sparse 2×2 block structure.

$$
P_{IO} = \begin{bmatrix}
e11w & 0 & 0 & 0 & 0 & 0 & 0 & e11n \\
0 & e21w & e21s & 0 & 0 & 0 & 0 & 0 \\
0 & n21w & n21s & 0 & 0 & 0 & 0 & 0 \\
0 & 0 & 0 & n22s & n22e & 0 & 0 & 0 \\
0 & 0 & 0 & w22s & w22e & 0 & 0 & 0 \\
0 & 0 & 0 & 0 & 0 & w12e & w12n & 0 \\
0 & 0 & 0 & 0 & 0 & s12e & s12n & 0 \\
s11w & 0 & 0 & 0 & 0 & 0 & 0 & s11n
\end{bmatrix}
$$

$$
P_{HH} = \begin{bmatrix}
0 & n11s & 0 & 0 & 0 & 0 & n11e & 0 \\
s21n & 0 & 0 & s21e & 0 & 0 & 0 & 0 \\
w21n & 0 & 0 & w21e & 0 & 0 & 0 & 0 \\
0 & 0 & e22w & 0 & 0 & e22n & 0 & 0 \\
0 & 0 & s22w & 0 & 0 & s22n & 0 & 0 \\
0 & 0 & 0 & 0 & n12s & 0 & 0 & n12w \\
0 & 0 & 0 & 0 & e12s & 0 & 0 & e12w \\
0 & w11s & 0 & 0 & 0 & 0 & w11e & 0
\end{bmatrix}
$$

$$
P_{HO} = \begin{bmatrix}
n11w & 0 & 0 & 0 & 0 & 0 & 0 & n11n \\
0 & s21w & s21s & 0 & 0 & 0 & 0 & 0 \\
0 & w21w & w21s & 0 & 0 & 0 & 0 & 0 \\
0 & 0 & 0 & e22s & e22e & 0 & 0 & 0 \\
0 & 0 & 0 & s22s & s22e & 0 & 0 & 0 \\
0 & 0 & 0 & 0 & 0 & n12e & n12n & 0 \\
0 & 0 & 0 & 0 & 0 & e12e & e12n & 0 \\
w11w & 0 & 0 & 0 & 0 & 0 & 0 & w11n
\end{bmatrix}
$$

$$
P_{IH} = \begin{bmatrix}
0 & e11s & 0 & 0 & 0 & 0 & e11e & 0 \\
e21n & 0 & 0 & e21e & 0 & 0 & 0 & 0 \\
n21n & 0 & 0 & n21e & 0 & 0 & 0 & 0 \\
0 & 0 & n22w & 0 & 0 & n22n & 0 & 0 \\
0 & 0 & w22w & 0 & 0 & w22n & 0 & 0 \\
0 & 0 & 0 & 0 & w12s & 0 & 0 & w12w \\
0 & 0 & 0 & 0 & s12s & 0 & 0 & s12w \\
0 & s11s & 0 & 0 & 0 & 0 & s11e & 0
\end{bmatrix}
$$

3 Solution to the direct problem

In terms of these matrices it is possible to express the relationship between the "object", i.e. the set of 64 parameters and the resulting set of 64 "data" by the relation

$$Q_{IO} = P_{IO} + P_{IH} \left(I - P_{HH} \right)^{-1} P_{HO} \tag{1}$$

which gives the input-output relation, i.e. the probability matrix of ever reaching an outgoing state (the site of a detector) from an incoming state (the site of a source) [1]. We define the direct map as the solution to the forward problem, calculating Q_{IO} from P_{IO}, P_{HH}, P_{HO}, and P_{IH}. If we were dealing with an $n \times n$ object we would have a total of 16 n^2 parameters describing the object and an equal number of data coming from the input-output relations. A natural question arises. Can one invert this (nonlinear) map?

4 Rank of the direct map

Before attempting to invert the direct map, we tried to determine the rank of the direct map. Using the symbol manipulators Macsyma and Maple, we computed the Jacobian of the direct map. Neither Macsyma nor Maple has been able to evaluate analytically the rank of this map as a function of the point in question. We have used both systems, however, to evaluate this Jacobian at many generic points and compute the rank of the resulting numerical matrices. The result has always been 48, which lead us to assume that this is indeed the rank at a generic point. At certain points the rank is lower.

5 A special model

Motivated by the observation that the rank of the direct map is 48, we looked for a model which has 48 independent parameters, distributed evenly among the four pixels. In order to reduce the number of independent parameters we made the following identifications:

$$eije = wijw, \; sijs = nijn, \; sijn = nijs, \; eijw = wije \text{ for all } i, j.$$

The first two constraints are fairly natural. They represent instances of the principle of "microscopic reversibility". Notice that the other two conditions are a bit less natural; they represent a certain type of "mirror symmetry". We have not made use of the concept of microscopic reversibility since that would lead to less than twelve free parameters per pixel and would reduce the rank of the direct map to 28. Finally notice that we have not made the assumption that the probability of being killed in a pixel is independent of the direction from which the photon entered the pixel.

Once the above identifications are made, the problem has the following features: the rank of the forward map remains 48, the main diagonals of P_{IO} and P_{HH} are common, and the off diagonals of P_{HO} and P_{IH} are common.

6 Rewrite the equations

After studying equation (1), one notices that by making several non-linear changes of variables, one may remove or "move" the non-linearities from (1) to the identifications. First define, (assuming that P_{HO} is invertible),

$$A = P_{HO}^{-1}$$

Equation (1) may be re-written as

$$(Q_{IO} - P_{IO})A(I - P_{HH}) - P_{IH} = 0 \tag{1a}$$

A few more changes of variables are required to make (1a) linear:

$$W = -AP_{HH} \tag{2a}$$

$$X = P_{IO}A \tag{2b}$$

$$Y = P_{IO}W + P_{IH}. \tag{2c}$$

Assuming that A is invertible we can recover, P_{HH}, P_{IO}, and P_{IH} in terms of A if we know W, X, and Y. Under these substitutions, the matrix equation (1) becomes

$$0 = Q_{IO}(A + W) - (X + Y). \tag{3}$$

Recall that Q_{IO} is the data, so the equation is linear in the unknown matrices, A, W, X,y and Y. $(A + W)$ and $(X + Y)$ are shown below. (Note their zero structures.)

$$A + W = \begin{bmatrix} a[1,1] & w[1,2] & 0 & 0 & 0 & 0 & w[1,7] & a[1,8] \\ w[2,1] & a[2,2] & a[2,3] & w[2,4] & 0 & 0 & 0 & 0 \\ w[3,1] & a[3,2] & a[3,3] & w[3,4] & 0 & 0 & 0 & 0 \\ 0 & 0 & w[4,3] & a[4,4] & a[4,5] & w[4,6] & 0 & 0 \\ 0 & 0 & w[5,3] & a[5,4] & a[5,5] & w[5,6] & 0 & 0 \\ 0 & 0 & 0 & 0 & w[6,5] & a[6,6] & a[6,7] & w[6,8] \\ 0 & 0 & 0 & 0 & w[7,5] & a[7,6] & a[7,7] & w[7,8] \\ a[8,1] & w[8,2] & 0 & 0 & 0 & 0 & w[8,7] & a[8,8] \end{bmatrix} \tag{4}$$

$$X + Y = \begin{bmatrix} x[1,1] & y[1,2] & 0 & 0 & 0 & 0 & y[1,7] & x[1,8] \\ y[2,1] & x[2,2] & x[2,3] & y[2,4] & 0 & 0 & 0 & 0 \\ y[3,1] & x[3,2] & x[3,3] & y[3,4] & 0 & 0 & 0 & 0 \\ 0 & 0 & y[4,3] & x[4,4] & x[4,5] & y[4,6] & 0 & 0 \\ 0 & 0 & y[5,3] & x[5,4] & x[5,5] & y[5,6] & 0 & 0 \\ 0 & 0 & 0 & 0 & y[6,5] & x[6,6] & x[6,7] & y[6,8] \\ 0 & 0 & 0 & 0 & y[7,5] & x[7,6] & x[7,7] & y[7,8] \\ x[8,1] & y[8,2] & 0 & 0 & 0 & 0 & y[8,7] & x[8,8] \end{bmatrix} \tag{5}$$

Now the equations in column one of (3) are linear in the variables,

$$\{a[1,1], a[8,1], w[2,1], w[3,1], x[1,1], x[8,1], y[2,1], y[3,1]\}.$$

These eight equations can be written as a homogeneous matrix equation:

$$0 = \begin{bmatrix} Q[4,1] & Q[4,8] & Q[4,2] & Q[4,3] & 0 & 0 & 0 & 0 \\ Q[5,1] & Q[5,8] & Q[5,2] & Q[5,3] & 0 & 0 & 0 & 0 \\ Q[6,1] & Q[6,8] & Q[6,2] & Q[6,3] & 0 & 0 & 0 & 0 \\ Q[7,1] & Q[7,8] & Q[7,2] & Q[7,3] & 0 & 0 & 0 & 0 \\ Q[2,1] & Q[2,8] & Q[2,2] & Q[2,3] & -1 & 0 & 0 & 0 \\ Q[3,1] & Q[3,8] & Q[3,2] & Q[3,3] & 0 & -1 & 0 & 0 \\ Q[1,1] & Q[1,8] & Q[1,2] & Q[1,3] & 0 & 0 & -1 & 0 \\ Q[8,1] & Q[8,8] & Q[8,2] & Q[8,3] & 0 & 0 & 0 & -1 \end{bmatrix} \begin{bmatrix} a[1,1] \\ a[8,1] \\ w[2,1] \\ w[3,1] \\ y[2,1] \\ y[3,1] \\ x[1,1] \\ x[8,1] \end{bmatrix} \tag{6}$$

One can do the same for the equations in the other columns of (3). To each column of (3) there corresponds a system of eight linear equations in the variables which appear in the corresponding columns of (4) and (5). Note that as far as their zero structures are concerned, the columns of (4) and (5) come in pairs. The roles of the $a[i,j]'s$ and $w[i,j]'s$ are reversed in the first two columns of (4) as are the roles of the $x[i,j]'s$ and $y[i,j]'s$ in the first two columns of (5). Hence, one may re-order the variables to get the "same" matrix equation for the first and second columns. Similarly, one can represent the linear systems corresponding to the third and fourth columns of (1) with a single matrix equation; the fifth and sixth columns with a third matrix equation; and the seventh and eighth columns with a fourth matrix equation. One is left with four sets of homogeneous linear equations. One also notes that the upper left 4×4 matrix of the larger 8×8 matrices in (6) has rank 2. This gives rise to many consistency conditions amongst the data. Lack of space prohibits a detailed discussion of consistency conditions. Many of these consistency conditions were discussed in [5].

As one can see, the variables for each system of equations contains two each of the $a[i,j]'s$, $w[i,j]'s$, $x[i,j]'s$, and $y[i,j]'s$. One must not forget, however, that the $w[i,j]'s$, $x[i,j]'s$, and $y[i,j]'s$ are functions of $a[i,j]'s$ which correspond to other columns. Although the variables differ from column to column (exactly 64 variables total—no repeats between columns), the equations are only artificially decoupled.

Recall that only six equations per column are independent. Since the $w[i,j]'s$, $x[i,j]'s$, and $y[i,j]'s$ are already functions of $a[i,j]'s$, it seems natural to solve for them in terms of the $a[i,j]'s$. One can follow this procedure for all eight columns, reducing the number of variables from 64 to 16.

To solve (6) for the $w[i,j]'s$, $x[i,j]'s$, and $y[i,j]'s$ in terms of the $a[i,j]'s$, one solves the system:

$$\begin{bmatrix} Q[6,2] & Q[6,3] & 0 & 0 & 0 & 0 \\ Q[7,2] & Q[7,3] & 0 & 0 & 0 & 0 \\ Q[2,2] & Q[2,3] & -1 & 0 & 0 & 0 \\ Q[3,2] & Q[3,3] & 0 & -1 & 0 & 0 \\ Q[1,2] & Q[1,3] & 0 & 0 & -1 & 0 \\ Q[8,2] & Q[8,3] & 0 & 0 & 0 & -1 \end{bmatrix} \begin{bmatrix} w[2,1] \\ w[3,1] \\ y[2,1] \\ y[3,1] \\ x[1,1] \\ x[8,1] \end{bmatrix} = \begin{bmatrix} Q[6,1] & Q[6,8] \\ Q[7,1] & Q[7,8] \\ Q[2,1] & Q[2,8] \\ Q[3,1] & Q[3,8] \\ Q[1,1] & Q[1,8] \\ Q[8,1] & Q[8,8] \end{bmatrix} \begin{bmatrix} a[1,1] \\ a[8,1] \end{bmatrix} \tag{7}$$

For the sake of simplicity, the "wa" equations which did not have coefficients from rows 2, 3, 6, or 7 of the data matrix Q were omitted. Let $d[[a,b],[c,d]]$

denote the determinant of the minor taken from the rows a and b and columns c and d of the data matrix. Then the determinant of the left-hand matrix in (7) is $d[[6,7],[2,3]]$. Hence, (7) can be uniquely solved if and only if $d[[6,7],[2,3]] \neq 0$. One finds the same sort of requirement for each of the other columns of (5). The determinants which must be non-zero are listed below:

$$d[[6,7],[2,3]], \; d[[6,7],[1,8]], \; d[[6,7],[4,5]],$$
$$d[[2,3],[6,7]], \; d[[2,3],[1,8]], \; d[[2,3],[4,5]].$$

If the data satisfy these mild requirements then one can solve the 48 independent equations in 64 variables linearly in terms of the 16 variables in $A = P_{HO}^{-1}$. Unfortunately, that exhausts the supply of equations for the original model. But if one uses the 16 identifications described above for the 48 parameter model, one may solve the problem, (at a generic point). We discuss our solution in sections 8, 9, and 10.

7 Graßmannians and the Plücker embedding

Since equation (7) is a linear system of six equations in eight unknowns, one is led to study Graßmannians and the Plücker embedding. They will be used later to simplify the equations in (3).

7.1 Graßmannians

Given integers k and n, where $k < n$, $G(k,n)$ is defined as the set of all k-dimensional linear spaces in $\mathbb{C}^n$. Let Λ be an element of $G(k,n)$. Then there exists a set of k $1 \times n$ spanning vectors of Λ. Represent Λ as a $k \times n$ matrix whose rows are these spanning vectors. Because the choice of spanning vectors is not unique, a family of matrices represent Λ. Given any g in $GL(k)$, define $\Lambda' = g\Lambda$. The rows of Λ' span the same space as the rows of Λ, so we identify Λ' and Λ.

Under these identifications, it is easy to construct a bijection between a dense, open subset of $G(k,n)$ and $\mathbb{C}^{k(n-k)}$. Let O be the set of all points in $G(k,n)$ which may be represented by a $k \times n$ matrix whose first k columns are independent. O is a dense open set in $G(k,n)$. Given any representation for Λ in O, one can easily find the representation $\tilde{\Lambda}$ such that the first k columns of $\tilde{\Lambda}$ are the identity matrix. (Simply take g^{-1} to be the first k columns of Λ. The rows of Λ are independent so g^{-1} is invertible. Define $\tilde{\Lambda} = g\Lambda$.) Only the entries of the $k \times (n-k)_{th}$ submatrix of $\tilde{\Lambda}$ are unconstrained.

More generally, let $I = (i_1, i_2, i_3, \ldots, i_k)$ index k independent columns of the original representation for Λ. Then given any Λ, one can define the map ϕ_I such that $\phi_I(\Lambda)$ satisfies the following: column ij of $\phi_I(\Lambda) = e_j$, where ij is the j_{th} index in I and e_j is the j_{th} canonical vector. In order to define ϕ_I, first set g^{-1} equal to the matrix with columns $i_1, i_2, i_3, \ldots, i_k$ of Λ. Once again, g^{-1} is of rank k so g exists. Then $\tilde{\Lambda} = g\Lambda$ is has column ij equal to e_j. Clearly, the inverse of ϕ_I is always uniquely defined.

Let I' denote another set of linearly independent columns. Define $\Lambda_I = \phi_I(\Lambda)$ and $\Lambda_{I'} = \phi_{I'}(\Lambda)$. Then Λ_I and $\Lambda_{I'}$ satisfy $\Lambda_I = g\Lambda$ and $\Lambda_{I'} = g'\Lambda$ for some g, g' in $GL(k)$. Then $\Lambda_{I'} = h\Lambda_I$ for some matrix h in $GL(k)$. A moment's reflection reveals that h must be the inverse of the matrix composed of the columns of Λ_I which are indexed by I'. Note that the entries of h are analytic functions of the entries of Λ_I, so $G(k, n)$ has the structure of a complex manifold.

7.2 The Plücker Embedding

Any Graßmannian, $G(k, n)$, may be embedded into $\mathbb{P}^{\binom{n}{k}-1}$. $\mathbb{P}^N$ is N dimensional projective space over the complex numbers; one can think of $\mathbb{P}^N$ as an N dimensional sphere lying in $N + 1$ dimensional space with antipodal points identified. A point, P, in $\mathbb{P}^N$ may denoted by $(p_0, p_1, p_2, \ldots, p_N)$. This point is identified with all other points $\alpha \times (p_0, p_1, p_2, \ldots, p_N)$ for any non-zero scalar α.

In the simplest case, $G(1, n)$, any point of the Graßmannian is represented by a single row vector. Since one may multiply each element of the Graßmannian by a non-zero scalar, one may canonically identify $\mathbb{P}^{n-1}$ and $G(1, n)$. Every element, Λ, of a $G(k, n)$ defines a k dimensional linear space in $\mathbb{C}^n$. The dual space corresponding to Λ is $\Lambda^\perp$, the $(n - k)$ dimensional linear space in $\mathbb{C}^n$ orthogonal to Λ. There is a $1 - 1$ correspondence between the Plücker coordinates of Λ and $\Lambda^\perp$. Since the set of $(n - 1)$ dimensional spaces in $\mathbb{C}^n$ is isomorphic to the space of one dimensional spaces, one can identify any Λ in $G(n - 1, n)$ with its dual, $\Lambda^\perp$. Hence $G(n - 1, n)$ and $\mathbb{P}^{n-1}$ are idenified.

The case k such that $1 < k < (n - 1)$ requires more complicated relations to embed $G(k, n)$ in some projective space. The Plücker coordinates of a Graßmannian Λ are simply the determinants of all $k \times k$ minors of any representation Λ of an element in $G(k, n)$. It is easy to check that Plücker coordinates are projectively invariant under representation of Λ. Let Λ and Λ' be equivalent representations for the same element of $G(k, n)$. Then $\Lambda = g\Lambda'$ for some g in $GL(k)$. Let I be any index of k columns. The minor of the I columns of Λ equals g times the minor of the I columns of Λ'. By the rules of determinants, $\det(AB) = \det(A)\det(B)$, and so the determinant of the I_{th} minor of Λ equals the determinant of g times the determinant of the I_{th} minor of Λ'. This holds for all I, and so if the $\binom{n}{k}$-tuple P are the Plücker coordinates of Λ then $\det(g)P$ are the Plücker coordinates of Λ'. Since there are $\binom{n}{k}$ minors of a $k \times n$ matrix, the Plücker map takes $G(k, n)$ into $\mathbb{P}^{\left(\binom{n}{k}-1\right)}$. (Note that the Plücker map is not onto.)

One can check that for $1 < k < n$, $\left(\binom{n}{k} - 1\right)$ is larger than or equal to $k(n - k)$. In order for the Plücker map to be an embedding, there must be some (i.e., $\left(\binom{n}{k} - 1\right) - k(n-k)$ independent) relations amongst the Plücker coordinates of a point Λ in $G(k,n)$. For $G(2,n)$ these are the Plücker relations. For general $G(k,n)$ they are called Graßmann relations. In either case the relations are quadratic in the Plücker coordinates for Λ. The Graßmann relations are easily derived.

7.3 Derivation of the Graßmann Relations

Let Λ be a matrix representation of an element of $G(k,n)$ where $\Lambda = (a)_{ij}$. Let $I = (i_1, i_2, i_3, \ldots, i_{(k-1)})$ index $(k-1)$ distinct columns of Λ. Let $J = (j_1, j_2, j_3, \ldots, j_{(k+1)})$ index $(k+1)$ distinct columns of Λ. Consider the sum,

$$\sum_{\lambda=1}^{k+1}(-1)^{\lambda+1}\begin{vmatrix} a_{1,i_1} & a_{1,i_2} & \cdots & a_{1,i_{k-1}} & a_{1,j_\lambda} \\ \vdots & & & \vdots & \vdots \\ a_{k,i_1} & \cdots & & \cdots & a_{k,i_{k-1}} & a_{k,j_\lambda} \end{vmatrix}\begin{vmatrix} a_{1,j_1} & \cdots & a_{1,j_{\lambda-1}} & a_{1,j_{\lambda+1}} & \cdots & a_{1,j_{k+1}} \\ \vdots & & \vdots & \vdots & & \vdots \\ a_{k,j_1} & \cdots & a_{k,j_{\lambda-1}} & a_{k,j_{\lambda+1}} & \cdots & a_{k,j_{k+1}} \end{vmatrix} \tag{8}$$

For each λ, expand the first determinant along the last column, i.e. use

$$\begin{vmatrix} a_{1,i_1} & a_{1,i_2} & \cdots & a_{1,i_{k-1}} & a_{1,j_\lambda} \\ \vdots & & & & \vdots \\ a_{k,i_1} & & & a_{k,i_{k-1}} & a_{k,j_\lambda} \end{vmatrix} = \sum_{\mu=1}^{k} a_{\mu,j_\lambda} CF_\mu \tag{9}$$

where CF_μ is the cofactor of the matrix on the left-hand side of (9) about the $(\mu, k)_{th}$ entry. Then (8) becomes

$$\sum_{\lambda=1}^{k+1}(-1)^{\lambda+1}\sum_{\mu=1}^{k} a_{\mu,j_\lambda} CF_\mu \begin{vmatrix} a_{1,j_1} & \cdots & a_{1,j_{\lambda-1}} & a_{1,j_{\lambda+1}} & \cdots & a_{1,j_{k+1}} \\ \vdots & & \vdots & \vdots & & \vdots \\ a_{k,j_1} & \cdots & a_{k,j_{\lambda-1}} & a_{k,j_{\lambda+1}} & \cdots & a_{k,j_{k+1}} \end{vmatrix}$$

$$= \sum_{\mu=1}^{k} CF_\mu \sum_{\lambda=1}^{k+1}(-1)^{\lambda+1}\begin{vmatrix} 0 & \cdots & 0 & a_{\mu,j_\lambda} & 0 & \cdots & 0 \\ a_{1,j_1} & \cdots & a_{1,j_{\lambda-1}} & a_{1,j_\lambda} & a_{1,j_{\lambda+1}} & \cdots & a_{1,j_{k+1}} \\ \vdots & & \vdots & \vdots & & & \vdots \\ a_{k,j_1} & \cdots & a_{k,j_{\lambda-1}} & a_{k,j_\lambda} & \cdots & & a_{k,j_{k+1}} \end{vmatrix} \tag{10}$$

$$= \sum_{\mu=1}^{k} CF_\mu \begin{vmatrix} a_{\mu,j_1} & a_{\mu,2} & \cdots & a_{\mu,j_{k+1}} \\ a_{1,j_1} & a_{1,j_2} & \cdots & a_{1,j_{k+1}} \\ \vdots & \vdots & & \vdots \\ a_{k,j_1} & a_{k,j_2} & \cdots & a_{k,j_{k+1}} \end{vmatrix} = \sum_{\mu=1}^{k} CF_\mu \cdot 0 = 0.$$

Denote by π_I the determinant of the minor whose columns are indexed by the multi-index I. Then

$$\sum_{\lambda=1}^{k+1} \pi_{(i_1,i_2,\ldots,i_{k-1},j_\lambda)} \pi_{(j_1,j_2,\ldots,j_{\lambda-1},j_{\lambda+1},\ldots,j_{k+1})} = 0. \tag{11}$$

Equation (11) defines the Graßmann relations.

Note that some of these Plücker relations are trivial. For there to be a non-trivial Plücker relation, we require that at least four of the "j_s" be distinct from the "i_s". (In fact, there may only be an even number of non-repeated indices.) Suppose for any $k < (n-1)$ and Λ in $G(k, n)$, there are exactly four indices, l_1, l_2, l_3, and l_4, which are not repeated. Suppose further that $I = (i_1, i_2, \ldots, i_{(k-2)}, i_{l_1})$ and $J = (j_1, j_2, \ldots, j_{(k-2)}, j_{l_2}, j_{l_3}, j_{l_4})$. One may analyse the Plücker relations for I and J as though one were looking at $G(2, 4)$ and $I = (l_1)$, $J = (l_2, l_3, l_4)$. The other indices don't "matter" because they contribute only zero terms to equation (11). For $G(2, 4)$ there is only one non-trivial Graßmann relation:

$$\pi_{l_1 l_2} \pi_{l_3 l_4} - \pi_{l_1 l_3} \pi_{l_2 l_4} + \pi_{l_1 l_4} \pi_{l_2 l_3} = 0 \tag{12}$$

Setting $I = (l_2)$ and $J = (l_1, l_2, l_3)$ one may reorder and reverse the signs of the terms in (12), but the results are equivalent. This makes sense, since $G(2, 4)$ is isomorphic to a dense, open subset of $\mathbb{C}^4$, and the Plücker map takes $G(2, 4)$ into $\mathbb{P}^5$. If one considers $G(2, 5)$, however, one finds a slightly more complicated situation. $G(2, 5)$ is isomorphic to a dense subset of $\mathbb{C}^6$, and there are $\binom{5}{4} = 5$ non-trivial Plücker relations. But the Plücker map takes $G(2, 5)$ into $\mathbb{P}^9$, so there must be only three independent Plücker relations amongst the five Plücker relations:

$$\pi_{l_1 l_2} \pi_{l_3 l_4} - \pi_{l_1 l_3} \pi_{l_2 l_4} + \pi_{l_1 l_4} \pi_{l_2 l_3} = 0 \tag{13a}$$

$$\pi_{l_1 l_2} \pi_{l_3 l_5} - \pi_{l_1 l_3} \pi_{l_2 l_5} + \pi_{l_1 l_5} \pi_{l_2 l_3} = 0 \tag{13b}$$

$$\pi_{l_1 l_2} \pi_{l_5 l_4} - \pi_{l_1 l_5} \pi_{l_2 l_4} + \pi_{l_1 l_4} \pi_{l_2 l_5} = 0 \tag{13c}$$

$$\pi_{l_1 l_5} \pi_{l_3 l_4} - \pi_{l_1 l_3} \pi_{l_5 l_4} + \pi_{l_1 l_4} \pi_{l_5 l_3} = 0 \tag{13d}$$

$$\pi_{l_5 l_2} \pi_{l_3 l_4} - \pi_{l_5 l_3} \pi_{l_2 l_4} + \pi_{l_5 l_4} \pi_{l_2 l_3} = 0. \tag{13e}$$

Similarly, if there are six non-repeated indices, then one can calculate the number of non-trivial Plücker relations as though one were working in $G(3, 6)$. In $G(3, 6)$, I and J have two and four components, respectively. So there are $\binom{6}{4} = 15$ ways to pick I and J. Hence there are 15 non-trivial Plücker relations. Note, however, that $G(3, 6)$ is isomorphic to a dense subset of $\mathbb{C}^{3(6-3)} = \mathbb{C}^9$ and that the Plücker map takes $G(3, 6)$ into $\mathbb{P}^{\binom{6}{3}-1} = \mathbb{P}^{19}$. Hence, among the 15 Plücker relations, only ten are independent. The Plücker relations will be used later to simplify the last 16 equations which were calculated in section 6.

8 Back to tomography

If one solves for all of the entries in P_{HH}, P_{IO}, P_{HO}, and P_{IH} in terms the $a[i, j]$'s, one may then substitute these solutions into the 16 identifications defined in section 5. Half of the resulting 16 equations are quadratic in the $a[i, j]$'s, the other half cubic in the $a[i,j]'s$. These "cubic and quadratic relations" are highly structured. The cubic equations have 13 terms each, whereas the quadratic equations have only seven.

8.1 Quadratics

Each of these quadratics is a polynomial in six of the $a[i,j]'s$. Furthermore, the $a[i,j]'s$ appear in a cyclic pattern. Recall that A has a 2×2 block structure. Four of the six variables come from one block of A; the other two variables come from one of A's adjacent blocks. The equation is linear in each of these variables separately. One of the quadratics is shown below:

$$
\begin{aligned}
&- d[[6,7],[2,3]]d[[3,6,7],[1,3,8]]d[[6,7],[4,5]]a[2,3]a[3,2] \\
&\quad - d[[6,7],[2,3]]d[[3,6,7],[3,4,5]]d[[6,7],[1,8]]a[2,2]a[3,3] \\
&\quad + d[[6,7],[1,8]]d[[6,7],[4,5]]d[[6,7],[1,3]]a[1,1]a[3,3] \\
&\quad - d[[6,7],[1,8]]d[[6,7],[4,5]]d[[6,7],[3,8]]a[8,1]a[3,3] \\
&\quad + d[[6,7],[1,8]]d[[6,7],[4,5]]d[[6,7],[1,2]]a[2,3]a[1,1] \\
&\quad - d[[6,7],[1,8]]d[[6,7],[4,5]]d[[6,7],[2,8]]a[2,3]a[8,1] \\
&\quad - d[[6,7],[2,3]](d[[3,6,7],[1,2,8]]d[[6,7],[4,5]] \\
&\quad + d[[3,6,7],[2,4,5]]d[[6,7],[1,8]])a[2,3]a[2,2]
\end{aligned}
\tag{14}
$$

(Hereafter this equation will be referred to as "quad".)

8.2 Cubics

As with the quadratics, each of the cubic equations is a polynomial in six of the $a[i,j]'s$. To each of the quadratics there exists a corresponding cubic equation in the same six variables. Each cubic is a sum of one linear and 12 cubic terms. The linear term is a multiple of one of the eight off diagonal entries in A. The block of four of the six variables contains the variable which appears linearly. As with the quadratics each cubic is linear in each of its variables. The following equation is the cubic corresponding to "quad":

$$- a[2,3]\%6^2\%2\%1 + a[3,3]\%7a[8,1]^2\%4\%2\%1$$

$$+ a[3,3]\%5a[1,1]^2\%3\%2\%1$$

$$- \%6\%1(d[[3,6,7],[1,2,8]]\%7 + d[[3,6,7],[2,3,8]]\%2)a[8,1]a[3,3]a[2,2]$$

$$+ \%6\%2(d[[3,6,7],[2,3,8]]\%1 - d[[3,6,7],[2,4,5]]\%7)a[3,2]a[2,3]a[8,1]$$

$$+ \%6\%2(d[[3,6,7],[1,2,3]]\%1 + d[[3,6,7],[2,4,5]]\%5)a[1,1]a[3,2]a[2,3]$$

$$- \%2\%1(\%5\%4 + \%3\%7)a[1,1]a[8,1]a[3,3]$$

$$+ \%6\%5(d[[3,6,7],[1,3,8]]\%1 + d[[3,6,7],[3,4,5]]\%2)a[3,3]a[1,1]a[3,2]$$

$$- \%6\%7(d[[3,6,7],[1,3,8]]\%1 + d[[3,6,7],[3,4,5]]\%2)a[3,3]a[8,1]a[3,2]$$

$$+ a[2,3]\%4^2a[8,1]^2\%2\%1 \tag{15}$$

$$- \%6\%1(d[[3,6,7],[1,2,3]]\%2 - d[[3,6,7],[1,2,8]]\%5)a[1,1]a[3,3]a[2,2]$$

$$- 2a[2,3]\%3a[1,1]\%4a[8,1]\%2\%1 + a[2,3]\%3^2a[1,1]^2\%2\%1$$

$\%1 :=$	$d[[6,7],[4,5]]$
$\%2 :=$	$d[[6,7],[1,8]]$
$\%3 :=$	$d[[6,7],[1,2]]$
$\%4 :=$	$d[[6,7],[2,8]]$
$\%5 :=$	$d[[6,7],[1,3]]$
$\%6 :=$	$d[[6,7],[2,3]]$
$\%7 :=$	$d[[6,7],[3,8]]$

(Hereafter this equation will be referred to as "lcub". This notation is a bit confusing. The percent signs are defined at the bottom of "lcub" and should be substituted into "lcub".)

9 Simplifying the equations

Adding to a given cubic a multiple of the corresponding quadratic yields an equivalent cubic with only nine terms. For example, multiplying "quad" by

$$(d[[6,7],[2,8]]a[8,1] - d[[6,7],[1,2]]a[1,1])$$

and adding "lcub" yields "scub":

$$
\begin{aligned}
&- (\%3d[[3,6,7],[1,2,3]]\%1 + \%1\%5d[[6,7],[1,3]] \\
&+ d[[6,7],[1,2]]\%4\%3)a[3,2]a[2,3]a[1,1] - (\%3d[[3,6,7],[2,3,8]])\%1 \\
&- \%1\%5d[[6,7],[3,8]] - d[[6,7],[2,8]]\%4\%3)a[3,2]a[2,3]a[8,1] \\
&+ (\%3d[[3,6,7],[1,2,3]]\%1 - \%3\%6d[[6,7],[1,3]] \\
&- d[[6,7],[1,2]]\%2\%1)a[3,3]a[2,2]a[1,1] + (\%3d[[3,6,7],[2,3,8]]\%1 \\
&+ \%3\%6d[[6,7],[3,8]] + d[[6,7],[2,8]]\%2\%1)a[8,1]a[3,3]a[2,2] \\
&- d[[6,7],[1,2]](\%6\%3 + \%5\%1)a[2,2]a[2,3]a[1,1] \\
&+ d[[6,7],[2,8]](\%6\%3 + \%5\%1)a[8,1]a[2,2]a[2,3] \\
&+ d[[6,7],[2,3]]a[2,3]\%1\%3 \\
&- d[[6,7],[1,3]](\%4\%3 + \%2\%1)a[3,2]a[1,1]a[3,3] \\
&+ d[[6,7],[3,8]](\%4\%3 + \%2\%1)a[3,2]a[8,1]a[3,3]
\end{aligned}
\tag{16}
$$

$\%1 :=$	$d[[6,7],[1,8]]$
$\%2 :=$	$d[[3,6,7],[3,4,5]]$
$\%3 :=$	$d[[6,7],[4,5]]$
$\%4 :=$	$d[[3,6,7],[1,3,8]]$
$\%5 :=$	$d[[3,6,7],[2,4,5]]$
$\%6 :=$	$d[[3,6,7],[1,2,8]]$

9.1 Simplifying cubics by adding Graßmann identities

In the shortened cubic above, one notices clusters of determinants of minors in the data. These clusters are either cubic or quadratic in the determinants. The cubic clusters are listed below:

$$
\begin{aligned}
&d[[6,7],[4,5]]d[[3,6,7],[1,2,3]]d[[6,7],[1,8]] \\
&+d[[6,7],[1,8]]d[[3,6,7],[2,4,5]]d[[6,7],[1,3]] \\
&+d[[6,7],[1,2]]d[[3,6,7],[1,3,8]]d[[6,7],[4,5]] \\
&= \quad d[[6,7],[4,5]](d[[3,6,7],[1,2,3]]d[[6,7],[1,8]] \\
&\qquad + d[[3,6,7],[1,3,8]]d[[6,7],[1,2]]) \\
&\qquad + d[[6,7],[1,8]]d[[3,6,7],[2,4,5]]d[[6,7],[1,3]]
\end{aligned}
\tag{17a}
$$

$$
\begin{aligned}
&d[[6,7],[4,5]]d[[3,6,7],[1,2,3]]d[[6,7],[1,8]] \\
&-d[[6,7],[4,5]]d[[3,6,7],[1,2,8]]d[[6,7],[1,3]] \\
&-d[[6,7],[1,2]]d[[3,6,7],[3,4,5]]d[[6,7],[1,8]] \\
&= \quad d[[6,7],[4,5]](d[[3,6,7],[1,2,3]]d[[6,7],[1,8]] \\
&\qquad - d[[3,6,7],[1,2,8]]d[[6,7],[1,3]]) \\
&\qquad - d[[6,7],[1,2]]d[[3,6,7],[3,4,5]]d[[6,7],[1,8]]
\end{aligned}
\tag{17b}
$$

$$d[[6,7],[4,5]]d[[3,6,7],[2,3,8]]d[[6,7],[1,8]]$$
$$-d[[6,7],[2,8]]d[[3,6,7],[1,3,8]]d[[6,7],[4,5]]$$
$$-d[[6,7],[1,8]]d[[3,6,7],[2,4,5]]d[[6,7],[3,8]]$$
$$\begin{aligned} = \quad & d[[6,7],[4,5]](d[[3,6,7],[2,3,8]]d[[6,7],[1,8]] \\ & - d[[3,6,7],[1,3,8]]d[[6,7],[2,8]]) \\ & - d[[6,7],[1,8]]d[[3,6,7],[2,4,5]]d[[6,7],[3,8]] \end{aligned} \tag{17c}$$

$$d[[6,7],[4,5]]d[[3,6,7],[2,3,8]]d[[6,7],[1,8]]$$
$$+d[[6,7],[4,5]]d[[3,6,7],[1,2,8]]d[[6,7],[3,8]]$$
$$+d[[6,7],[2,8]]d[[3,6,7],[3,4,5]]d[[6,7],[1,8]]$$
$$\begin{aligned} = \quad & d[[6,7],[4,5]](d[[3,6,7],[2,3,8]]d[[6,7],[1,8]] \\ & + d[[3,6,7],[1,2,8]]d[[6,7],[3,8]]) \\ & + d[[6,7],[2,8]]d[[3,6,7],[3,4,5]]d[[6,7],[1,8]] \end{aligned} \tag{17d}$$

The quadratic clusters are listed below:

$$\begin{aligned} & d[[3,6,7],[1,2,8]]d[[6,7],[4,5]] \\ & \quad + d[[3,6,7],[2,4,5]]d[[6,7],[1,8]] \end{aligned} \tag{18a}$$

$$\begin{aligned} & d[[3,6,7],[1,3,8]]d[[6,7],[4,5]] \\ & \quad + d[[3,6,7],[3,4,5]]d[[6,7],[1,8]] \end{aligned} \tag{18b}$$

Consider the element of the Graßmannian $G(3,7)$

$$\begin{bmatrix} Q[3,1] & Q[3,2] & Q[3,3] & Q[3,4] & Q[3,5] & Q[3,8] & 1 \\ Q[6,1] & Q[6,2] & Q[6,3] & Q[6,4] & Q[6,5] & Q[6,8] & 0 \\ Q[7,1] & Q[7,2] & Q[7,3] & Q[7,4] & Q[7,5] & Q[7,8] & 0 \end{bmatrix} \tag{19}$$

and recall the discussion of Graßmannians and the Plücker Embedding. Note that the Graßmann relations,

$$0 = \pi_{123}\pi_{167} - \pi_{126}\pi_{137} + \pi_{127}\pi_{136} \tag{20a}$$
$$0 = \pi_{136}\pi_{276} - \pi_{236}\pi_{176} + \pi_{736}\pi_{126} \tag{20b}$$

translate to

$$\begin{aligned} 0 = \; & d[[3,6,7],[1,2,3]]d[[6,7],[1,8]] - d[[3,6,7],[1,2,8]]d[[6,7],[1,3]] \\ & + d[[3,6,7],[1,3,8]]d[[6,7],[1,2]] \end{aligned} \tag{21a}$$

$$\begin{aligned} 0 = \; & -d[[3,6,7],[1,3,8]]d[[6,7],[2,8]] + d[[3,6,7],[2,3,8]]d[[6,7],[1,8]] \\ & + d[[3,6,7],[1,2,8]]d[[6,7],[3,8]] \end{aligned} \tag{21b}$$

respectively, when applied to the Graßmannian in (19). (21a) and (21b) can be used to simplify the cubic clusters above. The cubic clusters in equations (17a), (17b), (17c), (17d) become

$$d[[6,7],[1,3]](d[[3,6,7],[1,2,8]]d[[6,7],[4,5]]$$
$$+ d[[3,6,7],[2,4,5]]d[[6,7],[1,8]]), \qquad (22a)$$

$$-d[[6,7],[1,2]](d[[3,6,7],[1,3,8]]d[[6,7],[4,5]]$$
$$+ d[[3,6,7],[3,4,5]]d[[6,7],[1,8]]), \qquad (22b)$$

$$-d[[6,7],[3,8]](d[[3,6,7],[1,2,8]]d[[6,7],[4,5]]$$
$$+ d[[3,6,7],[2,4,5]]d[[6,7],[1,8]]), \qquad (22c)$$

$$d[[6,7],[2,8]](d[[3,6,7],[1,3,8]]d[[6,7],[4,5]]$$
$$+ d[[3,6,7],[3,4,5]]d[[6,7],[1,8]]), \qquad (22d)$$

respectively. Once these substitutions have been made "scub" becomes

$$d[[6,7],[2,3]]a[2,3]d[[6,7],[1,8]]d[[6,7],[4,5]]$$
$$+ d[[6,7],[2,8]]\%4a[8,1]a[2,2]a[2,3]$$
$$- d[[6,7],[1,2]]\%4a[2,2]a[2,3]a[1,1]$$
$$+ d[[6,7],[2,8]]\%3a[8,1]a[3,3]a[2,2]$$
$$- d[[6,7],[1,2]]\%3a[3,3]a[2,2]a[1,1]$$
$$+ d[[6,7],[3,8]]\%4a[3,2]a[2,3]a[8,1] \qquad (23)$$
$$- d[[6,7],[1,3]]\%4a[3,2]a[2,3]a[1,1]$$
$$+ d[[6,7],[3,8]]\%3a[3,2]a[8,1]a[3,3]$$
$$- d[[6,7],[1,3]]\%3a[3,2]a[1,1]a[3,3],$$

$$\%3 := d[[3,6,7],[1,3,8]]d[[6,7],[4,5]]$$
$$+ d[[3,6,7],[3,4,5]]d[[6,7],[1,8]] \qquad (23a)$$

$$\%4 := d[[3,6,7],[1,2,8]]d[[6,7],[4,5]]$$
$$+ d[[3,6,7],[2,4,5]]d[[6,7],[1,8]] \qquad (23b)$$

(Hereafter this equation will be referred to as "scub2".)

Once these simplifications have been made in all of the "scubics", the co-efficients are functions of determinants of 54 minors of the data. The resulting "scubics" have two different types of coefficients. An example of the other type is

$$d[[2,3],[4,6]]\%4a[5,4]a[6,6]a[4,5]$$
$$+\,d[[2,3],[5,7]]\%3a[7,6]a[5,5]a[4,4]$$
$$-\,a[5,4]d[[2,3],[6,7]]d[[6,7],[2,3]]d[[2,3],[4,5]]$$
$$+\,d[[2,3],[4,7]]\%4a[4,5]a[7,6]a[5,4]$$
$$+\,d[[2,3],[4,7]]\%3a[4,4]a[7,6]a[4,5]$$
$$+\,d[[2,3],[5,7]]\%4a[5,4]a[7,6]a[5,5]$$
$$+\,d[[2,3],[4,6]]\%3a[4,4]a[6,6]a[4,5]$$
$$+\,d[[2,3],[5,6]]\%4a[5,4]a[6,6]a[5,5]$$
$$+\,d[[2,3],[5,6]]\%3a[6,6]a[5,5]a[4,4]$$
(24)

$$\%3 := -d[[2,3,4],[4,6,7]]d[[6,7],[2,3]]$$
$$+\,d[[4,6,7],[2,3,4]]d[[2,3],[6,7]]$$
(24a)

$$\%4 := -d[[2,3,4],[5,6,7]]d[[6,7],[2,3]]$$
$$+\,d[[4,6,7],[2,3,5]]d[[2,3],[6,7]]$$
(24b)

Note that the quadratic factors of coefficients in (23) look like they might be part of some Graßmann relations. Consider the Graßmann relations

$$0 = \pi_{162}\pi_{457} - \pi_{164}\pi_{257} + \pi_{165}\pi_{247} - \pi_{167}\pi_{245} \tag{25a}$$
$$0 = \pi_{451}\pi_{367} - \pi_{453}\pi_{167} + \pi_{456}\pi_{137} - \pi_{457}\pi_{136} \tag{25b}$$

which, for the matrix in (19) become

$$0 = d[[3,6,7],[1,4,5]]d[[6,7],[3,8]] - d[[3,6,7],[3,4,5]]d[[6,7],[1,8]]$$
$$+\,d[[3,6,7],[4,5,8]]d[[6,7],[1,3]] - d[[3,6,7],[1,3,8]]d[[6,7],[4,5]] \tag{26a}$$

$$0 = -d[[3,6,7],[1,2,8]]d[[6,7],[4,5]] + d[[3,6,7],[1,4,8]]d[[6,7],[2,5]]$$
$$+\,d[[3,6,7],[1,5,8]]d[[6,7],[2,4]] - d[[3,6,7],[2,4,5]]d[[6,7],[1,8]] \tag{26b}$$

respectively. Naively, there seems little hope of using these relations to simplify the quadratics clusters in (23) as the cubic clusters were simplified. The hope of simplification dwindles further when one notices that after all cubic clusters have been simplified, the coefficients in the cubic and quadratic equations are funtions of determinants in exactly 54 minors of the data matrix, and not all of the terms in (26a) and (26b) are amongst these 54 minors. The clusters in (24) look even less promising; they are not part of a Plücker relation for any submatrix of Q_{IO}.

These observations lead one to make the following substitutions:

halfplücksubs = {

$d[[3, 6, 7], [1, 2, 8]]d[[6, 7], [4, 5]] + d[[3, 6, 7], [2, 4, 5]]d[[6, 7], [1, 8]] =$ hpa1,

$d[[6, 7], [1, 8]]d[[3, 6, 7], [3, 4, 5]] + d[[3, 6, 7], [1, 3, 8]]d[[6, 7], [4, 5]] =$ hpb1,

$d[[2, 3], [4, 5]]d[[2, 3, 7], [1, 7, 8]] + d[[2, 3, 7], [4, 5, 7]]d[[2, 3], [1, 8]] =$ hpa2,

$d[[2, 3, 7], [1, 6, 8]]d[[2, 3], [4, 5]] + d[[2, 3, 7], [4, 5, 6]]d[[2, 3], [1, 8]] =$ hpb2, (27)

$d[[2, 6, 7], [1, 3, 8]]d[[6, 7], [4, 5]] + d[[2, 6, 7], [3, 4, 5]]d[[6, 7], [1, 8]] =$ hpa3,

$d[[2, 6, 7], [1, 2, 8]]d[[6, 7], [4, 5]] + d[[6, 7], [1, 8]]d[[2, 6, 7], [2, 4, 5]] =$ hpb3,

$d[[2, 3, 6], [1, 6, 8]]d[[2, 3], [4, 5]] + d[[2, 3], [1, 8]]d[[2, 3, 6], [4, 5, 6]] =$ hpa4,

$d[[2, 3, 6], [1, 7, 8]]d[[2, 3], [4, 5]] + d[[2, 3, 6], [4, 5, 7]]d[[2, 3], [1, 8]] =$ hpb4}

and

othersubs = {

$-d[[2, 3, 4], [5, 6, 7]]d[[6, 7], [2, 3]] + d[[4, 6, 7], [2, 3, 5]]d[[2, 3], [6, 7]] =$ os2,

$d[[6, 7, 8], [2, 3, 8]]d[[2, 3], [6, 7]] - d[[6, 7], [2, 3]]d[[2, 3, 8], [6, 7, 8]] =$ os3,

$-d[[2, 3, 5], [5, 6, 7]]d[[6, 7], [2, 3]] + d[[5, 6, 7], [2, 3, 5]]d[[2, 3], [6, 7]] =$ os4,

$-d[[2, 3, 5], [4, 6, 7]]d[[6, 7], [2, 3]] + d[[5, 6, 7], [2, 3, 4]]d[[2, 3], [6, 7]] =$ os5, (28)

$-d[[2, 3, 8], [1, 6, 7]]d[[6, 7], [2, 3]] + d[[6, 7, 8], [1, 2, 3]]d[[2, 3], [6, 7]] =$ os6,

$d[[1, 6, 7], [1, 2, 3]]d[[2, 3], [6, 7]] - d[[6, 7], [2, 3]]d[[1, 2, 3], [1, 6, 7]] =$ os7,

$d[[1, 6, 7], [2, 3, 8]]d[[2, 3], [6, 7]] - d[[1, 2, 3], [6, 7, 8]]d[[6, 7], [2, 3]] =$ os8,

$d[[4, 6, 7], [2, 3, 4]]d[[2, 3], [6, 7]] - d[[2, 3, 4], [4, 6, 7]]d[[6, 7], [2, 3]] =$ os1}

Once these substitutions have been made "quad" becomes

$$- d[[6, 7], [2, 3]]d[[3, 6, 7], [1, 3, 8]]d[[6, 7], [4, 5]]a[2, 3]a[3, 2]$$
$$- d[[6, 7], [2, 3]]d[[3, 6, 7], [3, 4, 5]]d[[6, 7], [1, 8]]a[2, 2]a[3, 3]$$
$$+ d[[6, 7], [1, 8]]d[[6, 7], [4, 5]]a[3, 3]d[[6, 7], [1, 3]]a[1, 1]$$
$$- d[[6, 7], [1, 8]]d[[6, 7], [4, 5]]a[3, 3]d[[6, 7], [3, 8]]a[8, 1] \qquad (29)$$
$$+ d[[6, 7], [1, 8]]d[[6, 7], [4, 5]]a[2, 3]d[[6, 7], [1, 2]]a[1, 1]$$
$$- d[[6, 7], [1, 8]]d[[6, 7], [4, 5]]a[2, 3]d[[6, 7], [2, 8]]a[8, 1]$$
$$- d[[6, 7], [2, 3]] \text{ hpa1 } a[2, 3]a[2, 2]$$

and "scub" can be written very neatly!

$$(\text{hpa1 } a[2, 3] + \text{ hpb1 } a[3, 3])(d[[6, 7], [2, 8]]a[8, 1]a[2, 2]$$
$$- d[[6, 7], [1, 2]]a[2, 2]a[1, 1]$$
$$+ d[[6, 7], [3, 8]]a[3, 2]a[8, 1] - d[[6, 7], [1, 3]]a[3, 2]a[1, 1]) \qquad (30)$$
$$+ d[[6, 7], [2, 3]]a[2, 3]d[[6, 7], [1, 8]]d[[6, 7], [4, 5]]$$

Even the cubics of the "other" type neaten up:

$$
\begin{aligned}
(\text{os2 } a[5,4] + \ & \text{os1 } a[4,4])(d[[2,3],[4,6]]a[6,6]a[4,5] \\
& + d[[2,3],[4,7]]a[4,5]a[7,6] \\
& + d[[2,3],[5,7]]a[7,6]a[5,5] + d[[2,3],[5,6]]a[6,6]a[5,5]) \\
& - a[5,4]d[[2,3],[6,7]]d[[6,7],[2,3]]d[[2,3],[4,5]]
\end{aligned}
\tag{31}
$$

10 Solving the equations

We observed that the variables in these equations appear in a very systematic form. Each equation is a polynomial in 6 variables: four from one of the 2×2 sub-blocks of A, the other two from one of the neighboring sub-blocks of A. Furthermore, for each cubics there exists a quadratic which is a function of the same six $a[i,j]$'s. Finally, we noticed that each of the equations is linear in the $a[i,j]$'s from the neighboring sub-block. The following chart shows the pairings of variables of the quadratics and the shortened cubics:

pair of equations	linear variable	block of 2 linear variables	four variables
1.	$a[2,3]$	$a[1,1], a[8,1]$	$a[2,2], a[2,3], a[3,2], a[3,3]$
2.	$a[3,2]$	$a[4,4], a[5,4]$	$a[2,2], a[2,3], a[3,2], a[3,3]$
3.	$a[1,8]$	$a[2,2], a[3,2]$	$a[1,1], a[1,8], a[8,1], a[8,8]$
4.	$a[8,1]$	$a[6,7], a[7,7]$	$a[1,1], a[1,8], a[8,1], a[8,8]$
5.	$a[4,5]$	$a[2,3], a[3,3]$	$a[4,4], a[4,5], a[5,4], a[5,5]$
6.	$a[5,4]$	$a[6,6], a[7,6]$	$a[4,4], a[4,5], a[5,4], a[5,5]$
7.	$a[6,7]$	$a[4,5], a[5,5]$	$a[6,6], a[6,7], a[7,6], a[7,7]$
8.	$a[7,6]$	$a[1,8], a[8,8]$	$a[6,6], a[6,7], a[7,6], a[7,7]$

(Note that the first column applies only to the cubics.)

From the above chart, one can see that one may either solve pairs $1, 2, 7$, and 8 for the sub-blocks containing $a[1,1], a[1,8], a[8,1], a[8,8], a[4,4], a[4,5], a[5,4]$, and $a[5,5]$, (or vice-versa). After making these linear solves and substituting into pairs $3, 4, 5$ and 6, we were left with 8 messy equations. They have one important feature, however: the remaining $a[i,j]$'s appear in pairs. To solve the last equations we make the substitutions

$$
\{a[3,2] = a[2,2]q_1, \ a[3,3] = a[2,3]q_2, \ a[7,7] = a[6,7]q_3, \ a[7,6] = a[6,6]q_4\}.
$$

We can solve the four (previously) cubic equations for $a[6,6]^2, a[6,7]^2, a[2,2]^2$, $a[2,3]^2$. (Note that the last two $a[i,j]$'s were mislabeled in references 5 and 6.) Substituting these solutions into the four (previously) quadratic equations we get four highly NONLINEAR EQUATIONS in q_1, q_2, q_3, q_4.

Each of these equations is a 12-term polynomial, involves only three of the variables, and is linear in one of them. The roles of the four variables repeat cyclically. Solving for two of the variables (linearly) and replacing the result into the remaining two equations we get two nonlinear equations of 35 terms each. All of this can be done with the general equations!! Thus far, our data have been the symbols $Q[i,j]$, where $i,j = 1,\ldots,8$. In the equations above, the Q's are rewritten in our determinant notation. The size of the polynomials (or rather the coefficients) prohibits further computation with a general data set. We are now forced to use the phantom, i.e. numerical data, to finish the computation. Taking the resultant of these equations gives us a huge polynomial equation in one variable, which we factor. In our numerical tests we have always observed that there is only one linear factor over the rationals. This linear factor allows us to determine the phantom.

11 Conclusions

In this paper we have given a brief description of a general, albeit tiny, problem in diffuse tomography. This is an inverse problem with 64 variables and a system of only 48 independent non-linear equations. We make 16 linear "identifications" of variables to close the system. Rather than tackle this imposing system directly, we make a change of variables. The transformed system is composed of 48 linear and 16 non-linear equations, which is easily reduced to a system of 8 cubic and 8 quadratic equations in 16 unknowns. The bulk of the paper discusses how one can simplify these equations. The quadratics may be used to reduce the number of terms in the cubics. Next, Graßmann relations may be used to simplify the coefficients of both the cubics and the quadratics. Finally, we give an algorithm for solving these last 16 equations.

Our next project is to extend our method to a larger problem. At this point, the best option seems to be a recursive algorithm whose "base case" is the 2×2 problem described here. The greatest obstacle to inverting the $n \times n$ problem will be solving highly non-linear consistency conditions analogous to the cubics and quadratics described above. If the two-dimensional $n \times n$ problem can be solved with a reasonable computational effort, then a similar method should carry over to the three-dimensional $n \times n \times n$ problem.

References

1. Grünbaum, F. A.: Tomography with Diffusion. In: Sabatier, P. (ed.): *Inverse Problems in Action*, pp. 16–21. Springer-Verlag, Berlin 1990.
2. Singer, J., Grünbaum, F. A., Kohn, P., Zubelli, J.: Image reconstruction of the interior of bodies that diffuse radiation. Science **248** (1990) 990–993.
3. Grünbaum, F. A: Backscattering comes to the rescue. In Contemporary Mathematics, E. Grinberg and E. T. Quinto editors, **113** (1990) 137–239.
4. Grünbaum, F. A.: An inverse problem in transport theory: diffuse tomography. In: Corones, J. (ed.): *Invariant Imbedding and Inverse Problems. In Honor of R. Kruger*, pp. 209–215. SIAM 1993.
5. Grünbaum, F. A., Patch, S. K.: Analytic inversion of a general model in diffuse tomography. In: Fiddy, M. A. (ed.): *Inverse Problems in Scattering and Imaging*. PROC S.P.I.E. Vol. 1767 44–54, (1992).

6. Grünbaum, F. A., Patch, S. K.: Simplification of a general model in diffuse tomography. Inverse Problems in Scattering and Imaging. To appear in PROC S.P.I.E.
7. Griffiths, P., Harris, J.: *Principles of Algebraic Geometry*, Wiley and Sons, Inc., 1978.
8. Hodge, W. V. D., Pedoe, D.: *Methods of Algebraic Geometry*, Cambridge University Press, 1968.

Generic Uniqueness and Stability in Some Inverse Parabolic Problem

Karl-Heinz Hoffmann[1], *Masahiro Yamamoto*[2]

[1] Institut für Angewandte Mathematik und Statistik, Technische Universität München, Dachauerstraße 9a, 8000 München 2, Germany
[2] Department of Mathematical Sciences, University of Tokyo, 3-8-1 Komaba, Meguro, Tokyo 153 Japan

1 Introduction

We consider an initial - boundary value problem for a parabolic equation:

$$\left\{\begin{array}{ll} u_t = \Delta u - p_0(x)u - q(x)f(x,t) & (x \in r\Omega, \, t > 0), \\ u(x,0) = 0 & (x \in r\Omega), \\ u(x,t) = 0 & (x \in \partial(r\Omega), \, t > 0). \end{array}\right\} \tag{1.1}$$

Here Δ is the Laplacian and $r > 0$ is a parameter. The problem (1.1) with $r = 1$ is considered as the original or unperturbed problem. The domain Ω is bounded in $\mathbb{R}^n$ with smooth boundary $\partial\Omega$. We set $r\Omega = \{rx; x \in \Omega\}$. Let us fix $p_0 \in L^\infty(r\Omega)$. Henceforth $Cl(D)$ denotes the closure of a set $D \subset \mathbb{R}^n$. For each $q \in L^\infty(r\Omega)$, we denote the strong solution to (1.1) by $u(q)(x,t)$ (e.g. Pazy [8]).

Our inverse parabolic problem is:

Inverse Problem (IP) Let $T > 0$ be given.
(I) (Uniqueness) Do the data

$$u(q)(x,T) = g(x) \qquad (x \in r\Omega) \tag{1.2}$$

determine q uniquely ?
(II) (Stability) Is the correspondence $g \Longrightarrow q$ continuous in an appropriate sense, provided that it is well-defined ?

Remark. Let $f(p)(x,t)$ satisfy

$$\left\{\begin{array}{ll} f_t = \Delta f - p(x)f & (x \in r\Omega, \, t > 0), \\ f(x,0) = \varphi(x) & (x \in r\Omega), \\ f(x,t) = \psi(x,t) & (x \in \partial(r\Omega), \, t > 0). \end{array}\right\} \tag{1.3}$$

Then (IP) corresponds to a linearized problem for another inverse problem (IP2): Determine p from $f(p)(x,T)$ $(x \in r\Omega)$.

Thus for studying (IP2), it is important to discuss uniqueness and stability in our linearized problem (IP). For (IP2), we can refer to Anikonov [1], Beznoshchenko [2], Bukhgeim and Klibanov [3], Bukhgeim and Yakhno [4], Isakov [5], Klibanov [7], Prilepko and Solov'ev [9], Prilepko and Vasin [10], Yakhno [11]. The paper [2] is probably first on this kind of inverse problem, although the formulation is a little bit different. Moreover in [2], Beznoshchenko considers also the existence of coefficients realizing a given state of the solution at time T. In [4] and [11], the authors apply the technique in [2] to (IP2), so that uniqueness and stability are established if the domain where the coefficients are unknown, is sufficiently small.

From [3] and [7], we can see the following: Let $\Omega = \mathbb{R}^n$. If p is known outside any subdomain $\Omega_0 \subset \mathbb{R}^n$, then we can uniquely determine p in Ω_0 from the values of the solution over Ω_0 at time T. In this case, the measure of Ω_0 need not be small for uniqueness.

The paper [5] gives a counterexample for uniqueness in the case of $\Omega = (0,1)$: There exists some function $f \in C^\infty([0,1] \times [0,T])$ with $f(x,t) > 0$ $(x \in [0,1], t \in [0,T])$ such that $u_t = u_{xx} + q(x)f(x,t)$ $(0 < x < 1, 0 < t < T)$, $u(x,t) = 0$ $(x = 0,1, 0 < t < T)$, $u(x,0) = 0$ $(0 < x < 1)$ and $u(x,T) = 0$ $(0 < x < 1)$ has a non-zero classical solution u for some non-zero $q \in C^0[0,1]$. In this counterexample, we should note that the coefficient q is unknown over the whole domain, and so this does not contradict the result in [3] and [7].

The purpose of this paper is to discuss uniqueness and stability for our linearized inverse problem (IP) in domains which are not small and to give some interpretation of Isakov's counterexample. As is seen from our main result below, uniqueness and stability break down "very rarely", and Isakov's counterexample shows that we actually have such a rare breakdown.

Our methodology is based on a theorem from the theory of analytic perturbations of linear operators (e.g. Kato [6]) and on uniqueness for (IP) with small $r > 0$. Thus our method is applicable mainly to parabolic equations where the associated semigroups are analytic in time t.

2 Main result

Throughout this paper, we assume:

(2.1) *The domain $\Omega \subset \mathbb{R}^n$ is bounded with smooth boundary $\partial\Omega$ and star-shaped with respect to 0.*

(2.2) *There exists some constant $\delta > 0$ such that $|f(x,T)| \geq \delta$ $(x \in Cl(r\Omega))$*

(2.3) *$f, \partial f/\partial t \in C^\theta([0,T]; L^\infty(r\Omega))$ for some $\theta \in (0,1)$.*

Here $C^\theta([0,T]; L^\infty(r\Omega))$ is the space of Hölder continuous functions with exponent θ on $[0,T]$ with values in $L^\infty(r\Omega)$.

Now we can state our main result:

Theorem 1. *Let γ and R be arbitrarily fixed such that $0 < \gamma < R$. There exists a finite set $\Lambda \subset [\gamma, R]$ such that for any $r \in [\gamma, R] \backslash \Lambda$, the data $u(q)(\cdot, T)$ uniquely determine $q \in L^\infty(r\Omega)$ in (1.1) and the stability estimate*

$$\|q\|_{L^2(r\Omega)} \leq M(r)\|u(q)(\cdot, T)\|_{H^2(r\Omega)}$$

holds. Here $M(r) > 0$ is a constant depending only on r and $H^2(r\Omega) = W^{2,2}(r\Omega)$ is the Sobolev space.

Remark. This theorem asserts that there exists at most a finite number of domain parameters r violating the uniqueness and stability property for (IP), which means that uniqueness and stability is a generic property. From the counterexample of Isakov [5], we can see that Λ is not empty. That is, there actually exists r violating uniqueness and stability.

Remark. For the set Λ being finite, Dirichlet boundary conditions are important. As it is seen from the following example, it may happen that for any $r > 0$, uniqueness and stability do not hold for (IP) in $r\Omega$ if we consider Neumann boundary conditions in (1.1). This comes from the fact that a spatial uniform solution may exist.

Example 1. Let $\Omega = (0, 1)$ and $f(x, t) = 2t - T$ $(t > 0)$. However $q \equiv 1$ and $u(x, t) = t^2 - Tt$ $(x \in (0, r))$ satisfy

$$u_t = u_{xx} + qf(x, t)   \qquad (x \in r\Omega, t > 0),$$

$$\frac{\partial u}{\partial n} = 0   \qquad (x \in \partial(r\Omega), t > 0),$$

$$u(x, 0) = u(x, T) = 0   \qquad (x \in r\Omega)$$

for any $r > 0$.

From this theorem, we can get

Corollary 2. *Let us set $\Omega_\epsilon = \{x \in \Omega; dist(x, \partial\Omega) > \epsilon\}$. In addition to (2.1) and (2.2) – (2.3) with $r = 1$, we assume that the values of $u(q)(x, t)$ $(x \in \Omega \backslash \Omega_\epsilon, 0 < t < T)$ are known for any fixed $\epsilon > 0$. Then $u(q)(x, T)$ $(x \in \Omega_\epsilon)$ determines q uniquely and there exists a constant $M > 0$ such that*

$$\|q\|_{L^2(\Omega)} \leq M\|u(q)(\cdot, T)\|_{H^2(\Omega)}$$

for each $q \in L^\infty(\Omega)$.

3 Proof

We will give an outline of the proof in the following five steps.

1) By change of variables $x = ry(x \in r\Omega)$, we rewrite (1.1) and (1.2) in the varying domain $r\Omega$ as (3.1) and (3.2) in the fixed domain Ω:

$$\left\{ \begin{array}{ll} U_t = r^{-2}\Delta_y U - P_0(y)U - Q(y)F(y,t) & (y \in \Omega, t > 0), \\ U(y,0) = 0 & (y \in \Omega), \\ U(y,t) = 0 & (y \in \partial\Omega, t > 0) \end{array} \right\} \qquad (3.1)$$

$$U(y,T) = G(y) \qquad (y \in \Omega). \tag{3.2}$$

Here we set $U(y,t) \equiv U(Q)(y,t) = u(x,t)$, $P_0(y) = p_0(x)$, $Q(y) = q(x)$, $F(y,t) = f(x,t)$, $G(y) = g(x)$ and $\Delta_y U = \sum_{i=1}^n U_{y_i y_i}$ $(y = (y_1, ..., y_n) \in \Omega)$. Since $0 \in \Omega$ and Ω is a star-shaped domain with respect to 0, the problem (3.1) and (3.2) is equivalent to (1.1) and (1.2).

Henceforth let $-A$ be an operator in $L^2(\Omega)$ defined by $-Au(y) = \Delta_y u(y)$ $(y \in \Omega)$ and $\mathcal{D}(A) = \{u \in H^2(\Omega) ; u_{|\partial\Omega} = 0\}$ and let us consider (3.1) as an evolution equation in the Hilbert space $X = L^2(\Omega)$ with norm $\| \cdot \|$.

2) By the condition (2.3), we can prove that $U \in C^2((0,\infty);X)$ and $(AU)_t = AU_t$. Therefore setting $V(Q)(t) = \frac{dU(Q)}{dt}(t)$, we see that

$$\left\{ \begin{array}{ll} V_t = -r^{-2}AV - P_0V - QF_t(t) & (t > 0), \\ V(0) = -QF(0) \end{array} \right\} \tag{3.3}$$

and

$$V(Q)(T) = -r^{-2}AG - P_0G - QF(T). \tag{3.4}$$

3) We rewrite (3.3) as

$$V(Q)(t) = -\int_0^t \exp\left(-\frac{t-s}{r^2}A\right) P_0V(Q)(s)ds - \int_0^t \exp\left(-\frac{t-s}{r^2}A\right) QF_s(s)ds$$

$$- \exp\left(-\frac{t}{r^2}A\right) QF(0).$$

Therefore, by (2.2), the equality (3.4) can be reduced to

$$Q = F(T)^{-1} \int_0^T \exp\left(-\frac{T-s}{r^2}A\right) P_0V(Q)(s)ds$$

$$+ F(T)^{-1} \int_0^T \exp\left(-\frac{T-s}{r^2}A\right) QF_s(s)ds$$

$$+ F(T)^{-1} \exp\left(-\frac{T}{r^2}A\right) QF(0) - F(T)^{-1}(r^{-2}AG + P_0G)$$

$$\equiv L_1(r)Q + L_2(r)Q + L_3(r)Q - F(T)^{-1}(r^{-2}AG + P_0G).$$

We consider $L_i(r)\,(1 \le i \le 3)$ as linear operators from X to X.

4) Since the real part of any point in the spectrum of A is positive and A^{-1} is compact on X to X, we can see that (e.g. Pazy [8]):

We can take a $\lambda > 0$ satisfying the following estimate: for any $\alpha \geq 0$, there exists $M_1(\alpha) > 0$ such that

$$\|A^\alpha \exp(-tA)\|_{X \to X} \leq M_1(\alpha) \exp(-\lambda t) t^{-\alpha} \qquad (t > 0). \qquad (3.5)$$

For any $\epsilon > 0$, the operator $A^{-\epsilon}$ is compact from X to X. \qquad (3.6)

Moreover the operator $-A$ generates a compact analytic semigroup on X. Therefore by (3.5) and (3.6) we obtain

Lemma 1 *For a given $T > 0$, there exists $M_2 > 0$ independent of r such that*
$\|V(Q)(t)\| \leq M_2\|Q\| \qquad (0 \leq t \leq T)$.

Lemma 2 *The linear operators $L_i(r)\,(1 \leq i \leq 3)$ from X to X are analytic in $r > 0$.*

Lemma 3 $L_i(r)\,(1 \leq i \leq 3)$ *are compact from X to X.*

By Lemma 3 and the Riesz-Schauder theorem, it is sufficient to prove that 1 is not an eigenvalue of $L_1(r) + L_2(r) + L_3(r)$ for obtaining uniqueness and stability. On the other hand, by Lemmas 2 and 3, we can apply a result of analytic perturbation theory (e.g. [6], p.370), so that we have the alternative:

(i) 1 is an eigenvalue of $L_1(r) + L_2(r) + L_3(r)$ for any $r > 0$.

(ii) For any $0 < \gamma < R$, there exists a finite set $\Lambda \subset [\gamma, R]$ such that 1 is not an eigenvalue of $L_1(r) + L_2(r) + L_3(r)$ for any $r \in [\gamma, R] \setminus \Lambda$.

5) In order to complete the proof, we have only to exclude the first case (i). By (3.5) and (2.2) we can easily estimate $\|L_2(r)\|_{X \to X}$ to obtain

$$\|L_2(r)\|_{X \to X} \leq \frac{M_1(0)r^2}{\lambda \delta}(1 - \exp(-\lambda T r^{-2})) \times \max_{0 \leq s \leq T} \|F_s(s)\|_{L^\infty(\Omega)}.$$

As for the estimation of $\|L_1(r)\|_{X \to X}$, we have only to use Lemma 1 besides (3.5) and (2.2) to get a similar inequality. As for $L_3(r)$, we can proceed similarly. Thus, for a sufficiently small $r > 0$, we see that $\|L_1(r) + L_2(r) + L_3(r)\|_{X \to X} < 1$. This implies that 1 is not an eigenvalue of $L_1(r) + L_2(r) + L_3(r)$ by the Neumann series. Thus the case (i) is excluded.

Finally we will prove the Corollary. It is sufficient to verify uniqueness. That is, assume that

$$u_t = \Delta u - p_0(x)u - q(x)f(x,t) \qquad (x \in \Omega,\, t > 0)$$

$$u(x,0) = 0 \qquad (x \in \Omega)$$

$$u(x,T) = 0 \qquad (x \in \Omega)$$

and
$$u(x,t) = 0 \qquad (x \in \Omega \setminus \Omega_\epsilon, \, 0 < t < T).$$

By the Theorem, there exists a small $\beta > 0$ such that $(1 - \beta)\Omega \supset \Omega_\epsilon$ and the inverse problem (IP) has at most one solution in $(1-\beta)\Omega$. Since $\partial((1-\beta)\Omega) \subset \Omega \setminus \Omega_\epsilon$, we have $u(x,t) = 0$ $(x \in \partial((1-\beta)\Omega), \, t > 0)$. Thus from uniqueness of (IP) in the domain $(1-\beta)\Omega$, we can get $q(x) = 0$ $(x \in (1-\beta)\Omega)$. On the other hand, $u(x,t) = 0 \, (x \in \Omega \setminus \Omega_\epsilon, \, 0 < t < T)$ satisfies $u_t = \Delta u - p_0(x)u - q(x)f(x,t)$ and so $q(x) = 0 \, (x \in \Omega \setminus \Omega_\epsilon)$. Thus we can get $q(x) = 0$ for $x \in (1-\beta)\Omega \cup (\Omega \setminus \Omega_\epsilon)(\supset \Omega)$.

Acknowledgements A part of this paper has been written during the stay of the second author at Technische Universität München and the stay has been supported by the Alexander von Humboldt Foundation. The second author thanks the Foundation for the support.

References

1. Anikonov, Yu. E.: Formulas in multidimensional inverse problems for evolution equations. (English translation) Soviet Math. Dokl. **41** (1990) 385–388.
2. Beznoshchenko, N. Ya.: Determination of coefficients of higher terms in a parabolic equation. (English translation) Siberian Math. J. **16** (1975) 360–367.
3. Bukhgeim, A. L., Klibanov, M. V.: Global uniqueness of a class of multidimensional inverse problems. (English translation) Soviet Math. Dokl. **24** (1981) 244–247.
4. Bukhgeim A. L., Yakhno, V. G.: On two inverse problems for differential equations. (English translation) Soviet Math. Dokl. **17** (1976) 1083–1085.
5. Isakov, V.: Inverse parabolic problems with the final overdetermination. Comm. Pure Appl. Math. **44** (1991) 185–209.
6. Kato, T.: *Perturbation Theory for Linear Operators.* Second edition, Springer Verlag, Berlin, 1976.
7. Klibanov, M. V.: Inverse problems in the "large" and Carleman bounds. (English translation) Differential Equations **20** (1984) 755–760.
8. Pazy, A.: *Semigroups of Linear Operators and Applications to Partial Differential Equations.* Springer Verlag, Berlin, 1983.
9. Prilepko, A. I., Solov'ev, V. V.: Solvability of the inverse boundary-value problem of finding a coefficient of a lower-order derivative in a parabolic equation. (English translation) Differential Equations **23** (1987) 101–107.
10. Prilepko, A. I., Vasin, I. A.: On a non-linear non-stationary inverse problem of hydrodynamics. Inverse Problems **7** (1991) L13–L16.
11. Yakhno, V. G.: An inverse problem for a parabolic system. (English translation) Differential Equations **15** (1979) 398–400.

Regularization - Analytic and Stochastic Aspects

Bernd Hofmann

Technische Hochschule Zittau, Department of Mathematics, Theodor-Körner-Allee 16, D-O-8800 Zittau, Germany

Abstract *There are presented two studies on descriptive regularization. In a first study for a Volterra integral equation of the first kind the influence of ill-posedness effects and random data errors are compared. The higher degree of ill-posedness for a Fredholm integral equation of the first kind with smooth kernel becomes evident in a second study on the determination of monotone representatives of profile functions solving a specific Urysohn integral equation.*

1 A study on ill-posedness and stochastic data errors

In this section we are going to consider a specific aspect of solving a discretized Volterra integral equation of the first kind with experimental data for both the kernel and the right-hand side. The integral equation

$$\int_0^s a(s-t)\,x(t)\,dt \; = \; b(s) \quad (0 \le s \le T) \tag{1.1}$$

with

$$a(t) = \frac{d}{dt}\left(\frac{v(t)}{p(t)}\right), \quad b(s) = p(s) - p(0) \tag{1.2}$$

occurs in reservoir mechanics, where only the physical quantities $v(t) \ge 0$ and $p(t) > 0$ (volume and pressure) are observable in a certain sense (for details see [4]). In the sequel we assume that the continuous response function $x(t)$ ($0 \le t \le T$) solving (1.1) has "memory" character and is *non-negative, non-increasing* and *convex*.

Now we discretize the integral equation in a simple manner (cf. e.g. the ideas of [7]) by using the vectors $\underline{v} = (v_1, ..., v_n)^T, \underline{p} = (p_1, ..., p_n)^T, \underline{b} = (b_1, ..., b_n)^T$ and $\underline{a} = (a_1, ..., a_n)^T$ with $h = \frac{T}{n}$,

$$t_i = ih, \; v_i = v(t_i), \; p_i = p(t_i) \quad (i = 0, 1, ..., n);$$

$$a_i = \frac{v_i}{p_i} - \frac{v_{i-1}}{p_{i-1}}, \quad b_i = p_i - p_0 \quad (i = 1, 2, ..., n).$$

As a discrete version of (1.1) we thus have

$$\underline{A}\,\underline{x} = \underline{b} \tag{1.3}$$

with a semicirculant matrix

$$\underline{A} = \begin{pmatrix} a_1 & & & 0 \\ a_2 & a_1 & & \\ \vdots & \ddots & \ddots & \\ a_n & \dots & a_2 & a_1 \end{pmatrix},$$

where the components x_i of the vector $\underline{x} = (x_1, ..., x_n)^T$ should approximate average values $x_i^* = \frac{1}{h} \int\limits_{t_{i-1}}^{t_i} x(t)\, dt$ of the solution to (1.1). Under some smoothness assumptions it can be shown that $\max\limits_{1 \le i \le n} |x_i - x_i^*| = O(h^2)$ (see [5]), otherwise at least $O(h)$ is observed in most cases. Of course, solution vectors $\underline{x}$ of (1.3) are only interpretable as skeleton solutions of the integral equation (1.1) in our context if they belong to the set

$$\mathbf{S} = \{\underline{x} \in \mathbf{R}^n : x_1 \ge x_2 \ge ... \ge x_n \ge 0;\ x_{i-1} + x_{i+1} \ge 2x_i\ (i = 2, 3, ..., n-1)\}. \tag{1.4}$$

For solving the discrete ill-posed problem under real assumptions we are going to consider an experimental data variant: It is assumed that the noisy data vectors $\underline{\hat{v}} = (\hat{v}_1, ..., \hat{v}_n)^T$ and $\underline{\hat{p}} = (\hat{p}_1, ..., \hat{p}_n)^T$ with

$$\hat{v}_i = v_i + \sum_{j=1}^{i} \eta_j\ > 0 \quad (i = 1, ..., n) \tag{1.5}$$

and

$$\hat{p}_i = p_i + \xi_i\ > 0 \qquad (i = 1, ..., n) \tag{1.6}$$

yield the observations. In this context let $\underline{\eta} = (\eta_1, ..., \eta_n)^T$ and $\underline{\xi} = (\xi_1, ..., \xi_n)^T$ with

$$\mathbf{E}\underline{\eta} = \underline{0}, \quad \mathbf{E}\underline{\xi} = \underline{0}, \quad \mathrm{cov}(\underline{\eta}) = \sigma_v^2 \underline{I}, \quad \mathrm{cov}(\underline{\xi}) = \sigma_p^2 \underline{I} \tag{1.7}$$

be stochastically independent continuous random vectors. The different structure of the formulae (1.4) and (1.5) is motivated by the fact that only the increments $v_i - v_{i-1}$ are observable with respect to v, whereas the perturbed values of p_i are measured directly.

Even if the noise vectors $\underline{\eta}$ and $\underline{\xi}$ are nearly normally distributed, the linear system with random data

$$\underline{\hat{A}}\,\underline{\hat{x}} = \underline{\hat{b}} \tag{1.8}$$

with

$$\underline{\hat{A}} = \begin{pmatrix} \hat{a}_1 & & & 0 \\ \hat{a}_2 & \hat{a}_1 & & \\ \vdots & \ddots & \ddots & \\ \hat{a}_n & \dots & \hat{a}_2 & \hat{a}_1 \end{pmatrix}, \quad \underline{\hat{b}} = (\hat{p}_1 - p_0, ... \hat{p}_n - p_0)^T$$

$$\hat{a}_1 = \frac{\hat{v}_1}{\hat{p}_1} - \frac{v_0}{p_0}, \quad \hat{a}_i = \frac{\hat{v}_i}{\hat{p}_i} - \frac{\hat{v}_{i-1}}{\hat{p}_{i-1}} \quad (i = 2, 3, ..., n)$$

may be far away from the normal case $\hat{A} = \underline{A} + \Delta\underline{A}$ with $\mathrm{E}\Delta\underline{A} = \underline{0}$ and normally distributed elements in the noise matrix $\Delta\underline{A}$.

In order to get a rough idea of the deviations $\|x - \hat{x}\|_\infty$ (cf. (1.3) and (1.7)) arising from random errors in the data as well as from ill-posedness effects of the integral equation, we should simplify the problem under consideration. For the sake of simplicity we thus restrict our attention to the expected mean square error

$$e_1 = \sqrt{\mathrm{E}(x_1 - \hat{x}_1)^2}$$

of the first component x_1 and to the particular case $v_1 > v_0 = 0$, $p_1 > p_0 > 0$. Moreover, let $\hat{v}_1$ and $\hat{p}_1$ be lognormally distributed with

$$\mathrm{E}\hat{v}_1 = v_1, \quad \mathrm{D}^2\hat{v}_1 = \sigma_v^2, \quad \mathrm{E}\hat{p}_1 = p_1, \quad \mathrm{D}^2\hat{p}_1 = \sigma_p^2.$$

Then one can express e_1 in an explicit manner:

Theorem 1.1 *For $f_p = 1 + \frac{\sigma_p^2}{p_1^2}$ and $f_v = 1 + \frac{\sigma_v^2}{v_1^2}$ we have*

$$e_1 = \frac{p_1}{v_1}\{(p_1 - p_0)^2 + 2(p_1 - p_0)(p_0 - p_1 f_p)f_v + f_v^3(p_1^2 f_p^6 - 2p_0 p_1 f_p^3 + p_0^2 f_p^2)\}^{\frac{1}{2}}.$$
$$(1.9)$$

The theorem presented above gives a chance to separate the ill-posedness effects expressed by the first factor $\frac{p_1}{v_1}$ and the random data effects expressed by the terms within braces. If n is large, then $p_1 \approx p_0$, $v_1 = v(\frac{T}{n}) \approx v_0 = 0$ and $\frac{p_1}{v_1} \to \infty$ as $n \to \infty$.

For a tutorial example with $T = 1$, $p_0 = 1$, $x(t) = \exp(-t)$, $p(t) = 1 + 2(t - 1 + \exp(-t))$ and $v(t) = 2t^3 - t^2 + 2t^2 \exp(-t)$ Table 1 gives some impression of the influence of both error factors. From top to bottom of the table the influence of the ill-posedness grows with the number n of used nodes, whereas from the left to the right the results of a growing variance σ_p^2 are observable. Note that a linear approximation of formula (1.9) in the marginal case $\sigma_v = 0$ and for fixed n yields $e_1 = O(\sigma_p)$. This rate approximately occurs also in Table 1.

Unfortunately, an analytic separation of error factors is impossible for the whole error

$$e_2 = \mathrm{E}\|\underline{x} - \underline{x}^*\|_\infty.$$

In [5] a Monte-Carlo simulation study fo this error had been performed, where x expressed a solution of the constrained least-squares problem

$$\|\hat{\underline{A}}\tilde{\underline{x}} - \hat{\underline{b}}\|_2 \longrightarrow \min, \quad \text{subject to} \quad \tilde{\underline{x}} \in \mathbf{S} \qquad (1.10)$$

(see (1.4)). As the numerical experience shows, because of the a priori information about the solution collected in the set $\mathbf{S}$ there is an essential reduction of the error e_2 in comparision with e_1. For the case $\sigma_v = 0$, a rate of about $e_2 = O(\sigma_p^{0.6})$ with a small constant of proportionality was observed. For the discretized Volterra equation under consideration the descriptive regularization by monotonicity and convexity according to (1.10) works sufficiently well. In particular, it is important that in view of the stabilizing character of those constraints the accuracy of the results does not sensitively depend on the discretization level h.

n	$\|x_1 - x_1^*\|$	$e_1(\sigma_p = 10^{-3})$	$e_1(\sigma_p = 10^{-2})$	$e_1(\sigma_p = 10^{-1})$
16	0.0101	0.2572	2.5713	26.8409
32	0.0051	1.0252	10.2560	107.1024
64	0.0026	4.0976	40.9945	428.1442
128	0.0013	16.3874	163.9483	1712.3096
256	0.0006	65.5466	655.7632	6848.9704

Table 1:   First component errors e_1 for $\sigma_v/v_1 = 0.01$

2 A study on regularization by monotonicity

In the paper [1] there has been developed a decreasing rearrangement approach for the class

$$\int_0^1 \kappa(s, p(t))\, dt = r(s), \quad (0 \le s \le 1) \tag{2.1}$$

of Urysohn integral equations with continuous kernel $\kappa(s, p)$. Nonlinear integral equations of this ill-posed form arise for example in optics (cf. e.g. [6]). Since their solutions are never unique, decreasing rearrangements of the profiles $p(t) \ge 0$ may serve as representatives of the family of possible solutions.

One can rewrite (2.1) as a linear Fredholm integral equation of the first kind

$$\int_a^b k(s, t)\, x(t)\, dt = y(s) \quad (0 \le s \le 1) \tag{2.2}$$

and only search for the distribution function

$$x(t) = \lambda(\{\tau \in [0, 1] : p(\tau) > t\}) \quad (0 \le t < \infty) \tag{2.3}$$

instead of the profile p itself. Here, λ denotes the Lesbegue measure. Provided that real numbers a and b are known with

$$0 \le a \le \mathrm{essmin}_{0 \le t \le 1}\, p(t) < \mathrm{essmax}_{0 \le t \le 1}\, p(t) \le b < \infty$$

and that the partial derivative $\kappa_p'(s, p)$ of the kernel $\kappa(s, p)$ is continuous, we have

$$k(s, t) = \kappa_p'(s, t), \quad y(s) = r(s) - \kappa(s, a). \tag{2.4}$$

Moreover, since the distribution function is right-continuous and non-increasing by definition, we have to determine a monotone L^∞- function x with $0 \leq x(t) \leq 1$ $(a \leq t \leq b)$. The compactness of the set of non-increasing and uniformly bounded functions in L^p $(1 \leq p < \infty)$ offers the chance to use regularization by monotonicity to solve the problem (2.1) approximately in a stable manner.

Assumption 2.1 *Let for $(s, p) \in [0, 1] \times [a, b]$*
(i) $\kappa(s, p)$ *be non-negative and continuous;*
(ii) $|\kappa_p'(s, p)| \leq H < \infty;$
(iii) $|\kappa_p'(s_1, p) - \kappa_p'(s_2, p)| \leq L_1 |s_1 - s_2|,\ s_1, s_2 \in [0, 1],\ p \in [a, b];$
(iv) $|\kappa_p'(s, p_1) - \kappa_p'(s, p_2)| \leq L_2 |p_1 - p_2|,\ s \in [0, 1], p_1, p_2 \in [a, b].$

Then (cf. [1, Sec. 4]) for measurable functions $p(t)$ $(0 \leq t \leq 1)$ that solve the non-linear integral equation (2.1) for given $r(s)$ $(0 \leq s \leq 1)$, the associated distribution functions $x(t)$ $(a \leq t \leq b)$ (cf. (2.3)) fulfil the linear integral equation (2.2). We furthermore assume that there is at most one distribution function x solving (2.2) for any given right-hand side function y.

It can be shown that the distribution function x to be determined may be approximated arbitrarily well by a least-squares collocation approach. We denote by n and m the numbers of nodes for the discretization of the t and s direction, respectively, i.e. $h_t = \frac{b-a}{n}$, $h_s = \frac{1}{m}$ are the stepsizes, $t_j = a + \frac{h_t}{2} + (j-1)h_t$, $x_j = x(t_j)$ $(j = 1, 2, ..., n)$ and $s_i = \frac{h_s}{2} + (i-1)h_s$, $y_i = y(s_i)$ $(i = 1, 2, ..., m)$ midpoints and corresponding function values of the subintervals. The data will be given in a perturbed form

$$\hat{y}_i \quad \text{with} \quad |\hat{y}_i - y_i| \leq \delta \quad (i = 1, 2, ..., m) \tag{2.5}$$

for a given deviation bound δ.

Now we replace (2.2) by the algebraic linear system

$$\sum_{j=1}^{n} h_t\, k(s_i, t_j)\, x_j = \hat{y}_i \quad (i = 1, 2, ..., m) \tag{2.6}$$

written as $\underline{A}\,\underline{x} = \underline{\hat{y}}$ in a more compact form, where we search for an optimal vector $\underline{x}^{opt} = (x_1^{opt}, ..., x_n^{opt})^T$ of

$$\|\underline{A}\,\underline{\tilde{x}} - \underline{\hat{y}}\| \longrightarrow \min, \quad \text{subject to } \underline{\tilde{x}} \in \mathbf{M} \tag{2.7}$$

with

$$\mathbf{M} = \{\underline{x} \in \mathbb{R}^n : 0 \leq x_n \leq x_{n-1} \leq ... \leq x_2 \leq x_1 \leq 1\}. \tag{2.8}$$

For the parameter $\eta = \max(\delta, h_s, h_t)$ we can thus introduce a piecewise constant approximate function

$$x_\eta(t) = x_j^{opt} \quad \text{if} \quad a + (j-1)h_t \leq t < a + jh_t \tag{2.9}$$

of $x(t)$ $(a \leq t \leq b)$ and prove the following convergence result (for details see [2]):

Theorem 2.2 *Under the assumptions stated above we have for $1 \leq q < \infty$*

$$\|x_\eta - x^*\|_{L^q[a,b]} \longrightarrow 0 \quad as \quad \eta \to 0, \qquad (2.10)$$

where x^* denotes the uniquely determined distribution function of the profiles p solving the non-linear integral equation (2.1).

Note that the condition $\eta \to 0$ does not require any intercorrelation between the values h_s, h_t and δ tending to zero. However, under that weak assumptions no rate of convergence can be won. This missing rate of convergence leads to some pessimistic results for the determination of x^* in practice.

In [2] the behaviour of solutions x_η was studied for the case $\kappa(s,p) = \sqrt{p^2 - s}$, $p(t) = \sqrt{25 - 21t}$ and $x^*(t) = \frac{25-t^2}{21}$ $(a = 2 \leq t \leq 5 = b)$. The monotonicity constraints gave rise to staircase functions as optimal approximations x_η. The low rate of convergence for $n, m \to \infty$ was refelcted by a slight decrease of errors and by a slight increase of the number of steps in the graphs of approximate functions x_η (cf. for $n = m = 20$ the broken line in Figure 1). Even if Tikhonov regularization subject to monotonicity

$$\|\underline{A}\tilde{x} - \underline{\hat{y}}\|_2^2 + \alpha^2 \|\underline{L}\,\tilde{x}\|_2^2 \to \min, \quad \text{subject to} \quad \tilde{x} \in M \qquad (2.11)$$

is applied with $\underline{L}\underline{x}$ as the discretized second derivative of the function x, the results x_α of (2.11) are not very satisfactory for a realistic noise-to-signal ratio. This is a typical situation for Fredholm equations with smooth kernels. Only for an extremely small noise-to-signal ratio (10^{-7} in Figure 1) the best possible regularization parameter α_{opt} yields acceptable solutions (dotted line in Fig. 1).

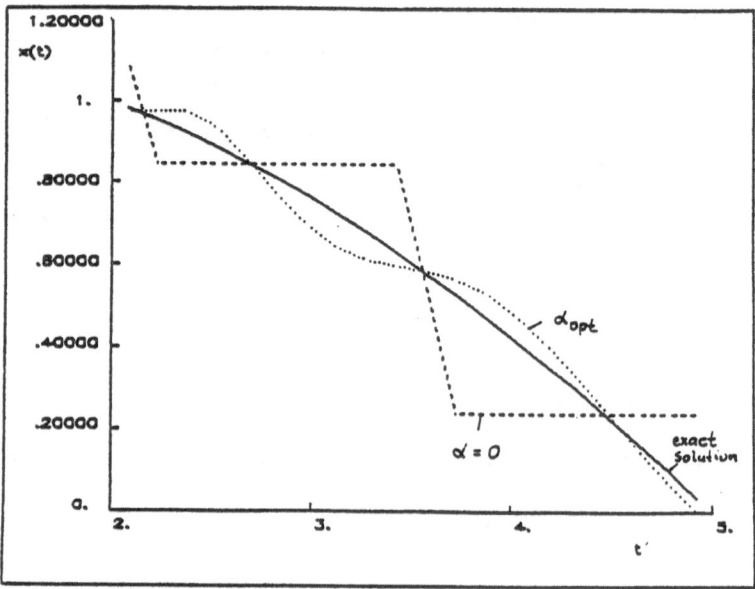

Fig. 1. Regularized solutions subject to monotonicity

The numerical experience of this study indicates that Hansen's L-curve method (cf. [3]) frequently provides a good approximation of this best possible regularization parameter.

References

1. Engl, H.W., Hofmann, B., Zeisel, H.: A decreasing rearrangement approach for a class of ill-posed integral equations. Institutsbericht Nr. 447. Johannes Kepler Universität Linz (Austria), Inst. f. Mathematik, 1991.
2. Gellrich, C., Hofmann, B.: A study of regularization by monotonicity. Paper submitted.
3. Hansen, P.C., O'Leary, D.P.: The use of the L-curve in the regularization of discrete ill-posed problems. UMIACS-TR-91-142. Danish Computing Center for Research and Education, Lyngby, 1991.
4. Hofmann, B.: On the analysis of a particular Volterra-Stieltjes convolution integral equation. Zeitschrift f. Analysis und Anwendungen 7 (1988) 247–257.
5. Hofmann, B., Hausding, R., Wolke, R.: Regularization of a Volterra integral equation by linear inequalities. Computing 43 (1990) 361–375.
6. Hofmann, B., Schachtzabel, H.: Uniqueness of monotone solutions to a non-linear Fredholm integral equation. Optimization 22 (1991) 765–774.
7. Linz, P.: *Analytical and Numerical Methods for Volterra Equations*. SIAM, Philadelphia 1985.

Problems in Impedance Imaging

David Isaacson[1], Margaret Cheney[1] and Jonathan Newell[2]

[1] Department of Mathematical Sciences, Rensselaer Polytechnic Institute, Troy, NY 12180
[2] Department of Biomedical Engineering, Rensselaer Polytechnic Institute, Troy, NY 12180

One of our long-term goals is to understand how to use electromagnetic fields to improve the diagnosis and treatment of disease, especially heart disease. An example in which information about the body's electromagnetic field is presently used is the electrocardiogram, which records the time-varying voltage difference between points on a person's body. These voltages on the body's surface, which are due to currents within the heart, can be used to diagnose some kinds of heart disease.

Our current project is constructing and improving a device to monitor patients for disease. We have built devices, which we call Adaptive Current Tomograph (ACT) systems, that apply small currents to the body through electrodes stuck to the skin. The ACT system measures the induced voltages, and sends all the current and voltage information to an algorithm, which processes the data and reconstructs approximate images of the conductivity and permittivity in the interior.

The equations governing electromagnetic phenomena are Maxwell's equations:

$$\nabla \wedge E = -\partial_t B \tag{1}$$

$$\nabla \wedge H = J + \partial_t D. \tag{2}$$

Here $\nabla \wedge$ is the curl operator, E is the electric field, H is the magnetic field, J is the current density, and D and B are related to E and H as defined below. We assume that $J = J^{\text{Ohmic}} + J^{\text{applied}}$, where, moreover, J^{applied} is time harmonic. We therefore look for time-harmonic solutions, which satisfy (with an abuse of notation)

$$E = E e^{i\omega t}, \qquad D = D e^{i\omega t}, \qquad \text{etc.,}$$

then Maxwell's equations reduce to

$$\nabla \wedge E = -i\omega B \tag{3}$$

$$\nabla \wedge H = J + i\omega D. \tag{4}$$

Since the fields involved are small in magnitude, we assume the linear constitutive relations $D = \epsilon(x,\omega)E$, $B = \mu(x,\omega)H$, $J^{\text{Ohmic}} = \sigma(x,\omega)E$ hold inside the body. Here σ is the electric conductivity, ϵ is the electric permittivity, and μ is the magnetic permeability. We will also use the quantity $1/\sigma = \rho$, the resistivity.

To obtain a simpler equation governing our experiments, we take the divergence of (4) to obtain

$$\nabla \cdot (\sigma + i\omega\epsilon)E = -\nabla \cdot J^{\mathrm{appl}}. \tag{5}$$

In our case, the right-hand side of (3) is negligible, because $\omega/(2\pi) \approx 10$ kHz, $\mu \approx \mu_0$, the permeability of free space, and $|\sigma + i\omega\epsilon| \approx .2(\Omega\text{-m})^{-1}$. If we take the curl of E to be zero, then we can write E as the gradient of a potential:

$$E = -\nabla u, \tag{6}$$

where u is the electric potential. Thus we have the equation

$$\nabla \cdot (\sigma(x,\omega) + i\omega\epsilon(x,\omega)) \nabla u = \nabla \cdot J^{\mathrm{appl}}. \tag{7}$$

To obtain the appropriate boundary conditions, we consider a small pillbox-shaped region that intersects the body, positioned so that the top and bottom of the pillbox are roughly parallel to the surface of the body. We integrate (5) over this volume and use the divergence theorem. By letting the height of the pillbox tend to zero so that the top and bottom of the pillbox coincide with the body's surface, we obtain the jump condition

$$[\![(\sigma + i\omega\epsilon)E \cdot \nu]\!] = -[\![J^{\mathrm{appl}} \cdot \nu]\!], \tag{8}$$

where ν denotes the outward unit normal to the body and the brackets denote the jump at the boundary. Because we apply currents only to the outside surface of the body, J^{appl} is zero on the inside of the body. We assume that the electric field outside the body is small, so that we obtain the condition

$$-(\sigma + i\omega\epsilon)E \cdot \nu\big|_{\mathrm{inside}} = -J^{\mathrm{appl}} \cdot \nu\big|_{\mathrm{outside}}. \tag{9}$$

If we denote by j the applied current density flowing into the body, then we have the boundary condition

$$(\sigma + i\omega\epsilon)\partial_\nu u = -J^{\mathrm{appl}} \cdot \nu \equiv j. \tag{10}$$

To simplify the notation, we write σ instead of $\sigma + i\epsilon$, with the understanding that σ may be complex. We thus study the "standard model"

$$\nabla \cdot \sigma\nabla u = 0 \qquad \text{in the body B.} \tag{11}$$

$$\sigma\partial_\nu u = j \qquad \text{on the surface S.} \tag{12}$$

A solution to this boundary value problem exists only if j satisfies the conservation of charge condition

$$\int_S j = 0, \tag{13}$$

and the solution is unique only if a ground or reference potential is specified, which we do by the condition

$$\int_S u = 0. \tag{14}$$

If, for example, the body is a cylinder of radius r_0, σ is a constant σ_0, and $j(\theta, z) = \cos n\theta$, then

$$u(r_0, \theta, z) = \frac{r_0}{\sigma_0 |n|} \cos n\theta. \tag{15}$$

If we denote the eigenvalue $r_0/(\sigma_0 |n|)$ by ρ_n, then we note that $n\rho_n = r_0/\sigma_0 = $ constant.

The mathematical problem is to find σ in B from knowledge of the map $R(\sigma) : j \mapsto u|_S$. For this problem, there are many interesting results, among them [KV, Sy, SUb, N, R, I, Su, and A]. A nice survey of some of this work appears in [SUa]. In order to use these results, we must solve the following problem: Suppose we know only the currents that are applied to the electrodes. Then what is j?

Suppose our system has L electrodes, which we denote by $e_1, e_2, \ldots, e_L$. Suppose I_l is the current applied to the lth electrode, and V_l is the voltage measured there. From the knowledge of these currents for all l, we want to find j. Knowing j, we can solve (11) and (12) to find u. Once we know u, we need to know how to predict the voltages. To do this, there are a number of models [CING], of which we will discuss the "nogap" or "continuum" model, the "gap" model, and the "complete" or "shunt plus surface impedance" model. To explain these models, we will use p_l to denote the center of electrode e_l. The area of the lth electrode we denote by A_l.

The "nogap" or "continuum" model merely takes j to be a smooth extrapolant of I_l/A_l, and takes V_l to be the potential $u(p_l)$ at the center of the lth electrode.

The "gap" model takes j to be constant on the electrodes and zero in the gaps:

$$j(p) = \begin{cases} I_l/A_l & \text{for } p \text{ on electrode } e_l, \ l = 1, 2, \ldots, L; \\ 0 & \text{if } p \text{ is in a gap,} \end{cases} \tag{16}$$

and V_l is again taken to be the potential $u(p_l)$ at the center of the lth electrode. Alternatively, V_l can be taken to be the average of $u(p)$ over the lth electrode, a model we refer to as the "avgap" model.

The "complete" or "shunt plus surface impedance" model is more complicated. In this model we replace the boundary condition (12) with the conditions

$$\int_{e_l} \sigma \partial_\nu u \, ds = I_l, \qquad l = 1, 2, \ldots, L, \tag{17}$$

$$\sigma \partial_\nu u = 0 \qquad \text{in the gaps} \tag{18}$$

and the constraints

$$u + z_l \sigma \partial_\nu u = \text{constant} = \text{"}V_l\text{"} \qquad \text{on } e_l. \tag{19}$$

Here the V_l are enclosed in quotes to remind us that in the forward problem, these constants are not known in advance, but are determined by solving the problem. It has been shown [SCI] that the forward problem has a unique solution, provided that the conservation of charge condition $\sum I_l = 0$ holds and that a ground has been chosen, for example by requiring that $\sum V_l = 0$.

Determining which of the above models is most accurate requires experimental tests. Our group built a cylindrical test tank in which each electrode extends the entire height, so the tank can be used to check two-dimensional calculations. This tank was filled with saline solution with about the same conductivity as the human body. A full set of linearly independent current patterns $I^1, I^2, \ldots, I^{L-1}$ was applied, and the corresponding voltages $V^1, V^2, \ldots, V^{L-1}$ measured. This results in $(L-1)L$ numbers, which, for our 32-electrode tank, is 992 numbers. These 992 numbers can be organized as eigenvalues and eigenvectors of the map $\Re : I^k \mapsto V^k$. These eigenvectors, moreover, can be found explicitly, because of the symmetry of the tank.

In particular, $\Re$ is a real, symmetric operator [SCI] that commutes with discrete rotations. These facts imply that the eigenvectors of $\Re$ are the trigonometric vectors $T^k = (\exp ik\theta_1, \exp ik\theta_2, \ldots, \exp ik\theta_L)$, $k = 1, 2, \ldots, L-1$, where $\theta_l = 2\pi l/L$. All but one of these eigenvectors are doubly degenerate, which means that only the $L/2$ eigenvalues are needed to recover all 992 numbers.

When these $L/2$ eigenvalues are plotted for the different models, we see that only the complete model is able to predict the experimental results to within the precision of the system.

The next problem is how to find a clinically useful approximation to σ. The methods that have been proposed fall into three categories: iterative optimization [Y, KM, D, K], linearization [C, BB, VS, AS, D], and direct methods [N, Sy, SCII]. Here we discuss the linearized optimization approach we have used.

This approach is based on least squares attempts to choose a resistivity ρ to minimize the error functional

$$E(\rho) = \sum_{k=1}^{L-1} \|U^k(\rho) - V^k\|^2, \tag{20}$$

where V^k denotes the measured voltage due to the kth current pattern and $U^k(\rho)$ denotes the computed voltage corresponding to the resistivity ρ, due to the kth current pattern. This functional can be rewritten in terms of the boundary maps as

$$E(\rho) = \sum_{k=1}^{L-1} \|[R(\rho) - R(\rho^*)]I^k\|^2, \tag{21}$$

where ρ^* denotes the true resistivity. Thus the least squares optimization method can be written as

$$\min_{\rho} E(\rho, I). \tag{22}$$

Note that this error functional depends on the current patterns. These current patterns, however, might not be able to distinguish ρ from ρ^*. This suggests the adaptive least squares procedure

$$\min_{\rho} \max_{I} E(\rho, I). \tag{23}$$

A flow chart outlining an algorithm to carry out this procedure is shown in Fig. 1.

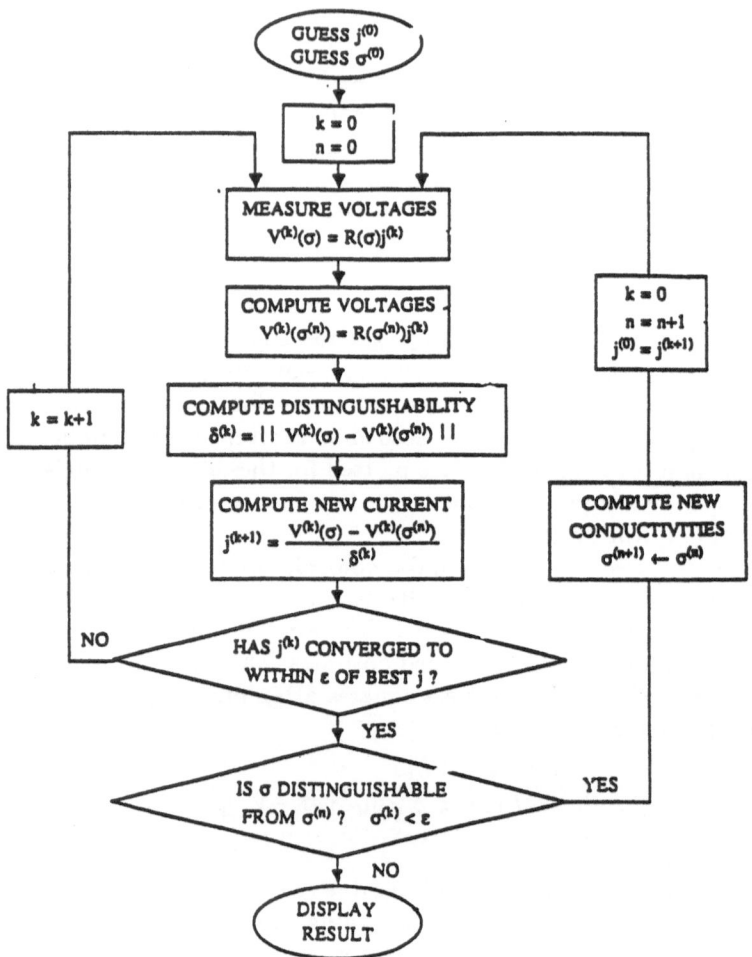

Fig. 1. This is a flow chart for a possible adaptive reconstruction algorithm.

The problem of finding $\max_I E(\rho, I)$ can be solved experimentally by an adaptive process [GIN].

To find $\min_{\rho} E(\rho)$ in practice, we first choose a basis for the resistivity distribution. A convenient basis is the set of characteristic functions of the elements of some mesh:

$$\rho = \sum_{n=1}^{N} \rho_n \chi_n(p), \tag{24}$$

where χ_n denotes the characteristic function of the nth mesh element. The minimization is then done with respect to the N variables $\rho_1, \rho_2, \ldots, \rho_N$. To carry out this minimization, we differentiate with respect to each of the variables, and set each derivative equal to zero:

$$0 = \frac{\partial E}{\partial \rho_n} \equiv F_n(\rho) = 2 \sum_{k=1}^{L-1} \langle U^k - V^k, \frac{\partial U^k}{\partial \rho_n} \rangle \qquad (25)$$

We can solve this system of nonlinear equations by Newton's method. If we denote an initial guess by $\rho^0 = (\rho^1, \rho^2, \ldots, \rho^N)$, then successive iterations can be computed by the formula

$$\rho^{i+1} = \rho^i - [F'(\rho^i)]^{-1} F(\rho^i), \qquad (26)$$

where

$$F'_{n,m}(\rho^i) = \frac{\partial}{\partial \rho_m} \frac{\partial E(\rho^i)}{\partial \rho_n} = 2 \sum_{k=1}^{L-1} \langle \frac{\partial U^k}{\partial \rho_n}, \frac{\partial U^k}{\partial \rho_m} \rangle + 2 \sum_{k=1}^{L-1} \langle U^k - V^k, \frac{\partial^2 U^k}{\partial \rho_n \partial \rho_m} \rangle \quad (27)$$

The next problem is to do the reconstruction rapidly. This we do by means of our code NOSER, which stands for Newton One-Step Error Reconstructor [CINGS]. This produces an approximate reconstruction by the following steps.

The first step is to find the best scalar resistivity ρ^0 by minimizing the error functional in the case when ρ is assumed to be constant. In this case, the linearity of the problem implies that $U(\rho) = \rho U(1)$. Thus the error functional can be written

$$E(\rho) = \sum_{k=1}^{L-1} \|\rho U^k(1) - V^k\|^2 = \sum \rho^2 \langle U^k, U^k \rangle - 2\rho \langle U^k, V^k \rangle + \langle V^k, V^k \rangle. \quad (28)$$

This quadratic equation can be solved for ρ:

$$\rho^0 = \frac{\sum_{k=1}^{L-1} \langle U^k(1), V^k \rangle}{\sum_{k=1}^{L-1} \langle U^k(1), U^k(1) \rangle}. \qquad (29)$$

The second stage is to take one (regularized) step of Newton's method. To do this rapidly, we organize the computation as follows. First we choose an orthonormal basis T^k, $k = 1, 2, \ldots, L-1$. We expand formula (25) in terms of this basis as

$$F_n = 2 \sum \langle U^k - V^k, T^\kappa \rangle \langle T^\kappa, \frac{\partial U^k}{\partial \rho_n} \rangle. \qquad (30)$$

In this formula, the measured data appears as the quantity $v_{k,\kappa} = \langle V^k, T^\kappa \rangle$. We must compute the quantities $\langle U^k, T^\kappa \rangle = \sum U_l^k T_l^\kappa$ and $\langle T^\kappa, \partial U^k / \partial \rho_n \rangle$.

The first of these quantities, $\langle U^k, T^\kappa \rangle$, can be computed as follows. Denote by u the solution to (11) together with the conditions

$$\int_{e_l} \sigma \partial_\nu u = I_l^k \qquad (31)$$

$$u + z\sigma\partial_\nu u = U_l^k, \tag{32}$$

and $\sigma\partial_\nu u = 0$ in the gaps. Denote by v the solution to (11) together with the condition

$$\sigma\partial_\nu v = \begin{cases} T_l^\kappa/A & \text{for } l = 1, 2, \ldots, L; \\ 0 & \text{in the gaps.} \end{cases} \tag{33}$$

We multiply the differential equation for v by u, integrate over the body, and use the divergence theorem to obtain

$$\int_B \sigma\nabla u \cdot \nabla v = \int_S u\sigma\partial_\nu v. \tag{34}$$

On the right side we use the boundary conditions for u to obtain

$$\int_B \sigma\nabla u \cdot \nabla v = \sum_{l=1}^{L} \int_{e_l} (U_l^k - z_l\sigma\partial_\nu u)\sigma\partial_\nu v\, ds = \sum_{l=1}^{L} U_l^k T_l^\kappa - \sum_l z_l T_l^k T_l^\kappa. \tag{35}$$

This gives us a formula for $\langle U^k, T^\kappa \rangle$.

The other needed quantities can be computed in a similar way. For more details in the two-dimensional case, see [CINGS]; for the three-dimensional case, see [G, GIC]

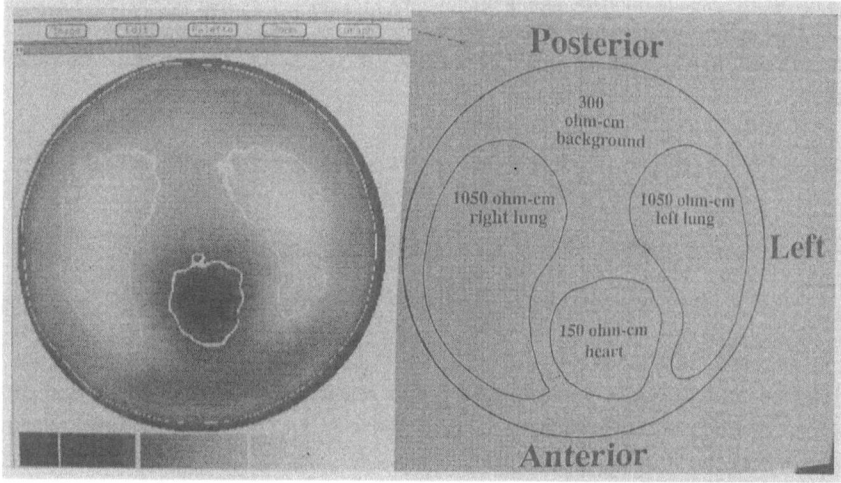

Fig. 2. The right side of the figure shows a diagram of a test tank containing a saline solution with resistivity 300 Ohm-cm. Into this saline solution were placed blocks of agar which had been molded to form "lungs" and a "heart". These blocks were constructed to have resistivities of 1050 Ohm-cm and 150 Ohm-cm, respectively. The test tank has 32 stainless steel electrodes imbedded around its circumference. These electrodes were connected to the ACT 3 system, which applied currents and measured the corresponding voltages. The left side of the figure shows a reconstruction from the NOSER algorithm. The grey scale ranges from 250 (black) to 350 Ohm-cm (white). Three white contour lines, which delineate the locations of the resistivity values indicated on the grey scale, have also been added.

Because the NOSER algorithm is based on a linearization, we expect it to provide a quantitatively accurate reconstruction only when the conductivity distribution is nearly constant. However, even when the conductivity distribution deviates a great deal from a constant, our experience is that NOSER still provides useful qualitative information [II, CII]. Fig. 2 shows an example of such a reconstruction.

To obtain quantitatively accurate reconstructions when the conductivity is not a small perturbation of a constant, one needs to develop nonlinear reconstruction algorithms. A possible nonlinear algorithm is the one based on layer-stripping, which is explained in a separate paper in this volume.

Acknowledgements We are grateful to Ray Cook, Peter Edic, John Goble, David Gisser, Gary Saulnier, and the rest of the Rensselaer impedance imaging group for their ongoing collaboration in building impedance imaging systems.

References

[A]   Alessandrini, G.: Stable determination of conductivity by boundary measurements. Applicable Analysis **27** (1988) 153–172.

[AS]   Allers, A., Santosa, F.: Stability and resolution analysis of a linearized problem in electrical impedance tomography. Inverse Problems **7** (1991) 515–533.

[BAG]   Berntsen, S., Andersen, J. B., Gross, E.: A general formulation of applied potential tomography. Preprint.

[BB]   Barber, D. C., Brown, B. H.: Applied potential tomography. J. Phys. E. Sci. Instrum. **17** (1984) 723–733.

[BBS]   Brown, B. H., Barber, D. C., Seagar, A. D.: Applied potential tomography: possible clinical applications. Clin. Phys. Physiol. Meas. **6** (1985) 109–121.

[C]   Calderón, A. P.: On an inverse boundary value problem. In: Meyer, W. H., Raupp, M. A. (eds.): *Seminar on Numerical Analysis and its Applications to Continuum Physics*, Brazilian Math. Society, Rio de Janeiro, 1980, 65–73.

[CII]   Cheney, M., Isaacson, D., Isaacson, E. L.: Exact solutions to a linearized inverse boundary value problem. Inverse Problems **6** (1990) 923–934.

[CING]   Cheng, K.-S., Isaacson, D., Newell, J. C., Gisser, D. G.: Electrode models for electric current computed tomography. IEEE Trans. Biomed. Engr. **36** (1989) 918–924.

[CINGS] Cheney, M., Isaacson, D., Newell, J.C., Goble, J., Simske, S.: NOSER: An algorithm for solving the inverse conductivity problem. Internat. J. Imaging Systems and Technology **2** (1990) 66–75.

[CW]   Connolly, T. J., Wall, D. J. N.: On an inverse problem, with boundary measurements for the steady state diffusion equation. Inverse Problems **4** (1988) 995–1012.

[D]   Dobson, D.: Estimates on resolution and stabilization for the linearized inverse conductivity problem. Inverse Problems **8** (1992) 71–81; Convergence of a reconstruction method for the inverse conductivity problem. SIAM J. Appl. Math. **52** (1992) 442–458.

[FV]   Friedman A., Vogelius, M.: Identification of small inhomogeneities of extreme conductivity by boundary measurements: a continuous dependence result. To appear in Archive Rat. Mech. Anal.

[G]   Goble, J.: The three-dimensional inverse problem in electric current comput-
          ed tomography. Ph.D. thesis, Rensselaer Polytechnic Institute (1990).

[GIC]   Goble, J., Isaacson, D., Cheney, M.: Impedance imaging in three dimensions.
          To appear in the Journal of the Applied Computational Electromagnetics
          Society.

[GIN]   Gisser, D.G., Isaacson, D., Newell, J. C.: Theory and performance of an
          adaptive current tomography system. Clin. Phys. Physiol. Meas. 9, Suppl.
          A (1988) 35–41; Current topics in impedance imaging. Clin. Phys. Physiol.
          Meas. 8 (1987) 39–46.

[I]   Isakov, V.: On uniqueness of recovery of a discontinuous conductivity coeffi-
          cient. Comm. Pure Appl. Math. 41 (1988) 865–877.

[IC]   Isaacson D., Cheney, M.: Current problems in impedance imaging. In: Colton,
          D., Ewing, R., Rundell, W. (eds.): *Inverse Problems in Partial Differential
          Equations*, SIAM, Philadelphia 1990.

[II]   Isaacson D., Isaacson, E. L.: Comment on Calderón's paper: 'On an inverse
          boundary value problem'. Math. Comput. 52 (1989) 553–559.

[K]   Klibanov, M.: Newton-Kantorovich method for impedance imaging. Preprint.

[KM]   Kohn, R. V., McKenney, A.: Numerical implementation of a variational
          method for electrical impedance imaging. Preprint.

[KV]   Kohn R. V., Vogelius, M.: Determining conductivity by boundary measure-
          ments. Comm. Pure Appl. Math. 37 (1984) 113–123.

[N]   Nachman, A. I.:Reconstructions from boundary measurements. Annals of
          Math. 128 (1988) 539–557.

[R]   Ramm, A. G.: Completeness of the products of solutions to PDE and unique-
          ness theormes in inverse scattering. Inverse Problems 3 (1987) L77–L87.

[SCI]   Somersalo, E., Cheney, M., Isaacson, D.: Existence and uniqueness for elec-
          trode models for electric current computed tomography, SIAM J. Appl. Math.
          52 (1992) 1023–1040.

[SCII]   Somersalo, E., Cheney, M., Isaacson, D., Isaacson, E. L.: Layer stripping: A
          direct numerical method for impedance imaging. Inverse Problems 7 (1991)
          899–926.

[Su]   Sun, Z.: On an inverse boundary value problem in two dimensions, Comm.
          PDE 14 (1989) 1101–1113.

[SUa]   Sylvester J., Uhlmann, G.: The Dirichlet to Neumann map and applications.
          In: Colton, D., Ewing, R., Rundell, W. (eds.):*Inverse Problems in Partial
          Differential Equations*, SIAM, Philadelphia 1990.

[SUb]   Sylvester, J., Uhlmann, G.: A uniqueness theorem for an inverse boundary
          value problem in electrical prospection. Comm. Pure Appl. Math. 39 (1986)
          91–112; A global uniqueness theorem for an inverse boundary value problem.
          Ann. of Math., 125 (1987) 153–169; Inverse boundary value problems at the
          boundary — continuous dependence. Comm. Pure Appl. Math. 41 (1988)
          197–221.

[SV]   Santosa, F., Vogelius, M.: A backprojection algorithm for electrical
          impedance imaging. SIAM J. Appl. Math. 50 (1990) 216–243.

[Sy]   Sylvester, J.: A convergent layer stripping algorithm for the radially sym-
          metric impedance tomography problem. To appear in Comm. Partial Diff.
          Eqs.

[Y]   Yorkey, T. J.: Comparing reconstruction methods for electrical impedance
          tomography, Ph.D. dissertation, Univ. of Wisconsin, August 1986.

Uniqueness for Inverse Problems in Quasilinear Differential Equations

*Victor Isakov**

Department of Mathematics and Statistics, Wicita State University, Wichita, KS 67208, USA

In this paper we consider identification problems for quasilinear elliptic and parabolic equations of second order. We are looking for terms $a(x,u)$ of these equations when in addition to classical boundary value data some other boundary data are prescribed. Such problems are quite important in applications where they model recovery of parameters in heat conduction, chemical cinetics, population dynamics etc. We refer for example to the paper of Ewing, Tao Lin and Falk [ETF] about oil search. We will describe a contemporary situation concerning uniqueness of identification. We will outline proofs and emphasize difficulties and problems.

1. Elliptic equations

Let Ω be a bounded domain in the euclidean space $\mathbf{R}^n$ with the C^2-boundary Γ. Consider the following Dirichlet Problem

(1) $$- \Delta u + a(x,u) = 0 \quad \text{in } \Omega$$

(2) $$u = g \quad \text{on } \Gamma$$

It can be shown that under the assumptions

---
* This research was supported in part by NSF grant #DMS–9101421

$$a, \dots, a_{uu} \in L_\infty(\Omega \times [- U, U]) \text{ for any } U < +\infty$$

(3)

$$0 \le a_u$$

for any function $g \in C^2(\Gamma)$ there is a unique solution u to the boundary value problem (1),(2) which is contained in the space $C^2(\Omega) \cap C^1(\Omega)$. So we have the well-posed direct problem (1),(2).

We are interested in the inverse (or identification) problem: determine a given additional boundary value data. We consider two types of additional information: the Neumann data

(4) $$\partial u / \partial N = h \quad \text{on } \Gamma$$

for one g (single boundary measurement) and the Neumann data for all g (many boundary measurements, or the Dirichlet-to-Neumann map). Accordingly, we assume that $a = a(u)$ or $a = a(x,u)$ and we will give uniqueness results in the both cases.

1.1 Single Boundary Measurements

In this case one can not determine a function $a(x,u)$ of $n+1$ variables from the additional Neumann data which is a function of n variables so it is natural to assume that

(5) $$a = a(u) + f(x), \qquad a(0) = 0$$

where f is a given function and a is unknown.

Let $n = 2$. We assume that Γ is the union of two connected arcs Γ_1, Γ_2 and their endpoints x^1 and x^2.

We give a local uniqueness result due to Pilant and Rundell [PR2].

Assume that

(6) $$g, h \in C^3(\Gamma), \qquad f \in C^2(\bar{\Omega})$$

that the solutions u_0, u_+, u_- to the boundary value problem (1),(2) (with $a = 0, -C$, $+C$) satisfy the inequalities

(7) $\qquad u_+(x^1) \leq u_+(x) \leq u_+(x^2) \qquad$ when $x \in \Omega$

and the boundary data

(8) $\qquad\qquad\qquad$ g is an increasing function of the arc length parameter on Γ_2

$\qquad$ Theorem 1. Let the conditions (6)-(8) be satisfied, then for sufficiently small $\epsilon > 0$ there is an unique solution (u, a), $u \in C^2(\bar{\Omega})$, $|a|_{2}(I) < \epsilon$, to the inverse problem provided $|\Delta_\tau(g - u_0)|_{2}(\Gamma) < \epsilon$.

$\qquad$ Here I is the range of u ,Δ_τ is the tangential Laplace Operator and we claim uniqueness of u only on I. Remind that $|\ |_k(\Omega)$ is the norm in the space $C^k(\bar{\Omega})$.

$\qquad$ The basic idea of the proof in [PR2] is to reduce the inverse problem to the equation

(9) $\qquad a(g(x)) = -f(x) + \Delta_\tau g(x) + (\Delta - \Delta_\tau)(u(x)), \quad x \in \Gamma_2$

(where $u = u(\ ; a)$ via the boundary value problem (1),(2)) and to solve this equation (with respect to a) by contractions arguments.
$\qquad$ Observe that we do not know global uniqueness results in this problem.
$\qquad$ It is interesting that for the equation

(10) $\qquad div(\ a\ \nabla u\ ) = 0 \qquad$ in Ω

there are global uniqueness results. We consider the Dirichlet Problem (10),(2) and prescribe the additional Neumann data

(11) $\qquad\qquad a\ \partial u/\partial N = h \quad$ on Γ_2

We assume that $a \in C^1(\bar{I})$, $a \geq \epsilon > 0$ on I.

Theorem 2. (Cannon [C], Pilant,Rundell [PR3]) If the function $g \in C^2(\Gamma)$ is monotone with respect to the arc length parameter on Γ_1 and on Γ_2 ,then a solution (u,a) to the inverse problem (10),(2),(11) is unique.

We outline a proof.

Let (u_1,a_1), (u_2,a_2) be two possible solutions. Introduce the functions

$$v_j(x) = \int_{u_j(x^1)}^{u_j(x)} a_j(r)\, dr$$

Let $v = v_2 - v_1$, $a = a_2 - a_1$. Then using (10) we obtain

(12) $$\Delta v = \Delta v_2 - \Delta v_1 = f - f = 0 \quad \text{in } \Omega$$

and using (11) we get

(13) $$\partial v / \partial N = h - h = 0 \quad \text{on } \Gamma_2$$

From (2) and from the definition of v_j we have

(14) $$v(x) = \int_{g(x^1)}^{g(x)} a(r)\, dr \qquad\qquad \text{when } x \in \Gamma$$

From (13) according to the Giraud's extremum principle ([I1,Theorem 1.3.1) we obtain that v can not achieve maximum or minimum over Ω neither on Γ_2 or in Ω unless v is constant. So maximum and minimum are achieved on Γ_1. But due to our assumptions on g range of g on Γ_2 is equal to range on Γ_1, so from (14) it follows that if maximum is achieved on Γ_1 it is achieved also on Γ_2 which is a contradiction to what we discussed above. Therefore v is constant and a is zero on I_1.

This completes the proof.

1.2 Many boundary measurements.

Now we turn to all posssible boundary measurements. As observed at the beginning of this section, for any smooth data g there is an unique solution u to the Dirichlet Problem (1),(2), so we have the well-defined Dirichlet - to Neumann map $\Lambda_a: g \to \partial u/\partial N$ on Γ. Consider a_1, a_2 and let $\Lambda_j = \Lambda_{a_j}$.

Theorem 3 (Isakov, Sylvester [IS]) Let $n \geq 3$. Let a_j satisfy the assumptions (3) and moreover

(15)
$$a_j(x,0) = 0 \quad \text{when} \quad x \in \Omega$$

(16)
$$\partial a_j/\partial u \text{ is bounded on } \Omega \times \mathbb{R}$$

If $\Lambda_1 = \Lambda_2$ (on $C^2(\Gamma)$), then

(17)
$$a_1 = a_2 \quad \text{on } \Omega \times \mathbb{R}$$

This Theorem generalizes the fundamental result of Sylvester and Uhlmann [SU] where $a(x,u) = a(x)\, u$.

We outline its proof referring for details to the forthcoming paper [IS].

1) Linearization.

Let $a^*_j(x,g) = \partial\, a_j/\partial u(\, x;\, u_j(x,g))$. Denote by Λ^*_j the map $g^* \to \partial v/\partial N$ on Γ which corresponds to the linear elliptic equation

(18)
$$-\Delta v + a^*_j v = 0 \quad \text{in } \Omega$$

Lemma 1.2.1 Under the conditions of Theorem 3 we have $\Lambda^*_1 = \Lambda^*_2$.

To prove this Lemma we let $g^* \in C^2(\Gamma)$ and consider the finite differences $v_j(\ ;\tau) = (u_j(\ ;g+\tau g^*) - u_j(\ ;g))/\tau$. By subtracting the equations (1) for u_2 and for u_1 and using the Mean Value Theorem we conclude that

$$-\Delta v_j(\ ;\tau) + a^*_j(\ ;g)\, v_j(\ ;\tau) + o\,(\tau)\, v_j(\ ;\tau) = 0 \qquad \text{in } \Omega$$

So by applying standard elliptic estimates and passing to the limit as $\tau \to 0$ we obtain that $v = v_j(\ ;0)$ solves the equation (18). It is clear that $v_j = g^*$ on Γ.

Since by the conditions of Theorem 3 we have $\partial u_1(\ ;g)/\partial N = \partial u_2(\ ;g)/\partial N$ on Γ we get $\partial v_1(\ ;\tau)/\partial N = \partial\, v_2(\ ;\tau)/\partial N$ there and passing to the limit we will have $\partial v_1(\ ;0)/\partial N = \partial v_2(\ ;0)/\partial N$ on Γ. This completes the proof.

From Lemma 1.2.1 and from the results of the papers [I2],[SU] on coincidence of linear elliptic operators (18) with the same Dirichlet-to-Neumann maps when $n \geq 3$ it follows that $a^*_1 = a^*_2$ and therefore

$$(19) \qquad \partial a_1/\partial u\ (\ x,\ u_1(x,g)) = \partial a_2/\partial u\ (x,\ u_2(x,g)) \quad \text{when } x \in \Omega$$

2) Range of u.

We need also the following result.

Lemma 1.2.2. Under the assumptions (3), (16) we have $u_j(x,\theta) \to +\infty$ as $\theta \to +\infty$ and $u_j(x,\theta) \to -\infty$ as $\theta \to -\infty$.

To prove this Lemma we differentiate the equalities (1),(2) with $a = a_j$ and $g = \theta$ with respect to θ to obtain

$$(20) \qquad -\,\Delta\,w + \,a_{ju}(x,u_j(x,\theta))\, w\, = 0 \quad \text{on } \Omega, \qquad w\, = 1 \text{ on } \Gamma$$

where $w = u_{j\theta}$. By the condition (16) we have $0 \leq a_{ju} \leq A$ where A is a positive number. Considering the solution w_A to the problem (20) with A replacing a_{ju} and applying comparison theorems we conclude that $w_A \leq w \leq 1$. Also observe that

$\epsilon(A) < w_A$ for some positive number $\epsilon(A)$ by extremum principles.

So $\epsilon(A) < u_{j\theta} \leq 1$ which completes the proof.

3) The basic step.

To complete the proof of Theorem 3 we assume that (17) is not true. We may assume then that the number

$$(21) \quad \theta^* = \sup \{ \theta: a_1(x, u_1(x,\theta)) = a_1(x, u_1(x,\theta)) \quad \text{for all } x \in \Omega \} < +\infty$$

and we will obtain a contradiction as soon as we make use of Lemma 1.2.2 and of the following result.

Lemma 1.2.3 If $\delta > 0$ is small then the relation (19) for all θ implies that $a_1(x,u_1(x,\theta)) = a_2(x,u_1(x,\theta))$ when $x \in \Omega$ and $\theta < \theta^* + \delta$.

To prove we rewrite (19) as follows

$$a_{2u}(x,u_1(x,\theta)) - a_{1u}(x,u_1(x,\theta)) = a_{2u}(x,u_1(x,\theta)) - a_{2u}(x,u_2(x,\theta)) =$$

$$(22) \quad (\int_0^1 a_{2uu}(x,u_2+\tau u) \, dt) \, u$$

where $u = u_2 - u_1$, $u_j = u_j(x,\theta)$. By subtracting the equations (1),(2) for u_1, u_2 we obtain

$$(23) \quad - \Delta u + bu = a_1(x,u_1) - a_2(x,u_2) \quad \text{on } \Omega , \quad u = 0 \text{ on } \Gamma$$

where $b(x)$ is the integral of $a_{2u}(x,(u_2+\tau u))$ over $(0,1)$.

By the definition of θ^* we have $a_1(x,u_1) = a_2(x,u_1)$ when $x \in \Omega$ and $\theta \leq \theta^*$. Using the Mean Value Theorem we get

$$| a_1(x,u_1) - a_2(x,u_1)) | \leq \sup|(a_{1u}-a_{2u})(x,u_1(,\xi))| \, |u_{1\theta}| \, (\theta-\theta^*)$$

where sup is taken over $\xi \in (\theta,\theta^*)$ and $\theta^* < \theta$. Combining this inequality with (22) and

(23) we conclude that $-\Delta u + bu = f$ in Ω where $\|f\|_\infty(\Omega) \leq C_1 (\theta - \theta^*) \|u\|_\infty(\Omega)$ with C depending only on θ^*. Since $\|b\|_\infty(\Omega)$ is bounded by a similar constant we have from elliptic estimates that $\|u\|_\infty(\Omega) \leq C\|u\|_\infty(\Omega)$. Choosing $\delta > 0$ so that $C_1(\theta - \theta^*) < 1$ we obtain that $u = 0$ or $u_1(x, \theta) = u_2(x, \theta)$. Hence, from (19) we derive that $\partial a_1/\partial u(x, u_1(x,\theta)) = \partial a_2/\partial u(x, u_1(x,\theta))$ when $x \in \Omega$, $\theta < \theta^* + \delta$. Now from the condition (15) we get the claim of Lemma 1.2.3.

We observe that the condition (15) is somehow essential, in fact one can not determine the right hand side of the Poisson equation given results of all possible boundary measurements. We refer for positive results and for counterexamples to the book [I1], sections 3.3,3.4.

The condition (16) can be removed but then we will be able to show only that $a_1 = a_2$ on the set $E = \{(x,u) : u_{1*}(x) < u < u^{1*}(x)\}$ where $u_1^*(x) = \sup \{u_1(x,\theta)$ over $\theta \in \mathbf{R}\}$ and $u_{1*}(x) = \inf \{u_1(x,\theta)$ over $\theta \in \mathbf{R}\}$. There are examples of equations (i.g., $-\Delta u + u^3 = 0$) when $E \neq \Omega \times \mathbf{R}$. For complete proofs we refer to the forthcoming paper [IS].

2. Parabolic equations.

Here we describe similar results for the parabolic initial boundary value problem

$$(24) \qquad u_t - \Delta u + a(x, t; u) = 0 \qquad \text{in } Q$$

where $Q = \Omega \times (0, T)$ with the initial data

$$(25) \qquad u = u_0 \qquad \text{on } \Omega \times \{0\}$$

and with the Dirichlet lateral boundary condition

$$(26) \qquad u = g \text{ on } \Gamma \times (0, T)$$

or with the Neumann condition

(27) $\partial u/\partial N = h$ on $\Gamma \times (0,\ T)$

Later on we will make assumptions which guarantee existence and uniqueness of
a solution $u(x,t\ ;\ a)$ to the boundary value problem (24),(25),(26) with given a. In the
inverse problem a is unknown and to find it we have to use additional data as above
for elliptic equations.

2.1 Single Measurements.

Here we consider the problem (24),(25),(27) and prescribe additional data g for
one single set of data h, so we have to restrict a and we assume that

(28) $a\ =\ b(u) + f(x,t),$ $b(0) = 0$

where f is given and b is to be determined.

As an additional data for determination of b we use

(29) $u(x^o,t) = g(t),$ $0 < t < T$

where x^o is a point at $\partial\Omega$.

We give the following uniqueness and existence result due to Pilant and Rundell
[PR3].

Assume that $\Gamma \in C^3$ and Ω is bounded. Also assume that

(30) $u_0 = 0,$ $f \in C^1(\overline{Q}),$ $f > 0$ on Q

(31) $h, h_t \in C^1(\Gamma \times [0,T]),$ $h(x,t) \leq h(x^o,t)$
                     $0 < h_t$ on $\Gamma \times [0,T],$ $g(x,\ 0) = 0$

The additional data (29) are assumed to be satisfying the following conditons

$$g \in C^2[0,T], \quad g(0) = 0, \ 0 < g' \ \text{on} \ [0,T]$$

(32)

$$| g - u(x^o, ;0)|_1 ([0,T]) < C, \quad g'(0) = f(x^o, 0)$$

Theorem 4. If the data g,h of the inverse problem (24),(25),(27),(29) satisfy the conditions (30),(31),and (32) then for some T depending only on C and Ω there is an unique solution (u,a) to the inverse problem with $u \in C^{2+\lambda, 1+\lambda/2}(\overline{Q})$, $a \in C^\lambda[\overline{I}]$.

This result is local (T is small) and we do not know a global result (even on uniqueness).

An idea of proof is similar to that one for elliptic equations. One considers the equation (24) at the point (x^o, t) and reduces the inverse problem to the following non-linear (integral) equation with respect to a

$$g'(t) - \Delta u \ (x^o, t; \ a) - f(x^o, t) = b((u(x^o, t; a))$$

where one represents $u(x^o, t; \ a)$ by Green's functions as a solution to the parabolic problem (24),(25), (27). For small T this equation can be solved by contraction argument.

We mention also the recent preprint of Klibanov [K] where there are references to some earlier works on quasilinear parabolic equations in somehow different formulations. In this preprint he studies the equation (24) with $a = a (x_2, ..., x_n, u)$ under the assumptions that a_u is bounded and nonnegative, $\partial u_0 / \partial x_1 < -\epsilon$, $h < -\epsilon$, $\partial g / \partial t > \epsilon$, and that a is given in some region depending on the initial data u_0, in his paper $\Omega = \{ x_1 > 0 \}$. He is using the hodograph transformation to reduce the nonlinear equation (24) to a linear one where one can apply the method of Carleman estimates suggested by Buchgeim and Klibanov in 1981 and exposed, for example, in the book [I1], section 2.6.

2.2 Many Measurements

As for elliptic equations we assume that

(33) $a,\ a_u,\ a_{uu} \in L_\infty\ (Q \times (-U,U))$ for any $U < +\infty$

and that either

(34) $0 \leq a_u$

or

(35) $a_u \in L_\infty(\ Q \times \mathbf{R})$

It is known that under the conditions (33),(34) or (33),(35) for any function
$g \in C^2(\Gamma \times [0,T])$ and for any function $u_0 \in C^1(\overline{\Omega} \times \{0\})$, $g = u_0$ on $\Gamma \times \{0\}$
there is an unique solution u to the direct problem (24),(25),(26) with second order x-
derivatives and first order t-derivative in $L_2(Q)$. Define u_T as u on $\Omega \times \{T\}$. We
are interested in the following inverse problem with all possible boundary
measurements. Given the map $\Lambda_a : (u_0,g) \to (u_T,h)$, from the space of functions
described above, find the term $a(x,t;\ u)$ of the differential equation (24).

   Theorem 5. Assume that the functions $a_j = a,\ j = 1,2$, satisfy the
conditions (33),(34) and the condition

(36) $a_j(x,t,0) = 0$

   Let $\Lambda_{a_1} = \Lambda_{a_2}$.
   i)If in addition the condition (35) is satisfied as well then $a_1 = a_2$ on $Q \times \mathbf{R}$.
   ii) If in addition $\partial a_j/\partial t = 0$ then $a_1 = a_2$ on $\Omega \times \mathbf{R}$.

   Observe that in general case (when a depends on x,t,u) we can not expect uniqueness of a without the condition (35) which is of course quite restrictive. A reason has
been explained after Theorem 3. In part ii) we assume that a_j is t-independent and

then we can drop this restrictive condition. This part is proven in th paper [I3] .The part i) is new and can be proven similarly to Theorem 3. We outline a scheme of a proof.

Let $a_j^*(x,t; u_o,g)$ be $\partial a_j/\partial u(x,t,u_j(x,t; u_o,g))$.

Lemma 2.2.1. Under the conditions of Theorem 5 the Dirichlet-to-Neumann maps $(u_o^*,g^*) \to (u_{jT}^*,h_j^*)$ which correspond to the linear parabolic equations $v_t - \Delta v + a_j^* \, v = 0$ coincide.

The proof is similar to the proof of Lemma 1.2.1.

From Lemma 2.2.1 and from results of paper [I2] on linear parabolic equations we have

$$(37) \qquad \partial a_1/\partial u(x,t,u_1(x,t; u_o,g)) = \partial a_2/\partial u(x,t,u_2(x,t;u_o,g))$$

when $(x,t) \in Q$. To prove the part ii) we let $t = 0$ and observe that then $u_1 = u_2 = u_o$. Since we can prescribe arbitrary smooth u_o we conclude that $\partial a_1/\partial u \, (x,u) = \partial a_2/\partial u \, (x,u)$ and from the assumption (36) we derive that $a_1 = a_2$.

To prove i) we make use of the following Lemmas.

Lemma 2.2.2 Under the assumption (35) we have $\lim u_j(x,t;\theta,\theta) = +\infty$ as $\theta \to +\infty$ and $\lim u_j(x,t;\theta,\theta) = -\infty$ as $\theta \to -\infty$.

Proof is quite similar to the proof of Lemma 1.2.2 since comparison theorems and extremum principles are valid for second order parabolic equations as well.

To complete the proof we introduce the number θ^* as in the proof of Theorem 3 and assuming that $a_1 \neq a_2$ we conclude that this number is finite. To obtain a contradiction it suffices to use the following result.

Lemma 2.2.3 If positive δ is small then the equality (37) implies that $a_1(x,t;u_1(x,t;\theta,\theta)) = a_2(x,t;u_1(x,t;\theta,\theta))$ when $(x,t) \in Q$ and $\theta < \theta^* + \delta$.

Now we consider a more difficult case when only results of lateral boundary measurements are available and we are looking for $a = a(x,u)$. We will impose some additional constraints which guarantee uniqueness.

We assume that $T = +\infty$, $u_0 = 0$ and we introduce the lateral Dirichlet-to-Neumann map $\Lambda_a^l : g \to h$. The assumptions (33),(34) guarantee that for any function g in $C_0^3(\Gamma \times (0,+\infty))$ there is an unique solution u to the problem (24)-(26) with second order x - derivatives and first order t-derivative continuous in Q.

Theorem 6. Let the functions $a = a_j$ satisfy the conditions (33),(34),(35) and in addition the condition

(38) $$a_j(x,u) = a_j^{\#}(x) \, u \qquad \text{when } |u| < \epsilon_0$$

If $\Lambda_{a_1}^l = \Lambda_{a_2}^l$ then $a_1 = a_2$ on $\Omega \times \mathbf{R}$.

The proof is based on the ideas explained above but now we have an additional difficulty connected with zero initial data and unbounded domain Q. To "cut off" zero and infinity we will use the condition (38) and the following asymptotic behavior of a solution $|u_j(x,t)| \leq C \, e^{-\epsilon t}$. We modify the proof of Theorems 3,5, in particular the construction (21).

Let $g_0(t)$ be a $C^\infty(\mathbf{R})$-function, $0 < g_0 < 1$ when $1/4 < t < 3/4$ and $g_1 = 0$ otherwise. Let $u_j(\ ;\theta)$ be a solution to the first boundary value problem (24),(25),(26) with $a = a_j$, $u_0 = 0$ and $g = \theta g_0$. Accordingly we change the definition of θ^* in (21).

A central new point of the proof is the following result

Lemma 2.2.4 Under the condition of Theorem 6 we have

(39) $$\int_Q (a_1^* - a_2^*) \, v_1 \, v_2^* \; = \; 0$$

for all solutions v_1 to the equation

(40) $$v_{1t} - \Delta \, v_1 + a_1^* \, v_1 = 0 \quad \text{near } \overline{\Omega} \times [0,T]$$

and for all solutions v_2^* to the equation

(41) $- v_{2t}{}^* - \Delta\, v_2{}^* + a_2{}^* v_2{}^* = 0$ near $\overline{\Omega} \times [0, T]$

provided T is sufficiently large.

A proof is given in [I3]. The main idea is to subtract two linear equations for v_2 and v_1 obtaining for their difference v the equation $v_t - \Delta v + a_2{}^* v = (a_1{}^* - a_2{}^*) v_1$ and the boundary conditions $v = 0$ on $\Omega \times \{0\}$ and the lateral conditions $v = \partial v / \partial N = 0$ on $\Gamma \times (0, +\infty)$. Then we multiply the both parts of the differential equation for v by a solution $v_2{}^*$ to (41) and integrate by parts. All boundary integrals for the left side will be zero because in fact $v = 0$ for small and large t. So we obtain the realtion (39) for all $v_2{}^*$ mentioned in Lemma and for all v_1 which are solutions to (40) in Q and satisfy zero initial conditions. It can be shown that any solution v_1 to (40) can be approximated by such solutions on $\Omega \times [1/T, T]$ for large T (a variant of Runge property) and this will complate the proof.

To show that $v = 0$ outside of $\Omega \times (1/T, T)$ we first prove that $a_1{}^* = a_2{}^*$ on $\Omega \times (0, 1/T)$, here we make use of the assumption (38) and identify the x- dependent coefficient of the linear parabolic equation from results of all lateral boundary measurements. Then we rememeber the asymptotic behavior of u_j for large t we conclude that $a_1{}^* = a_2{}^*$ also for $t > T$. Henceforth v solves a homogeneous linear parabolic equation on $\Omega \times (T, +\infty)$, in addition it has zero Cauchy data on the lateral surface $\Gamma \times (T, +\infty)$, so by uniqueness of the continuation we have $v = 0$ there.

By using Lemma 2.2.4 we can derive theorem 6 from theorem5.

We feel that the condition (38) can be removed if one integrates by parts in the proof of Lemma 2.2.4 over the domain $\Omega \times (0, T)$, let $T \to +\infty$ and makes use of asymptotic behavior to eliminate boundary integrals over $\Omega \times \{T\}$.

3. Open problems

We give some unsolved and challenging problems which at our opinion are of mathematical and applied importance.

Problem 3.1) Prove uniqueness of $a(u)$ of the equation (1) with given Dirichlet-to-Neumann map when $n=2$.

We think that one might try to use localization and parameter θ as it was done in the proof of Theorem 3.

<u>Problem 3.2)</u> Prove uniqueness of the coeffiecient $a(x,u)$ of the equation $div\,(a\,\nabla u)$ $= 0$ with given Dirichlet-to-Neumann map

Probably it is a difficult question, and at the timewe do not know how to attack it.

<u>Problem 3.3)</u> Is Theorem 6 true for a finite domain $\Omega \times (0, T)$?

<u>Problem 3.4)</u> Prove uniqueness of the coefficient $a(x,u)$ of the quasilinear parabolic equation $a\,u_t - \Delta u\, = 0$ with given Dirichlet-to-Neumann (lateral) map.

We'd like to emphasize that this problem is quite important for application to oil search. Some results for terms $a = a(u)$ entering the principal part of a quasi-linear parabolic equation were obtained by Cannon and DuChateau [CD] and by Lorenzi and Lunardi [LL].

We think it is possible to obtain similar results for hyperbolic equations.

Certainly, it is extremely important to analyse stability and to develop appropriate numerical algorithms. We observe that in such algorithms one can use the basic constructive idea of the proof of Theorem 3: locality with respect to parameter θ.

REFERENCES

[C]   Cannon, J.R., Determination of the unknown coefficient $k(u)$ in the equation
        $\nabla \cdot k(u)\nabla u = 0$ from overspecified boundary data, J. Math.Anal.Appl.,
        18 (1967), 112-114.

[CD]   Cannon, J.R., DuChateau, P., An inverse problem for a nonlinear diffusion
        equation, SIAM J.Appl.Math.,39 (1980), 272-289.

[ETF]   Ewing, R., Tao Lin, Falk, R., Inverse and Ill-Posed Problems in Resevoir
        Simulation, in "Inverse and Ill-Posed Problems", Academic Press, 1990,
        pp.483-492

[I1]   Isakov, V., Inverse Source Problems, AMS Math. Monographs and Surveys,
        Vol.24, Providence, RI, 1990

[I2]   Isakov, V., Completeness of products and some inverse problems for PDE,
        J. of Diff. Equat., 92 (19910, 305-317.

[I3]   Isakov, V., On Uniqueness in Inverse Problems for Quasilinear Parabolic
        Equations, Preprint, 1992.

[IS]   Isakov, V., Sylvester, J., Global Uniqueness for a Semilinear Elliptic
        Inverse Problem, Preprint, 1992.

[K]   Klibanov, M.V., Uniqueness of one multidimensional inverse problem for
        reaction-diffusion equation, preprint, 1992.

[LL]   Lorenzi, A., Lunardi, A., An identification problem in the theory of heat
        conduction, Differential and Integral Equations, 3 (1990), 237-252.

[PR1]   Pilant, M.,Rundell,W., An inverse problem for a nonlinear parabolic equa-
        tion, Comm. Part.Diff.Equat.; 11 (1986), 445-457.

[PR2]   Pilant, M., Rundell,W., An Inverse Problem for a Nonlinear Elliptic Diffe-
        rential Equation, SIAM J. of math.Anal., 18 (1987), 1801-1809.

[PR3]   Pilant, M.,Rundell,W., A Uniqueness Theorem for Determining Conductivity
        from Overspecified Boundary Data, J. Math.Anal.Appl., 136 (1988),20-28.

[SU]   Sylvester, J., Uhlmann, G., A global uniqueness theorem for an inverse boundary
        value problem, Ann.of Math., 125 (1987), 153-169.

Diffraction by Periodic Structures

Andreas Kirsch

Institut für Angewandte Mathematik, Universität Erlangen-Nürnberg, 8520 Erlangen, Germany

1 Introduction

In this paper we will study some aspects of the scattering of time harmonic plane waves by an acoustically soft periodic surface. We restrict ourselves to the two dimensional problem. The next section recalls some basic results including the Rayleigh expansion, uniqueness and existence, the latter by using the well known integral equation method (cf. [15]). The Green's function on which the integral equation method relies is not defined on an envelope of curves in the (k, θ)−plane. Here, $k > 0$ denotes the wave number and $\theta \in (0, \pi)$ the angle of the incident plane wave.

For this reason − and for the study of the behavior of the solution in the neighborhood of these exceptional values − we introduce in Section 3 a variational formulation which allows it to prove existence for all values of k and θ. It turns out that, in contrast to the scattering by bounded obstacles, the field does not depend analytically on k and θ at these values but admits square root like singularities.

Section 4 is devoted to the sensitivity of the solution with respect to variations of the boundary curve. First, the continuous dependence is shown and, second, the Fréchet differentiability of the field with respect to the curve is shown and a characterization of the derivative is given. In a future paper we plan to use this derivative for the construction of an efficient algorithm for solving the inverse scattering problem (cf. [10] for the case of a bounded obstacle).

Diffraction by gratings has a long history. We refer to the monograph edited by R. Petit [15] for some of the physical and mathematical background and to Wilcox' Lecture Notes [17] on the spectral theory. In particular, we want to direct the attention to a the paper by Millar [11] where a good overview on results until 1973 is given. For more recent papers we refer to [1, 5, 14].

2 Repetition of Some Basic Results

Let $k > 0$ the wavenumber and $\hat{\theta} = (\cos \theta, -\sin \theta)^{\top} \in \mathbb{R}^2$ a unit vector with second component $\hat{\theta}_2 < 0$ (i.e. $\theta \in (0, \pi)$). The incident plane wave is given by

$u^i(x) = \exp(ik\hat{\theta} \cdot x)$, $x \in \mathbb{R}^2$ and the scattering object by the curve $x_2 = f(x_1)$ with periodic function $f \in C^2(\mathbb{R})$. By a simple change of variables we can assume the period to be 2π.

It is the aim to compute the scattered wave u^s in the region

$$\Omega := \{x \in \mathbb{R}^2 : x_2 > f(x_1), \; x_1 \in \mathbb{R}\}$$

such that

$$\Delta u^s + k^2 u^s = 0 \quad \text{in } \Omega \tag{2.1}$$
$$u^s + u^i = 0 \quad \text{on } \partial\Omega. \tag{2.2}$$

This boundary value problem is not yet completely specified. First, one needs a radiation condition for the scattered part u^s as x_2 tends to infinity, and, second, a periodicity condition is necessary to achieve uniqueness of the solution. It is readily seen that, if u^s is a solution of the problem, then also $v^s(x_1, x_2) := u^s(x_1 + 2\pi, x_2)e^{-ik\hat{\theta}_1 2\pi}$. Therefore, we search for **quasi-periodic solutions** (with $\alpha = k\hat{\theta}_1$):

Definition 1. A function u is called **quasi-periodic** with parameter $\alpha > 0$ if $u(x_1 + 2\pi, x_2) = e^{i\alpha 2\pi}u(x_1, x_2)$ for all x_1, x_2, i.e. if $x_1 \mapsto e^{-i\alpha x_1}u(x_1, x_2)$ is 2π−periodic for every x_2.

We remark that one usually formulates the scattering problem for the total field $u = u^i + u^s$:

$$\Delta u + k^2 u = 0 \quad \text{in } \Omega, \qquad u = 0 \quad \text{on } \partial\Omega,$$

and a radiation condition as well as the quasi periodicity condition for $u^s := u - u^i$. We introduce the following set for our conveniance:

$$Q(\alpha) := \left\{ u \in C^2(\Omega) \cap C(\overline{\Omega}) : \begin{array}{l} u \text{ quasi-periodic with parameter } \alpha, \text{ bounded} \\ \text{and} \quad \Delta u + k^2 u = 0 \text{ in } \Omega. \end{array} \right\}$$

For $x_2 > \|f\|_\infty := \max_{0 \le t \le 2\pi} |f(t)|$ every $u \in Q(k\hat{\theta}_1)$ can be expanded into a Fourier series of the form

$$u(x) = \sum_{n \in \mathbb{Z}} u_n(x_2)e^{i(k\hat{\theta}_1 + n)x_1} = \sum_{n \in \mathbb{Z}} u_n(x_2)e^{i\alpha_n x_1} \quad \text{with } \alpha_n := n + k\hat{\theta}_1, \; n \in \mathbb{Z}.$$

Plugging this into the Helmholtz equation yields $u_n'' + (k^2 - \alpha_n^2)u_n = 0$. With the notations

$$\alpha_n = n + k\hat{\theta}_1 \quad \text{and} \quad \beta_n = \begin{cases} \sqrt{k^2 - \alpha_n^2}, & |\alpha_n| \le k, \\ i\sqrt{\alpha_n^2 - k^2}, & |\alpha_n| > k, \end{cases} \tag{2.3}$$

and using also the boundedness condition we have the Rayleigh expansion of the form

$$u(x) = \sum_{|\alpha_n| \leq k} a_n e^{i(\alpha_n x_1 - \beta_n x_2)} \qquad \text{``incoming wave''}$$

$$+ \sum_{|\alpha_n| \leq k} u_n e^{i(\alpha_n x_1 + \beta_n x_2)} \qquad \text{``outgoing wave''}$$

$$+ \sum_{|\alpha_n| > k} u_n e^{i(\alpha_n x_1 + \beta_n x_2)}. \qquad \text{``surface wave''}$$

The famous Rayleigh hypothesis treats the problem when this expansion is valid in all of Ω, cf. Millar [12, 13].

Since the incident field is of the form $u^i(x) = e^{i(k\hat{\theta}_1 x_1 + k\hat{\theta}_2 x_2)} = e^{i(\alpha_0 x_1 - \beta_0 x_2)}$ (note that $\hat{\theta}_2 < 0$!) we require that u^s has an expansion of the form

$$u^s(x) = \sum_{n \in \mathbf{Z}} u_n e^{i(\alpha_n x_1 + \beta_n x_2)} \quad \text{for } x_2 > \|f\|_\infty, \qquad (2.4)$$

which is the sum of outgoing and surface waves. This is the form of a **radiation condition** we will use in the following.

For the Dirichlet boundary condition under investigation we have uniqueness (cf. Cadilhac [4]). First we formulate a simple application of Green's formula in the set
$\Omega_R := \{x \in \mathbb{R}^2 : f(x_1) < x_2 < R, \ 0 < x_1 < 2\pi\}$ for some $R > \|f\|_\infty$ using the fact that the contributions of the vertical line integrals cancel because of the periodicity:

Lemma 2. *Let* $u, v \in Q(k\hat{\theta}_1) \cap C^1(\bar{\Omega})$ *satisfy the radiation condition. Set* $\Gamma := \{x \in \partial\Omega : 0 < x_1 < 2\pi\}$. *Then with* α_n, β_n *from (2.3):*

$$\int_\Gamma \left[u \frac{\partial \bar{v}}{\partial n} - \bar{v} \frac{\partial u}{\partial n} \right] ds = -4\pi i \sum_{|\alpha_n| \leq k} \beta_n u_n \bar{v}_n$$

where the unit normal vector $n(x)$ *at* $x \in \Gamma$ *is directed into* Ω *and* u_n, v_n *are the expansion coefficients of* u *and* v *resp.* $\bar{w}$ *denotes the complex conjugate of* w.

Theorem 3. *There exists at most one solution* $u \in Q(k\hat{\theta}_1)$ *of the problem described by (2.1), (2.2) and (2.4).*

Proof. Let u be the difference of two solutions. Then u satisfies the radiation condition (2.4) and vanishes on Γ. We proceed in three steps:

(i) By standard regularity results (cf. [6]) we conclude that $u \in C^1(\overline{\Omega})$ so that we can apply the preceding lemma.

(ii) $v := u$ in the lemma yields $\sum_{|\alpha_n| \leq k} \beta_n |u_n|^2 = 0$, i.e. $\beta_n u_n = 0$ for $|\alpha_n| \leq k$.

(iii) $v := \partial u/\partial x_2$ in the lemma yields

$$0 = \int_\Gamma \frac{\partial \bar{u}}{\partial x_2} \frac{\partial u}{\partial n} ds = \int_\Gamma n_2 \left| \frac{\partial u}{\partial n} \right|^2 ds,$$

thus $\partial u/\partial n = 0$ on Γ. Together with $u = 0$ on Γ Holmgren's uniqueness theorem (cf. [6]) implies $u \equiv 0$ in Ω. $\square$

Remark: A similar uniqueness result holds for the Robin boundary condition $\partial u/\partial n + i\lambda u = 0$ on $\partial \Omega$ with constant $\lambda \in \mathbb{C}$ such that $\mathrm{Re}\,\lambda > 0$. Application of the lemma to $v = u$ and use of the boundary condition yields

$$\mathrm{Im}\,(i\bar{\lambda}) \int_\Gamma |u|^2 ds = -2\pi \sum_{|\alpha_n| \le k} \beta_n |u_n|^2 \le 0.$$

Since $\mathrm{Im}\,(i\bar{\lambda}) = \mathrm{Re}\,\lambda > 0$ the field u has to vanish on Γ. From the boundary condition also $\partial u/\partial n = 0$ which again implies $u \equiv 0$ in Ω.

Uniqueness can also been proven for the transmission problem with a nontrivial jump in either u or $\partial u/\partial n$, but to my knowledge not for the Neumann problem or the inhomogeneous medium case.

A well known method for proving existence is the integral equation method. We define the free space quasi periodic Green's function by

$$G(x, y) = \frac{i}{4\pi} \sum_{n \in \mathbb{Z}} \frac{1}{\beta_n} \exp\left[i\alpha_n(x_1 - y_1) + i\beta_n |x_2 - y_2| \right], \quad x \ne y,$$

provided $\beta_n \ne 0 \; \forall n$, i.e. $|n + k\hat{\theta}_1| \ne k \; \forall n \in \mathbb{Z}$. From the asymptotic behaviour $\beta_n = i|n + k\hat{\theta}_1| + \mathcal{O}(1/|n|)$ and $1/\beta_n = -i/|n| + \mathcal{O}(1/|n|^2)$ for $|n| \to \infty$ we conclude that G differs by a continuous function from

$$\tilde{G}(x, y) := \frac{1}{4\pi} \sum_{n=1}^\infty \frac{1}{n} e^{(n+k\hat{\theta}_1)(i(x_1-y_1)-|x_2-y_2|)}$$

$$+ \frac{1}{4\pi} \sum_{n=1}^\infty \frac{1}{n} e^{(-n+k\hat{\theta}_1)(i(x_1-y_1)+|x_2-y_2|)}$$

$$= \frac{1}{4\pi} e^{k\hat{\theta}_1 z} \sum_{n=1}^\infty \frac{1}{n} e^{nz} + \frac{1}{4\pi} e^{-k\hat{\theta}_1 \bar{z}} \sum_{n=1}^\infty \frac{1}{n} e^{n\bar{z}} = \frac{1}{2\pi} \mathrm{Re} \sum_{n=1}^\infty \frac{1}{n} e^{nz} + A(x, y)$$

$$= -\frac{1}{2\pi} \mathrm{Re} \ln(1 - e^z) + A(x, y) = -\frac{1}{\pi} \ln(|x - y|) + B(x, y),$$

with $z = i(x_1 - y_1) - |x_2 - y_2|$ and continuous functions A and B. This shows that G has the same singularity as the fundamental solution $\Phi(x, y) := \frac{i}{2} H_0^{(1)}(k|x-y|)$ of the two dimensional Helmholtz equation. Since G is also a solution of the

Helmholtz equation it follows from a general argument (cf. [4]) that the difference $\Phi - G$ is even analytic in $[(0, 2\pi) \times \mathbb{R}] \times [(0, 2\pi) \times \mathbb{R}]$.

Now we turn to the integral equation method and make an ansatz for the scattered wave in the form of a combined double- and simple layer first proposed by Brackhage and Werner [3]:

$$u^s(x) = \int_\Gamma \left(\frac{\partial}{\partial n(y)} - i\lambda \right) G(x, y)\, \phi(y)\, ds(y) \qquad (2.5)$$

$$= \int_0^{2\pi} \left(\frac{\partial}{\partial n(y)} - i\lambda \right) G(x, y)|_{y_2 = f(y_1)} \sqrt{1 + f'(y_1)^2}\, \phi(y_1, f(y_1))\, dy_1, \quad x \in \Omega,$$

with "quasi periodic" and continuous density ϕ (the notion of quasi periodicity carries over in a natural way to functions living on Γ). The coupling parameter $\lambda \in \mathbb{C}$ is arbitrary with $\operatorname{Re}\lambda > 0$. We note that the integrand is 2π-periodic with respect to y_1 since the factors $\exp(ik\hat{\theta}_1 y_1)$ in ϕ and $\exp(-ik\hat{\theta}_1 y_1)$ in G cancel out. The potential theoretic jump conditions for the double- and simple layer potentials (cf. [6]) for G are the same as for Φ and yield

Theorem 4. *Let be* $|n + k\hat{\theta}_1| \neq k$ *for all* $n \in \mathbb{Z}$. u^s *from (2.5) is a solution of the scattering problem if and only if the density* ϕ *solves the boundary integral equation*

$$\phi + D\phi + i\lambda\, S\phi = -u^i \quad \text{on } \Gamma, \qquad (2.6)$$

where D *and* S *are defined by*

$$D\phi(x) := \int_\Gamma \phi(y)\, \frac{\partial}{\partial n(y)} G(x, y)\, ds(y), \quad x \in \Gamma,$$

$$S\phi(x) := \int_\Gamma \phi(y)\, G(x, y)\, ds(y), \quad x \in \Gamma.$$

Theorem 5. *Let be* $|n + k\hat{\theta}_1| \neq k$ *for all* $n \in \mathbb{Z}$. *Then equation (2.6) is uniquely solvable. Therefore, also the scattering problem (2.1),(2.2),(2.4) admits a unique solution* $u \in Q(k\hat{\theta}_1)$.

Proof. We multiply equation (2.6) by $\exp(-ik\hat{\theta}_1 x_1)$ and use the parametrization $y_2 = f(y_1)$. Then equation (2.6) takes the form

$$\tilde{\phi}(x_1) + \int_0^{2\pi} \tilde{\phi}(y_1)\, K(x_1, y_1)\, dy_1 = -e^{ik\hat{\theta}_2 f(x_1)}, \quad x_1 \in (0, 2\pi),$$

with 2π-periodic function $\tilde{\phi}(x_1) = \exp(-ik\hat{\theta}_1 x_1)\phi(x_1, f(x_1))$ and 2π-periodic kernel

$$K(x_1, y_1) := e^{ik\hat{\theta}_1(y_1 - x_1)} \left(\frac{\partial}{\partial n(y)} - i\lambda \right) G(x_1, f(x_1), y_1, y_2)|_{y_2 = f(y_1)} \sqrt{1 + f'(y_1)^2}.$$

This is a Fredholm equation of the second kind with weakly singular kernel in the space of continuous and 2π-periodic functions. Due to the Fredholm theory it is sufficient to prove uniqueness of equation (2.6). Thereore, let ϕ be a quasi periodic solution of (2.6) for $u^i = 0$ and define u by the formula (2.5) for all $x \notin \Gamma$. Then, by the preceding theorem, u solves the scattering problem in Ω for $u^i = 0$ and, by the uniqueness result of Theorem 2.3, has to vanish in Ω. Now we turn to the lower region $\Omega^- := \{x \in \mathbb{R}^2 : x_2 < f(x_1), \ x_1 \in \mathbb{R}\}$. Denoting by $u(x)|_\pm$ the upper and lower limit of u for $x \to \Gamma$ resp. we have from the jump conditions again:

$$u|_- = u|_- - u|_+ = -2\phi \quad \text{and} \quad \left.\frac{\partial u}{\partial n}\right|_- = \left.\frac{\partial u}{\partial n}\right|_- - \left.\frac{\partial u}{\partial n}\right|_+ = -2i\lambda\phi.$$

Eliminating ϕ yields

$$\left.\frac{\partial u}{\partial n}\right|_- - i\lambda u|_- = 0 \text{ on } \Gamma.$$

Therefore, u solves the homogeneous Robin boundary value problem in Ω^-. Due to the remark after Theorem 2.3 (by reversing the normal for the region Ω^-) u has to vanish also in Ω^-. This proves $\phi = 0$ which ends the proof of the theorem.   □

3 A Variational Formulation And Analytic Dependence on the Wave Number And the Angle of the Incoming Wave

In this section we propose a different method which couples a variational equation in $\Omega_R := \{x \in \mathbb{R}^2 : 0 < x_1 < 2\pi, \ f(x_1) < x_2 < R\}$ with the Rayleigh expansion in $\{x \in \mathbb{R}^2 : 0 < x_1 < 2\pi, \ x_2 > R\}$ for some fixed $R > \|f\|_\infty$. Again, we assume that $f \in C^2(\mathbb{R})$ is $2\pi-$periodic. (The strong smoothness assumption is only necessary to assure uniqueness by Theorem 2.3.) Apply the first Green's formula in Ω_R to the total field u and some quasi periodic test function $\phi \in H^1(\Omega_R)$ with $\phi = 0$ on Γ. With $\Gamma_R := \{x \in \mathbb{R}^2 : 0 < x_1 < 2\pi, \ x_2 = R\}$ this yields

$$\int_{\Omega_R} [\nabla u \nabla \bar\phi - k^2 u \bar\phi] \, dx - \int_{\Gamma_R} \bar\phi \frac{\partial u}{\partial x_2} \, ds = 0.$$

Again, the contributions from the vertical line integrals cancel. Now we introduce the "Dirichlet-to-Neumann map" L, i.e. $L\psi$ is defined by $\partial v / \partial x_2$ on Γ_R where v is the (unique) quasi periodic solution of the Helmholtz equation in the region $\{x \in \mathbb{R}^2 : 0 < x_1 < 2\pi, \ x_2 > R\}$ and the radiation condition (2.4) which satisfies $v = \psi$ on Γ_R. By the Rayleigh expansion $L\psi$ exists and is given by

$$L\psi = i \sum_{n \in \mathbb{Z}} \beta_n \psi_n \exp(i\alpha_n \cdot) \quad \text{if } \psi \text{ has the expansion} \quad \psi = \sum_{n \in \mathbb{Z}} \psi_n \exp(i\alpha_n \cdot).$$

Replacing u by $u^i + u^s$, using Lu^s for $\partial u^s / \partial x_2$ and replacing u^s again by $u - u^i$ yields

$$\int_{\Omega_R} [\nabla u \nabla \bar\phi - k^2 u \bar\phi] \, dx - \int_{\Gamma_R} \bar\phi \, Lu \, ds = \int_{\Gamma_R} \left(\frac{\partial u^i}{\partial x_2} - Lu^i \right) \bar\phi \, ds$$

$$= 2i\,k\,\hat{\theta}_2\,e^{ik\hat{\theta}_2 R}\int_0^{2\pi} e^{ik\hat{\theta}_1 x_1}\bar{\phi}(x_1, R)\,dx_1. \tag{3.1}$$

The left hand side defines the sesquilinear form

$$a(u, \phi) := \int_{\Omega_R} [\nabla u \nabla \bar{\phi} - k^2 u \bar{\phi}]\,dx - \int_{\Gamma_R} \bar{\phi}\,Lu\,ds. \tag{3.2}$$

We consider the variational equation (3.1) in the space

$$\tilde{H}_{qp}^1 := \{u \in H^1(\Omega_R) : u \text{ quasi periodic with parameter } k\hat{\theta}_1,\ u = 0 \text{ on } \Gamma\}.$$

From the representation above and the asymptotic behavior $\beta_n = i|n + k\hat{\theta}_1| + \mathcal{O}(1/|n|)$, $|n| \to \infty$, we see that L is a bounded mapping from $H^{1/2}(\Gamma_R)$ into $H^{-1/2}(\Gamma_R)$ where $H^{\pm 1/2}(\Gamma_R)$ denote the usual Sobolev spaces of order $1/2$ and $-1/2$ resp. Also, L is bijective iff $\beta_n \neq 0$ for all $n \in \mathbb{Z}$. From now on we interpret the integral $\int_{\Gamma_R} \bar{\phi}\,Lu\,ds$ for $u, \phi \in \tilde{H}_{qp}^1$ as the dual bracket $\langle Lu, \phi \rangle_{1/2}$ (since $Lu \in H^{-1/2}(\Gamma_R)$ and $\phi|_{\Gamma_R} \in H^{1/2}(\Gamma_R)$ by the trace theorem).

We can split the bilinear form a from (3.2) into two parts $a = a_0 + a_K$ with

$$a_0(u, \phi) := \int_{\Omega_R} [\nabla u \nabla \bar{\phi} + u\bar{\phi}]\,dx - \int_{\Gamma_R} \bar{\phi}\,L_0 u\,ds \quad \text{and}$$

$$a_K(u, \phi) := -(k^2 + 1)\int_{\Omega_R} u\bar{\phi}\,dx + \int_{\Gamma_R} \bar{\phi}\,(L_0 - L)u\,ds, \quad \text{where}$$

$$L_0\psi := -\sum_{n \in \mathbb{Z}} |n + k\hat{\theta}_1|\,\psi_n\,\exp(i\alpha_n\cdot) \quad \text{for} \quad \psi = \sum_{n \in \mathbb{Z}} \psi_n\,\exp(i\alpha_n\cdot).$$

Then $-L_0$ is nonnegative and, therefore, a_0 coercive in $\tilde{H}_{qp}^1$. By the theorem of Lax and Milgram (cf. [7]) there exists an isomorphism T from $\tilde{H}_{qp}^1$ onto itself with $a_0(\psi, \phi) = (T\psi, \phi)_{H^1(\Omega_R)}$ for all $\psi, \phi \in \tilde{H}_{qp}^1$. By $(\psi, \phi)_{H^1(\Omega_R)}$ we denote the inner product in $H^1(\Omega_R)$. By the Riesz theorem there exists $r \in \tilde{H}_{qp}^1$ and a bounded operator K from $\tilde{H}_{qp}^1$ into itself with $a_K(\psi, \phi) = (K\psi, \phi)_{H^1(\Omega_R)}$ for all $\psi, \phi \in \tilde{H}_{qp}^1$ and $2ik\hat{\theta}_2 \exp(ik\hat{\theta}_2 R)\int_0^{2\pi} \exp(ik\hat{\theta}_1 x_1)\bar{\phi}(x_1, R)\,dx_1 = (r, \phi)_{H^1(\Omega_R)}$ for all $\phi \in \tilde{H}_{qp}^1$. Furthermore, the operator $L - L_0$ is compact since $\beta_n = i|n + k\hat{\theta}_1| + \mathcal{O}(1/|n|)$ for $|n| \to \infty$. This implies (simple arguments, cf. proof of Lemma 3.4) that also K is compact.

The variational equation (3.1) can then be written in the form $Tu + Ku = r$ with isomorphism T and compact K. By standard regularity results (cf. [7]) it is easily seen that any solution $u \in \tilde{H}_{qp}^1$ of $Tu + Ku = r$ is a classical solution of the scattering problem if it is extended by the solution of the Dirichlet problem with boundary data u into the region $(0, 2\pi) \times (R, \infty)$. From this and Theorem 2.3 it follows that $T + K$ is one-to-one. Fredholm's alternative then again yields the existence of a unique solution of the variational equation (3.1) in $\tilde{H}_{qp}^1$. This construction can be made for other boundary conditions as well (cf. [1]).

In this formulation the space $\tilde{H}^1_{qp}$ depends on k and θ via the quasi periodicity. We can eliminate this dependence at the cost of changing the bilinear form a by replacing $u(x)$ by $\exp(ik\hat{\theta}_1 x_1)\,\tilde{u}(x)$ and $\phi(x)$ by $\exp(ik\hat{\theta}_1 x_1)\,\phi(x)$, where $\tilde{u}$ and the new ϕ are now 2π−periodic with respect to x_1. The variational equation (3.1) transforms into

$$\int\limits_{\Omega_R}\left[\nabla\tilde{u}\nabla\bar{\phi}+ik\hat{\theta}_1\left(\tilde{u}\frac{\partial\bar{\phi}}{x_1}-\bar{\phi}\frac{\partial\tilde{u}}{\partial x_1}\right)-k^2\hat{\theta}_2^2\,\tilde{u}\,\bar{\phi}\right]dx-\int\limits_{\Gamma_R}\bar{\phi}\,\tilde{L}\tilde{u}\,ds$$

$$=2i\,k\,\hat{\theta}_2\,e^{ik\hat{\theta}_2 R}\int\limits_{\Gamma_R}\bar{\phi}\,ds,\qquad(3.3)$$

where

$$\tilde{L}\psi=i\sum_{n\in\mathbf{Z}}\beta_n\psi_n\exp(in\cdot)\quad\text{for}\quad\psi=\sum_{n\in\mathbf{Z}}\psi_n\exp(in\cdot).$$

We consider this variational equation in the space V of (w.r.t. x_1) 2π−periodic functions which vanish on Γ. In the following theorem we prove the continuous and even piecewise analytic dependence of the solution on k and θ:

Theorem a *The solution $u\in Q(k\hat{\theta}_1)$ of the scattering problem (2.1), (2.2) and (2.4) depends continuously on $(k,\theta)\in(0,\infty)\times(0,\pi)$ and analytically on (k,θ) in the set*

$$\Lambda:=\{(k,\theta)\in(0,\infty)\times(0,\pi):(n+k\cos\theta)^2\neq k^2\text{ for every }n\in\mathbf{Z}\}.$$

(b) Let be $(k_0,\theta_0)\in(0,\infty)\times(0,\pi)$ such that $(n_0+k_0\cos\theta_0)^2=k_0^2$ for some $n_0\in\mathbf{Z}$. Then there exists a neighborhood U of (k_0,θ_0) and quasi periodic functions $v,w\in\tilde{H}^1_{qp}$ which depend analytically on $(k,\theta)\in U$ and satisfy $u(x)=v(x)+\beta_{n_0}w(x)$ for $x\in\Omega_R$, i.e. u depends on (k,θ) in the same way as β_{n_0}.

Before we prove this theorem we want to recall some notions from the multivalued complex function theory (cf. Hörmander [8]) and transfer it to functions with values in a Banach space.

We call a mapping F from an open set $U\subset\mathbf{C}^N$ into a complex Banachspace X *analytic* in $z^0\in U$ if there exist $R>0$ and elements $f_j\in X$, $j\in\mathbf{N}_0^N:=(\mathbf{N}\cup\{0\})^N$, such that

$$\left\|F(z)-\sum_{\substack{j\in\mathbf{N}_0^N\\|j|\le m}}\frac{(z-z^0)^j}{j!}f_j\right\|_X\longrightarrow 0\quad(m\to\infty)\quad\text{uniformly in }z\in U\cap D(z^0,R),$$

$$(3.4)$$

where we have used the standard notations:

$$z^j=z_1^{j_1}\cdots z_N^{j_N}\text{ for }z\in\mathbf{C}^N\text{ and }j\in\mathbf{N}_0^N,$$

$$|j|=\sum_{l=1}^N j_l\quad\text{and}\quad j!=j_1!\cdots j_N!\quad\text{for }j\in\mathbf{N}_0^N,$$

$$D(z^0; R) = \{z \in \mathbb{C}^N : |z_l - z_l^0| < R \; \forall l = 1, \ldots, N\} \quad \text{for} \quad z^0 \in \mathbb{C}^N, \; R > 0$$

("polydisc").

We write condition (3.4) in short form as

$$F(z) = \sum_{j \in \mathbb{N}_0^N} \frac{(z - z^0)^j}{j!} \, f_j \quad \text{for } z \in U \cap D(z^0, R).$$

We want to show that the solution $\tilde{u}$ of (3.2) depends analytically on (k, θ). We use the fact that the inverse of an invertible operator in a Banachspace which depends analytically on a parameter also depends analytically on this parameter. We will take for X the space $\mathcal{LB}(H^s(0, 2\pi), H^{s+\tau}(0, 2\pi))$ of linear and bounded operators from $H^s(0, 2\pi)$ into $H^{s+\tau}(0, 2\pi)$ equipped with the operator norm or the (closed) subspace V of $H^1(\Omega_R)$ consisting of functions which vanish on Γ and are 2π−periodic w.r.t. x_1. We need the following three lemmas. Their proofs are left to the reader (cf. [9] for the case $N = 1$).

Lemma 6. *Let be* $L(z) : H^s(0, 2\pi) \longrightarrow H^{s+\tau}(0, 2\pi)$ *for all* $z \in U \subset \mathbb{C}^N$ *defined by*

$$L(z)\psi := \sum_{n \in \mathbb{Z}} \alpha_n(z) \psi_n \exp(in\cdot) \quad \text{for} \quad \psi = \sum_{n \in \mathbb{Z}} \psi_n \exp(in\cdot),$$

and $\alpha_n(z) \in \mathbb{C}$ *satisfies* $\alpha_n(z) = \mathcal{O}(1/|n|^\tau)$, $|n| \geq 1$, *uniformly in* $z \in U$. *Furthermore, let* α_n *be analytic in* U *for every* $n \in \mathbb{Z}$. *Then* $z \mapsto L(z)$ *from* U *into* $\mathcal{LB}(H^s(0, 2\pi), H^{s+\tau}(0, 2\pi))$ *is analytic.*

Remark: From the asymptotic behaviour of $\alpha_n(z)$ it follows that the operators $L(z)$ are well defined and bounded.

Results of the following type for $N = 1$ can be found in [9]. Let Ω_R and Γ_R be defined as above.

Lemma 7. *Let* $L(z) : H^{1/2}(\Gamma_R) \rightarrow H^{-1/2}(\Gamma_R)$ *be nonnegative (i.e.* $\int_{\Gamma_R} \bar{\phi} \, L(z) u \, ds \geq 0$ *for all* $u, \phi \in H^{1/2}(\Gamma_R)$) *and depend analytically on* $z \in U \subset \mathbb{C}^N$. *Define the sesquilinear form* $b(z)$ *in some closed subspace* $V \subset H^1(\Omega_R)$ *by*

$$b(z)(u, \phi) := \int_{\Omega_R} [\nabla u \, \nabla \bar{\phi} + u \, \bar{\phi}] \, dx + \int_{\Gamma_R} \bar{\phi} \, L(z) u \, ds, \quad u, \phi \in V.$$

Then there exists an isomorphism $T(z)$ *from* V *onto itself which depends analytically on* $z \in U$ *such that*

$$b(z)(u, \phi) = (T(z)u, \phi)_{H^1(\Omega_R)} \quad \text{for all } u, \phi \in V, \; z \in U.$$

Remark: The sesquilinear forms $b(z)$ are well defined, bounded and coercive in V. This follows from the properties of $L(z)$ and the trace theorem.

Lemma 8. *Let* $L(z) : H^{1/2}(\Gamma_R) \to H^{-1/2}(\Gamma_R)$ *be compact and depend analytically on* $z \in U \subset \mathbb{C}^N$. *Let* $V \subset H^1(\Omega_R)$ *be a closed subspace. Then there exists a a compact* $K(z)$ *from* V *into itself which depends analytically on* $z \in U$ *such that*

$$\int_{\Gamma_R} \bar{\phi} \, L(z) u \, ds = (K(z)u, \phi)_{H^1(\Omega_R)} \quad \text{for all } u, \phi \in V, \ z \in U.$$

Now we begin with the **proof of Theorem 3.1**: We restrict ourselves to the proof of the analytic dependence and will apply Lemma 3.3 and 3.4. First we study the analytic dependence of L on (k, θ). Let be $(k_0, \theta_0) \in \Lambda$, i.e. $k_0^2 \neq (n + k_0 \cos \theta_0)^2$ for all $n \in \mathbb{Z}$. We cut the complex plane $\mathbb{C}$ into $\mathbb{C}_- := \{w \in \mathbb{C} : iw \notin [0, \infty) \subset \mathbb{R}\}$. Then we choose an open polydisc $U \subset \mathbb{C}^2$ with $(k_0, \theta_0) \in U$ and $z_1^2 - (n + z_1 \cos z_2)^2 \in \mathbb{C}_-$ for $z = (z_1, z_2) \in U$, $n \in \mathbb{Z}$. Then $z \mapsto \sqrt{z_1^2 - (n + z_1 \cos z_2)^2}$ is analytic in U for all $n \in \mathbb{Z}$.

(a) Since $|\sqrt{z_1^2 - (n + z_1 \cos z_2)^2}| \leq c|n|$ for all $n \in \mathbb{Z}$ and all $z \in U$ we conclude from Lemma 3.2 with $s = 1/2$ and $\tau = -1$ that L depends analytically on $z = (k, \theta)$. Now we rewrite (3.3) in the form

$$\int_{\Omega_R} [\nabla \tilde{u} \nabla \bar{\phi} + \tilde{u} \bar{\phi}] \, dx - \int_{\Gamma_R} \bar{\phi} \, \tilde{L}_0 \tilde{u} \, ds + ik \cos \theta \int_{\Omega_R} \left(\tilde{u} \frac{\partial \bar{\phi}}{\partial x_1} - \bar{\phi} \frac{\partial \tilde{u}}{\partial x_1} \right) dx -$$

$$- (k^2 \sin^2 \theta + 1) \int_{\Omega_R} \tilde{u} \bar{\phi} \, dx + \int_{\Gamma_R} \bar{\phi} (\tilde{L}_0 - \tilde{L}) \tilde{u} \, ds$$

$$= -2ik \sin \theta \, e^{-ikR \sin \theta} \int_{\Gamma_R} \bar{\phi} \, ds, \quad \phi \in V.$$

Here we choose $\tilde{L}_0$ to be (for some $\hat{n} \in \mathbb{N}$ with $\hat{n} \geq 1 + \sup\{|z_1| : z \in U\}$):

$$\tilde{L}_0 \psi = - \sum_{|n| \geq \hat{n}} |n + k \cos \theta| \, \psi_n \exp(in \cdot) \quad \text{for} \quad \psi = \sum_{n \in \mathbb{Z}} \psi_n \exp(in \cdot).$$

Then $\tilde{L}_0$ is nonnegative, depends analytically on $(k, \theta) \in U$, and $\tilde{L}_0 - \tilde{L}$ is compact. Using Lemma 3.3 and 3.4 this variational equation is equivalent to

$$[T(k, \theta) + ik \cos \theta \, K_1 - (k^2 \sin^2 \theta + 1) K_2 + K_3(k, \theta)] \tilde{u}$$

$$= -2ik \sin \theta \exp(-ikR \sin \theta) \, r, \tag{3.5}$$

with isomorphism T and compact K_1, K_2, K_3 where T and K_3 depend analytically on (k, θ).

Then also the solution $\tilde{u}$ depends analytically on (k, θ) and thus also $u(x) = \exp(ikx_1 \cos \theta) \, \tilde{u}(x)$.

(b) We split $\tilde{L}$ into

$$\tilde{L} \psi = [\tilde{L} \psi - i\beta_{n_0} \psi_{n_0} \exp(in_0 \cdot)] + i\beta_{n_0} \psi_{n_0} \exp(in_0 \cdot).$$

The first part on the right hand side depends analytically on (k, θ) in a neighborhood U of (k_0, θ_0). The splitting of $\tilde{L}$ corresponds to a splitting of K_3 in the form $K_3(k, \theta) = K_4(k, \theta) + \beta_{n_0} K_5$ with analytic K_4. ¿From the Neumann series the decomposition of $\tilde{u}$ into $\tilde{v} + \beta_{n_0} \tilde{w}$ follows where $\tilde{v}$ and $\tilde{w}$ depend on $\beta_{n_0}^2$ and, therefore, analytically on (k, θ). The multiplication with $\exp(ik\hat{\theta}_1 x_1)$ yields the desired decomposition.

4 Continuous And Differentiable Dependence on the Boundary

In this section we study the dependence of the solution on the boundary curve f. We expect that small changes in f will produce small changes in the solution u of the scattering problem (for fixed wave number k and incident wave u^i). This is true (Theorem 4.1) if the change in f is measured with respect to the C^1-norm, i.e. small changes in amplitude and low oszillations of f will lead to small changes of u in the H^1-norm. If f is perturbed by a small (in amplitude) but rapidly oszillating function then one expects from the theory of homogenization (cf. [2]) that u converges to some different, i.e. homogenized, problem.

In Theorem 4.2 we prove that u depends differentiable on $f \in C^1$ and give a characterization of the Fréchet-derivative.

Theorem 9. *Let be $f \in C^2(\mathbb{R})$ $2\pi-$periodic and $u \in Q(k\hat{\theta}_1)$ be the unique solution of the scattering problem (2.1), (2.2) and (2.4). Let $K \subset \Omega$ be compact. Then there exists $\epsilon > 0$ and $c > 0$ both depending on k, u^i, f and K, such that for all $2\pi-$periodic $g \in C^1(\mathbb{R})$ with $\|g - f\|_{1,\infty} \leq \epsilon$ the unique solution u_g of the scattering problem corresponding to g satisfies*

$$\|u_g - u\|_{H^1(K)} \leq c \|f - g\|_{1,\infty}.$$

Here, $\|h\|_{1,\infty} := \|h\|_\infty + \|h'\|_\infty$ denotes the norm in $C^1[0, 2\pi]$.

Proof. We will use the variational equation (3.1) for the "reference" problem with quantities f, Ω_R, $\tilde{H}_{qp}^1$ and solution u and also for the "perturbed" problem, where we indicate the dependence on g by $\Omega_R(g)$, $\tilde{H}_{qp}^1(g)$ and solution u_g.

Let be $R > 0$ such that $K \subset \Omega_R$ and also $R \geq \|f\|_\infty + \|g\|_\infty + 2$. We will transform the volume integral in

$$\int_{\Omega_R(g)} [\nabla u_g \nabla \bar{\phi} - k^2 u_g \bar{\phi}] \, dx - \int_{\Gamma_R} \bar{\phi} \, L u_g \, ds = 2i k \hat{\theta}_2 e^{ik\hat{\theta}_2 R} \int_0^{2\pi} e^{ik\hat{\theta}_1 x_1} \bar{\phi}(x_1, R) \, dx_1$$

$$(4.1)$$

into an integral over Ω_R by a suitable change of the variable x. Choose $\epsilon_0 > 0$ such that
$\epsilon_0 < \min\{ \min\{x_2 : (x_1, x_2) \in K\} - f(x_1) : x_1 \in \mathbb{R} \}$ and a function $\alpha \in C^1(\mathbb{R})$ with $\alpha(t) = 1$ for $t \leq \epsilon_0/2$ and $\alpha(t) = 0$ for $t \geq \epsilon_0$. For g such that $\|f-g\|_{1,\infty} \leq \epsilon_0$ we define $\mathcal{F} : \Omega_R \to \Omega_R(g)$ by

$$\mathcal{F}(y) = y + \alpha(y_2 - f(y_1)) [g(y_1) - f(y_1)] \hat{e}_2, \quad y \in \Omega_R,$$

where $\hat{e}_2 = (0,1)^{\mathsf{T}}$. For $\|f - g\|_{1,\infty} \leq \epsilon$ and sufficiently small $\epsilon \leq \epsilon_0$ $\mathcal{F}$ is a diffeomorphism and $\mathcal{F} = id$ on K. The Jacobian $J_{\mathcal{F}}$ of $\mathcal{F}$ satisfies $J_{\mathcal{F}}(y) = id + \mathcal{O}(\|f - g\|_{1,\infty})$ uniformly in $y \in \Omega_R$, and for the inverse $\mathcal{G}$ of $\mathcal{F}$ we also have $J_{\mathcal{G}}(x) = id + \mathcal{O}(\|f - g\|_{1,\infty})$. The change of variable $x = \mathcal{F}(y)$, $y \in \Omega_R$, transforms (4.1) into

$$\int_{\Omega_R} \left[\sum_{i,j=1}^{2} b_{ij} \frac{\partial \hat{u}_g}{\partial y_i} \frac{\partial \overline{\hat{\phi}}}{\partial y_j} - k^2 \, \hat{u}_g \overline{\hat{\phi}} \right] \det J_{\mathcal{F}} \, dy - \int_{\Gamma_R} \overline{\hat{\phi}} \, L\hat{u}_g \, ds = \int_0^{2\pi} r \overline{\hat{\phi}} \, ds, \quad (4.2)$$

where $r(y_1) = 2i\,k\,\hat{\theta}_2 \exp(ik(y_1\hat{\theta}_1 + R\hat{\theta}_2))$, $\hat{u}_g = u_g \circ \mathcal{F}$, $\hat{\phi} = \phi \circ \mathcal{F}$ both in $\tilde{H}^1_{qp}$, and

$$b_{ij}(y) = \sum_{l=1}^{2} \frac{\partial \mathcal{G}_i(x)}{\partial x_l} \frac{\partial \mathcal{G}_j(x)}{\partial x_l}\bigg|_{x=\mathcal{F}(y)}, \quad i,j = 1,2. \quad (4.3)$$

The left hand side of (4.2) defines a sesquilinear form $a_g(u, \phi)$ on $\tilde{H}^1_{qp}$. Since $\det J_{\mathcal{F}} = 1 + \mathcal{O}(\|f - g\|_{1,\infty})$ and $b_{ij} = \delta_{ij} + \mathcal{O}(\|f - g\|_{1,\infty})$ we conclude that

$$|a_g(u, \phi) - a(u, \phi)| \leq c\, \|f - g\|_{1,\infty} \|u\|_{H^1(\Omega_R)} \|\phi\|_{H^1(\Omega_R)} \quad \text{for all } u, \phi \in \tilde{H}^1_{qp}.$$

Here, $a(u, \phi) := \int_{\Omega_R} [\nabla u \nabla \bar{\phi} - k^2 u\bar{\phi}] \, dx - \int_{\Gamma_R} \bar{\phi} \, Lu \, ds$ was defined in (3.2). Now we can apply the general perturbation theory of variational equations (cf. [9]) which yields

$$\|u_g \circ \mathcal{F} - u\|_{H^1(\Omega_R)} = \|\hat{u}_g - u\|_{H^1(\Omega_R)} \leq c\, \|f - g\|_{1,\infty}$$

and thus, since $\mathcal{F} = id$ on K,

$$\|u_g - u\|_{H^1(K)} \leq c\, \|f - g\|_{1,\infty}.$$

This ends the proof of the theorem. $\qquad\qquad\qquad\qquad\qquad\qquad\qquad\qquad\qquad$ □

Now we turn to the differential dependence on the boundary curve. For fixed 2π−periodic function $f \in C^2(\mathbb{R})$ let be Ω and u as before. We have seen in the previous theorem that the mapping $T : g \mapsto u_g|_K$ from a C^1−neighborhood U of f into $H^1(K)$ is continuous. The next theorem proves the differentiability of T at f and determines its Fréchet derivative.

Theorem 10. *Let be $f \in C^2(\mathbb{R})$ 2π−periodic and $u \in Q(k\hat{\theta}_1)$ be the solution of the scattering problem (2.1), (2.2) and (2.4). Let $K \subset \Omega$ be compact and the mapping T be defined as above. Then the Frechet derivative $T'(f; h)$ of T at f exists and is given by $u'_h|_K$ where $u'_h \in Q(k\hat{\theta}_1)$ solves the boundary value problem*

$$u'_h = -\frac{h}{\sqrt{1 + f'^2}} \frac{\partial u}{\partial n} \left(= -h \frac{\partial u}{\partial x_2} \right) \quad \text{on } \partial\Omega, \quad (4.4)$$

and u'_h satisfies (2.4).

Remark: The solution u is in C^1 since $f \in C^2(\mathbb{R})$ by well known regularity results (cf. [6]).

Proof. The first part follows exactly the proof of Theorem 4.1. We use the same notations, in particular $\mathcal{F}$ and α. Let g be close to f and set $h = g - f$. After the change of variables $x = \mathcal{F}(y)$ we arrive at the variational equation (4.2), i.e.

$$a_{f+h}(\hat{u}_{f+h}, \phi) = \int_{\Gamma_R} r\,\bar{\phi}\,ds \quad \text{for all } \phi \in V,$$

where again $V := \{\phi \in H^1(\Omega_R) : \phi|_\Gamma = 0, \ \phi(\cdot, x_2) \text{ is } 2\pi - \text{periodic }\}$. Let be u_h' the solution of (4.4) and define $v_h(x) := \alpha(x_2 - f(x_1))\,h(x_1)\,\partial u/\partial x_2$ for $x \in \Omega$. We will show that

$$\frac{1}{\|h\|_{1,\infty}}\|\hat{u}_{f+h} - u - (u_h' + v_h)\|_{H^1(\Omega_R)} \longrightarrow 0 \quad \text{for } \|h\|_{1,\infty} \to 0$$

which would prove the theorem since $v_h = 0$ on K.

With a from (3.2) we have for $\phi \in V \cap H^2(\Omega_R)$ (omitting the subscript $f+h$ in $\hat{u}_{f+h}$ in the following):

$$a(\hat{u} - u - u_h' - v_h, \phi) = -[a_{f+h}(\hat{u}, \phi) - a(\hat{u}, \phi)] - a(u_h' + v_h, \phi)]. \qquad (4.5)$$

Let us consider the first term on the right hand side:

$$a_{f+h}(\hat{u}, \phi) - a(\hat{u}, \phi) = \int_{\Omega_R} \left[\nabla\hat{u}\,(\det J_{\mathcal{F}}\,B - id)\,\nabla\bar{\phi} - k^2(\det J_{\mathcal{F}} - 1)\,\hat{u}\,\bar{\phi}\right] dy$$

where $B = (b_{ij})_{i,j=1,2}$ from (4.3). One easily computes: $\det J_{\mathcal{F}}(y) = 1 + \alpha'(y_2 - f(y_1))\,h(y_1)$ and

$$B(y) = \begin{pmatrix} 1 & \alpha'f'h - \alpha h' \\ \alpha'f'h - \alpha h' & 1 - 2\alpha'h - \alpha h' \end{pmatrix} + \mathcal{O}(\|h\|_{1,\infty}^2).$$

Here, we have omitted the arguments $y_2 - f(y_1)$ of α and y_1 of h, h' and f'. Therefore, $B \det J_{\mathcal{F}} - id = \tilde{B} + \mathcal{O}(\|h\|_{1,\infty}^2)$ with

$$\tilde{B} = \begin{pmatrix} \alpha'h & \alpha'f'h - \alpha h' \\ \alpha'f'h - \alpha h' & -\alpha'h \end{pmatrix}.$$

A simple but lengthy application of the product rule yields

$$\nabla\hat{u}\,\tilde{B}\,\nabla\bar{\phi} = -\text{div}\left\{ v_h\nabla\bar{\phi} + \alpha\,h\,\frac{\partial\bar{\phi}}{\partial y_2}\,\nabla\hat{u} - (\nabla\bar{\phi}\nabla\hat{u})\,\alpha\,h\,\hat{e}_2 \right\} + v_h\triangle\bar{\phi} + \alpha\,h\,\frac{\partial\bar{\phi}}{\partial y_2}\,\triangle\hat{u}.$$

Therefore, and since $\alpha(y_2 - f(y_1))$ vanishes on Γ_R, $\alpha(y_2 - f(y_1)) = 1$ on Γ, and $\phi = 0$ on Γ:

$$a_{f+h}(\hat{u}, \phi) - a(\hat{u}, \phi) = \int_\Gamma \left[v_h \frac{\partial \bar{\phi}}{\partial n} + \alpha h \frac{\partial \bar{\phi}}{\partial y_2} \frac{\partial \hat{u}}{\partial n} - (\nabla \bar{\phi} \nabla \hat{u}) \alpha h n_2 \right] ds$$

$$- \int_{\Omega_R} \nabla v_h \nabla \bar{\phi} \, dy - \int_\Gamma v_h \frac{\partial \bar{\phi}}{\partial n} \, ds - k^2 \int_{\Omega_R} \alpha h \frac{\partial \bar{\phi}}{\partial y_2} \, \hat{u} \, dy$$

$$- k^2 \int_{\Omega_R} \alpha' h \, \hat{u} \, \bar{\phi} \, dy + \mathcal{O}(\|h\|_{1,\infty}^2 \|\hat{u}\|_{H^1(\Omega_R)} \|\phi\|_{H^1(\Omega_R)}) =$$

$$= \int_{\Omega_R} \nabla v_h \nabla \bar{\phi} \, dy - k^2 \int_{\Omega_R} \alpha h \frac{\partial \bar{\phi}}{\partial y_2} \, \hat{u} \, dy - k^2 \int_{\Omega_R} \alpha' h \, \hat{u} \, \bar{\phi} \, dy +$$

$$+ \mathcal{O}(\|h\|_{1,\infty}^2 \|\hat{u}\|_{H^1(\Omega_R)} \|\phi\|_{H^1(\Omega_R)}).$$

The second term on the right hand side of (4.5) is

$$a(u_h' + v_h, \phi) = \int_{\Omega_R} \left[\nabla v_h \nabla \bar{\phi} - k^2 v_h \bar{\phi} \right] dy + \int_{\Gamma_R} \bar{\phi} \frac{\partial u_h'}{\partial n} \, ds - \int_{\Gamma_R} \bar{\phi} \, L u_h' \, ds,$$

where we used that $v = 0$ on Γ_R and applied Green's formula to u_h'. Altogether we get

$$a(\hat{u} - u - u_h' - v_h, \phi) = \int_{\Omega_R} \nabla v_h \nabla \bar{\phi} \, dy + k^2 \int_{\Omega_R} \alpha h \frac{\partial \bar{\phi}}{\partial y_2} \, \hat{u} \, dy + k^2 \int_{\Omega_R} \alpha' h \, \hat{u} \, \bar{\phi} \, dy -$$

$$- \int_{\Omega_R} \left[\nabla v_h \nabla \bar{\phi} - k^2 v_h \bar{\phi} \right] dy + \int_{\Gamma_R} \bar{\phi} \left[L u_h' - \frac{\partial u_h'}{\partial y_2} \right] ds +$$

$$+ \mathcal{O}(\|h\|_{1,\infty}^2 \|\hat{u}\|_{H^1(\Omega_R)} \|\phi\|_{H^1(\Omega_R)}).$$

The terms with factor k^2 are combined to

$$k^2 \int_{\Omega_R} h \frac{\partial}{\partial y_2} (\alpha \, \bar{\phi} \, \hat{u}) \, dy = k^2 \int_{\Omega_R} \operatorname{div}(h \, \alpha \, \bar{\phi} \, \hat{u} \, \hat{e}_2) \, dy = 0$$

since $h = h(y_1)$, $\alpha(y_2 - f(y_1)) = 0$ on Γ_R and $\phi = 0$ on Γ_R. Finally, since $L u_h' = \partial u_h' / \partial y_2$ we arrive at

$$a(\hat{u}_{f+h} - u - u_h' - v_h, \phi) = \mathcal{O}(\|h\|_{1,\infty}^2 \|\hat{u}\|_{H^1(\Omega_R)} \|\phi\|_{H^1(\Omega_R)}) \quad \text{for all } \phi \in V \cap H^2(\Omega_R).$$

Since $H^2(\Omega_R)$ is dense in $H^1(\Omega_R)$ this implies

$$\|\hat{u}_{f+h} - u - u_h' - v_h\|_{H^1(\Omega_R)} = \mathcal{O}(\|h\|_{1,\infty}^2)$$

which proves the theorem.   □

Closely related to the Fréchet differentiability of $T : f \mapsto u|_K$ is the notion of the "domain derivative" of the mapping $\tilde{T} : \partial\Omega \mapsto u|_K$ (cf. [16] for a detailed discussion of domain derivatives). Let $a \in C^1(\partial\Omega, \mathbb{R}^2)$ be a vector field on $\partial\Omega$ which is $2\pi-$periodic w.r.t. x_1, and define "parallel boundaries" by $\{x + \varepsilon a(x) : x \in \partial\Omega\}$. These are the boundaries $\partial\Omega_{\varepsilon a}$ of some domains $\Omega_{\varepsilon a}$ for sufficiently small $\varepsilon > 0$. Let $\tilde{T}$ map each boundary in a C^1- neighborhood of a fixed reference boundary $\partial\Omega$ to the solution $u|_K$ of the corresponding scattering problem. The the domain derivative of $\tilde{T}$ is defined by

$$\tilde{T}'(\partial\Omega; a) := \lim_{\varepsilon \to 0+} \frac{1}{\varepsilon} \left(\tilde{T}(\partial\Omega_{\varepsilon a}) - \tilde{T}(\partial\Omega) \right),$$

and the limit is understood in the $H^1(K)-$sense. We can prove:

Theorem 11. *Let be $f \in C^2(\mathbb{R})$ $2\pi-$periodic and $u \in Q(k\hat{\theta}_1)$ be the solution of the scattering problem (2.1), (2.2) and (2.4). Let $K \subset \Omega$ be compact and the mapping $\tilde{T}$ be defined as above. Then the domain derivative $\tilde{T}'(\partial\Omega; a)$ of $\tilde{T}$ at f in the direction of any $2\pi-$periodic vector field $a \in C^1(\partial\Omega)$ exists and is given by $u'_a|_K$ where $u'_a \in Q(k\hat{\theta}_1)$ solves the boundary value problem*

$$u'_a = -(a \cdot n) \frac{\partial u}{\partial n} \quad \text{on } \partial\Omega, \tag{4.6}$$

and u'_a satisfies (2.4).

Proof. For $\partial\Omega_{\varepsilon a} = \{(x_1 + \varepsilon a_1(x_1), f(x_1) + \varepsilon a_2(x_1)) : x_1 \in \mathbb{R}\}$ we choose a different parametrization by setting $y_1 := x_1 + \varepsilon a_1(x_1)$. Then $x_1 = \rho_\varepsilon(y_1) = y_1 - \varepsilon a_1(y_1) + \mathcal{O}(\varepsilon^2)$, thus $\partial\Omega_{\varepsilon a} = \{(y_1, f(y_1) + g_\varepsilon(y_1)) : y_1 \in \mathbb{R}\}$ with $g_\varepsilon(y_1) = f(\rho_\varepsilon(y_1)) - f(y_1) + \varepsilon a_2(\rho_\varepsilon(y_1))$, $y_1 \in \mathbb{R}$. The $g_\varepsilon(y_1) = \varepsilon(a_2(y_1) - f'(y_1)a_1(y_1)) + \mathcal{O}(\varepsilon^2)$. From Theorem 4.2 we conclude that

$$\lim_{\varepsilon \to 0+} \frac{1}{\|g_\varepsilon\|_{1,\infty}} \left\| \tilde{T}(\partial\Omega_{\varepsilon a}) - \tilde{T}(\partial\Omega) - u'_a \right\|_{H^1(K)} = 0,$$

where u'_a solves the boundary value problem with boundary data

$$v'_a = -\frac{a_2 - f'a_1}{\sqrt{1 + f'^2}} \frac{\partial u}{\partial n} = -(a \cdot n) \frac{\partial u}{\partial n}$$

which proves the theorem. $\qquad\square$

References

1. Bellout, H., Friedman, A.: Scattering by strip grating. J. Math. Anal. Applic. **147** (1990) 228–248.
2. Bensoussan, A., Lions, J. L., Papanicolaou, G.: Asymptotic analysis for periodic structures. Studies in Mathematics and its applications 5. North Holland, 1978.
3. Brackhage, H., Werner, P.: Über das Dirichletsche Aussenraumproblem für die Helmholtzsche Schwingungsgleichung. Arch. Math. **16** (1965) 325–329.

4. Cadilhac, M.: Some mathematical aspects of the grating theory. In: Petit, R. (ed.): *Electromagnetic theory of gratings*, Springer, 1980, 53–62.
5. Chen, X., Friedman, A.: Maxwell's equations in a periodic structure. Trans. Amer. Math. Soc. **323** (1991) 456–507.
6. Colton, D., Kress, R.: *Integral equation methods in scattering theory*. Wiley, New York 1983.
7. Gilbarg, D., Trudinger, N. S.: *Elliptic partial differential equations of second order*. Springer, Berlin 1983.
8. Hörmander, L.: *An introduction to complex analysis in several variables*. D. Van Nostrand, 1966.
9. Kato, T.: *Perturbation theory for linear operators*. Springer, Berlin 1976.
10. Kirsch, A.: The domain derivative and two applications in inverse scattering theory. To appear in Inverse Problems.
11. Millar, R. F.: The Rayleigh hypothesis and a related least-squares solution to scattering problems for periodic surfaces and other scatterers. Radio Science **8** (1973) 785–796.
12. Millar, R. F.: On the Rayleigh assumption in scattering by a periodic surface. Proc. Camb. Phil. Soc. **65** (1969) 773–791.
13. Millar, R. F.: On the Rayleigh assumption in scattering by a periodic surface: Part II. Proc. Camb. Phil. Soc. **69** (1971) 217–225.
14. Nedelec, J. C., Starling, F.: Integral equation methods in a quasi-periodic diffraction problem for the time harmonic Maxwell's equations. SIAM J. Math. Anal. **22** (1991) 1679–1701.
15. Petit, R.: *Electromagnetic theory of gratings*. Springer, Berlin 1980.
16. Pironneau, O.: *Optimal shape design for elliptic systems*. Springer, Berlin 1984.
17. Wilcox, C. H.: *Scattering Theory for Diffraction Gratings*. Lecture Notes in Mathematics, Springer, Berlin 1984.

On Uniqueness in Inverse Obstacle Scattering

Rainer Kress

Institut für Numerische und Angewandte Mathematik, Universität Göttingen, Lotzestr. 16–18, D3400 Göttingen, Germany

1 Introduction

The inverse problem we consider is to reconstruct the shape of an obstacle from the knowledge of the far field pattern for the scattering of incident time-harmonic acoustic or electromagnetic waves. It occurs in a variety of applications such as nondestructive testing, ultrasound tomography and seismic imaging and is difficult to solve since it is nonlinear and improperly posed. In this survey we outline the three different methods which are available in the literature for proving uniqueness theorems in inverse scattering problems for impenetrable and penetrable obstacles. At the end we will mention a few open uniqueness problems.

2 The inverse scattering problem

Consider time-harmonic acoustic wave propagation in a homogeneous isotropic medium in $\mathbb{R}^3$ with speed of sound c. Then the wave motion can be described by the velocity potential $u(x)e^{-i\omega t}$ with frequency $\omega > 0$ where the space dependent part u satisfies the reduced wave equation or *Helmholtz equation*

$$\triangle u + k^2 u = 0 \tag{2.1}$$

with the *positive wave number k* given by $k = \omega/c$.

The scattering of acoustic time-harmonic plane waves by an impenetrable bounded obstacle described by a domain D (with a connected boundary ∂D) surrounded by a homogeneous isotropic medium leads to exterior boundary value problems for the Helmholtz equation. The total wave u must satisfy the Helmholtz equation in $\mathbb{R}^3 \setminus \bar{D}$ and is decomposed $u = u^i + u^s$ into the given incident plane wave $u^i(x) = e^{ik\,d\cdot x}$, where d is a unit vector giving the direction of propagation, and the unknown scattered wave u^s which is required to fulfill the *Sommerfeld radiation condition*

$$\lim_{r\to\infty} r\left(\frac{\partial u^s}{\partial r} - iku^s\right) = 0, \quad r = |x|, \tag{2.2}$$

uniformly in all directions. Solutions to the Helmholtz equation satisfying the Sommerfeld radiation condition are called radiating solutions. For a *sound-soft* obstacle the total wave $u \in C^2(\mathbb{R}^3 \setminus \bar{D}) \cap C(\mathbb{R}^3 \setminus D)$ has to satisfy a Dirichlet boundary condition

$$u = 0 \quad \text{on } \partial D \tag{2.3}$$

whereas for a *sound-hard* obstacle $u \in C^2(\mathbb{R}^3 \setminus \bar{D}) \cap C^1(\mathbb{R}^3 \setminus D)$ has to satisfy a Neumann boundary condition

$$\frac{\partial u}{\partial \nu} = 0 \quad \text{on } \partial D \tag{2.4}$$

where ν denotes the unit outward normal to the boundary ∂D. More generally, allowing obstacles for which the normal velocity on the boundary is proportional to the excess pressure on the boundary leads to an *impedance* boundary condition of the form

$$\frac{\partial u}{\partial \nu} + i\lambda u = 0 \quad \text{on } \partial D \tag{2.5}$$

with a positive constant λ.

The radiation condition (2.2) ensures the uniqueness for the exterior boundary value problems and leads to an asymptotic behavior of the form

$$u^s(x) = \frac{e^{ik|x|}}{|x|} \left\{ u_\infty \left(\frac{x}{|x|} \right) + O \left(\frac{1}{|x|} \right) \right\}, \quad |x| \to \infty, \tag{2.6}$$

uniformly in all directions where the function u_∞, defined on the unit sphere Ω in $\mathbb{R}^3$, is known as the *far field pattern* or scattering amplitude of the scattered wave. A vanishing far field pattern $u_\infty = 0$ on the unit sphere implies $u^s = 0$ by Rellich's lemma (see [1] or [2]), i.e., there is a one-to-one correspondence between a radiating solution u^s of the Helmholtz equation and its far field pattern u_∞. Since the far field pattern u_∞ is an analytic function (see [1] or [2]) it is completely determined on the whole unit sphere by only knowing it on some surface patch. In the sequel, in order to indicate the dependence on the incident direction d and the wave number k we will write $u = u(\cdot; d, k)$, $u^s = u^s(\cdot; d, k)$ and $u_\infty = u_\infty(\cdot; d, k)$, respectively.

Throughout this paper we assume the boundary ∂D of the scatterer to be of class C^2. Then existence of a solution and well-posedness for the exterior boundary value problems can be based on boundary integral equations. For details we refer to [1] or [2].

The *inverse problem* we are concerned with in this survey is, given the far field pattern $u_\infty(\cdot; d, k)$ of the scattered wave $u^s(\cdot; d, k)$ for one or several incoming plane waves with incident directions d and wave numbers k, and the information on the nature of the scatterer, i.e., the type of the boundary condition, to determine the shape of the scatterer D.

3 Uniqueness Theorems

We want to address the question of uniqueness for these inverse scattering problems, i.e., we want to investigate under what conditions an obstacle is uniquely determined by a knowledge of its far field patterns for incident plane waves. This problem of uniqueness is of interest both for the theoretical study and the implementation of numerical algorithms. We will formulate two uniqueness theorems for the inverse Dirichlet problem, one for incident waves with a fixed wave number and one for a fixed incident direction. For both theorems we will indicate two different proofs and then we will discuss their limitations and extensions with respect to other boundary conditions. In the following theorem, since the far field patterns depend analytically on the incident direction d it suffices to assume that they coincide for an infinite number of incident plane waves with distinct directions.

Theorem 1. *Assume that D_1 and D_2 are two sound-soft scatterers such that the corresponding far field patterns*

$$u_{\infty,1}(\,\cdot\,;d,k) = u_{\infty,2}(\,\cdot\,;d,k) \tag{3.1}$$

coincide for all incident directions $d \in \Omega$ and one fixed wave number k. Then $D_1 = D_2$.

1. Proof. This proof is due to Schiffer (see [10]). Since the far field pattern uniquely determines the scattered field, the assumption (3.1) implies that the scattered waves for both obstacles coincide

$$u_1^s(\,\cdot\,;d,k) = u_2^s(\,\cdot\,;d,k) \quad \text{in } G \tag{3.2}$$

for all $d \in \Omega$ where G denotes the unbounded component of the complement of $\bar{D}_1 \cup \bar{D}_2$. Since ∂G consists only of points belonging either to ∂D_1 or ∂D_2 the homogeneous Dirichlet condition for the total fields implies that $u_1(\,\cdot\,;d,k) = u_2(\,\cdot\,;d,k) = 0$ on ∂G.

Assume that $D_1 \neq D_2$. Without loss of generality, we can assume that $D^* := (\mathbb{R}^3 \setminus \bar{G}) \setminus \bar{D}_1$ is nonempty. Then u_1^s is defined in D^* since it describes the scattered wave for D_1, that is, $u_1 = u^i + u_1^s$ satisfies the Helmholtz equation in D^* and fulfills homogeneous boundary conditions $u_1 = 0$ on ∂D^*. Hence, u_1 is a classical Dirichlet eigenfunction for the negative Laplacian in the domain D^* with eigenvalue k^2. Since any classical solution to the homogeneous Dirichlet boundary value problem can be shown to be also a weak solution without any regularity requirement on ∂D^* (see Gieseke [4] and Weck [12] and also [2]) we have that u_1 belongs to the Sobolev space $H_0^1(D^*)$. Finally, the total fields $u_1(\,\cdot\,;d,k)$ for distinct incoming plane waves and fixed wave number can be shown to be linearly independent since the incident fields are (for the details see [2]). This is a contradiction since by the Rellich selection theorem for a fixed wave number k and a fixed domain D^* there exist only finitely many linearly independent Dirichlet eigenfunctions in $H_0^1(D^*)$. Hence $D_1=D_2$. □

Difficulties arise in attempting to extend Schiffer's approach to the case of other boundary conditions, e.g., the Neumann or the impedance boundary condition. This is due to the fact that the validity of the Rellich selection theorem in the Sobolev space $H^1(D^*)$, i.e., without homogeneous Dirichlet boundary values, requires the boundary to be sufficiently smooth. Hence, using Schiffer's method, we cannot exclude the existence of two scatterers D_1 and D_2 for which the boundary of the difference set D^* is not sufficiently smooth, for example by having cusps.

Isakov [6] has obtained uniqueness results on inverse scattering problems for penetrable obstacles by using techniques different from Schiffer's method. Kirsch and Kress [9] were able to simplify Isakov's approach and extend it to the case of impenetrable obstacles. We will outline the proof which consists of two parts for the case of the Dirichlet boundary condition. In the first step, from the coincidence of the scattered waves (3.2) for all incident plane waves we will deduce that the scattered waves also coincide for all point sources in G as incident waves. These are described in terms of the fundamental solution

$$\Phi(x, y) := \frac{1}{4\pi} \frac{e^{ik|x-y|}}{|x-y|}, \quad x \neq y,$$

to the Helmholtz equation in $\mathbb{R}^3$. Then in the second step, assuming that $D_1 \neq D_2$ we will arrive at a contradiction by letting the point source tend to a boundary point of ∂D_1 which does not belong to $\bar{D}_2$. For the first part we will need the following lemma on the approximation of arbitrary solutions to the Helmholtz equation by plane waves.

Lemma 2. *Let D be a bounded domain with a connected C^2 boundary ∂D such that k^2 is not a Dirichlet eigenvalue of the negative Laplacian for D. Let $v \in C^2(D) \cap C^1(\bar{D})$ satisfy $\Delta v + k^2 v = 0$ in D. Then there exists a sequence (v_n) in the span of plane waves*

$$V := \text{span}\{e^{ik\,x\cdot d} : d \in \Omega\}$$

such that

$$v_n \to v, \quad n \to \infty,$$

uniformly on compact subsets of D (together with all derivatives).

Proof. We begin by showing that the restriction of V on ∂D is complete in $L^2(\partial D)$. Assume $\varphi \in L^2(\partial D)$ satisfies

$$\int_{\partial D} \varphi(y) e^{-ik\,y\cdot d} ds(y) = 0$$

for all $d \in \Omega$. Then the single-layer potential

$$u(x) := \int_{\partial D} \varphi(y) \Phi(x, y) ds(y), \quad x \in \mathbb{R}^3 \setminus \partial D,$$

has vanishing far field pattern $u_\infty = 0$ and therefore we have $u = 0$ in $\mathbb{R}^3 \setminus \bar{D}$. The L^2 jump relations for single-layer potentials (see Kersten [8]) now imply

$$\varphi(x) - 2 \int_{\partial D} \varphi(y) \frac{\partial \Phi(x,y)}{\partial \nu(y)} \, ds(y) = 0 \quad x \in \partial D. \tag{3.3}$$

This integral equation guarantees that $\varphi \in C(\partial D)$ and, therefore, the continuity of the single-layer potential implies that u is a classical solution the homogeneous Dirichlet problem in D. Thus, by our assumption on D we conclude that $u = 0$ in D and the jump relations for the single-layer potential imply $\varphi = 0$. Hence, completeness of V in $L^2(\partial D)$ is established.

This completeness now allows us to choose a sequence (v_n) in V such that

$$\|v_n - v\|_{L^2(\partial D)} \to 0, \quad n \to \infty. \tag{3.4}$$

We can represent $v_n - v$ as a double-layer potential

$$v_n(x) - v(x) = \int_{\partial D} \psi_n(y) \frac{\partial \Phi(x,y)}{\partial \nu(y)} \, ds(y), \quad x \in D, \tag{3.5}$$

where the density $\psi_n \in C(\partial D)$ is given by the solution of the integral equation

$$\psi_n(x) - 2 \int_{\partial D} \psi_n(y) \frac{\partial \Phi(x,y)}{\partial \nu(y)} \, ds(y) = 2v(x) - 2v_n(x), \quad x \in \partial D. \tag{3.6}$$

The integral equation (3.6) is the adjoint of (3.3). Hence, by the Riesz–Fredholm theory, there exists a unique solution ψ_n of (3.6) depending continuously on the right hand side. Therefore, from (3.4) we have that

$$\|\psi_n\|_{L^2(\partial D)} \to 0, \quad n \to \infty. \tag{3.7}$$

Finally, applying Schwarz' inequality to the double-layer potential (3.5), we see that the L^2 convergence (3.7) of the densities ψ_n implies that the v_n converge uniformly to v on compact subsets of D. Convergence of the derivatives follows in an analogous manner.   □

We are no ready to present the second proof for Theorem 1.

2. Proof of Theorem 1. As in the first proof we begin by noting that for incident plane waves the scattered waves coincide as stated in (3.2). Choose $x_0 \in G$ and consider the two exterior Dirichlet problems for radiating solutions to the Helmholtz equation

$$\triangle w_j^s + k^2 w_j^s = 0 \quad \text{in } \mathbb{R}^3 \setminus \bar{D}_j, \quad j = 1, 2, \tag{3.8}$$

with boundary condition

$$w_j^s + \Phi(\cdot, x_0) = 0 \quad \text{on } \partial D_j, \quad j = 1, 2. \tag{3.9}$$

We will show that

$$w_1^s = w_2^s \quad \text{in } G. \tag{3.10}$$

To this end, we choose a bounded domain D with connected C^2 boundary ∂D such that $\bar{D}_1 \cup \bar{D}_2 \subset D$, $x_0 \notin \bar{D}$ and k^2 is not a Dirichlet eigenvalue for D. The latter is possible because of the strong monotonicity property of the eigenvalues of the negative Laplacian with respect to the domain (see [11], Theorem 4.7). Then, by Lemma 2, there exists a sequence (v_n) of linear combinations of plane waves such that

$$v_n \to \Phi(\cdot, x_0), \quad n \to \infty, \tag{3.11}$$

uniformly on $\bar{D}_1 \cup \bar{D}_2$. Since the v_n are linear combinations of plane waves, as a consequence of (3.2), the corresponding scattered waves $v_{n,1}^s$ and $v_{n,2}^s$ for the obstacles D_1 and D_2 coincide in G, that is, for $v_n^s = v_{n,1}^s = v_{n,2}^s$ we have

$$v_n^s + v_n = 0 \quad \text{on } \partial D_j \cap \partial G, \quad j = 1, 2. \tag{3.12}$$

The well-posedness of the exterior Dirichlet problem, the boundary conditions (3.9) and (3.12) and the convergence (3.11) now imply that

$$v_n^s \to w_j^s, \quad n \to \infty,$$

uniformly on compact subsets of G for $j = 1, 2$, whence (3.10) follows.

Now assume that $D_1 \neq D_2$. Then, without loss of generality, there exists $x^* \in \partial G$ such that $x^* \in \partial D_1$ and $x^* \notin \bar{D}_2$. We can choose a positive null sequence (α_n) such that the sequence

$$x_n := x^* + \alpha_n \nu(x^*), \quad n = 1, 2, \ldots,$$

is contained in G. Consider the solutions $w_{n,j}^s$ to the exterior Dirichlet problems (3.8), (3.9) with x_0 replaced by x_n. By (3.10) we have $w_{n,1}^s = w_{n,2}^s$ in G. Considering $w_n^s = w_{n,2}^s$ as the scattered wave corresponding to the obstacle D_2, we observe that the Dirichlet data

$$w_n^s = -\Phi(\cdot, x_n) \quad \text{on } \partial D_2$$

are uniformly bounded with respect to the maximum norm on ∂D_2. Therefore, by the well-posedness of the exterior Dirichlet problem we have that the w_n^s are uniformly bounded with respect to the maximum norm on closed subsets of $\mathbb{R}^3 \setminus \bar{D}_2$. In particular, this implies that

$$|w_n^s(x^*)| \leq C$$

for all n and some positive constant C. On the other hand, considering $w_n^s = w_{n,1}^s$ as the scattered wave corresponding to the obstacle D_1, from the boundary condition we have

$$|w_n^s(x^*)| = |\Phi(x^*, x_n)| = \frac{1}{4\pi |x^* - x_n|} \to \infty, \quad n \to \infty.$$

This is a contradiction. Therefore $D_1 = D_2$. $\qquad\square$

The ideas used in this proof for the inverse Dirichlet problem can be immediately extended to other boundary conditions by making the obvious modifications in (3.9) and (3.12). In particular, uniqueness for the inverse Neumann problem (see [9]) and for the inverse impedance problem can be proven by means of this technique.

We now proceed to discuss a uniqueness theorem for fixed incident direction and varying wave number. Since the far field patterns depend analytically on the wave number (see [2]), in the following theorem it suffices to assume that the far field patterns coincide for an infinite set of distinct wave numbers with a finite accumulation point.

Theorem 3. *Assume that D_1 and D_2 are two sound-soft scatterers such that the corresponding far field patterns*

$$u_{\infty,1}(\,\cdot\,;d,k) = u_{\infty,2}(\,\cdot\,;d,k) \tag{3.13}$$

coincide for a fixed incident direction $d \in \Omega$ and all wave numbers k contained in some interval $[k_1, k_2]$ with $0 \leq k_1 < k_2 < \infty$. Then $D_1 = D_2$.

1. Proof. The first proof essentially is the same as for Theorem 1. Corresponding to the assumption (3.13) we now have coinciding scattered waves

$$u_1^s(\,\cdot\,;d,k) = u_2^s(\,\cdot\,;d,k) \quad \text{in } G \tag{3.14}$$

for all $k \in [k_1, k_2]$. Hence, the only difference to Theorem 1 is, that now the total fields $u_1(\,\cdot\,;d,k)$ for fixed incident direction and different wave numbers become Dirichlet eigenfunctions for the negative Laplacian in D^* for different eigenvalues. These are automatically linearly independent by orthonormality. Therefore, we arrive at the same contradiction as in the proof of Theorem 1.

2. Proof. This proof is due to Jones [7]. We will use the fact that the scattered wave depends analytically on the wave number k and that the derivatives with respect to the space variables and with respect to the wave number can be interchanged. Therefore, from the Helmholtz equation for the total field $u_1 = u^i + u_1^s$ we derive the inhomogeneous Helmholtz equation

$$\triangle w + k^2 w = -2k u_1$$

for the derivative

$$w := \frac{\partial u_1}{\partial k} \ .$$

Now assume that $D_1 \neq D_2$. As above, without loss of generality, we can assume that $D^* := (\mathbb{R}^3 \setminus \bar{G}) \setminus \bar{D}_1$ is nonempty. Then u_1^s is defined in D^* since it describes the scattered wave for D_1, that is, the total wave $u_1 = u^i + u_1^s$ satisfies the Helmholtz equation in D^* and fulfills homogeneous boundary conditions $u_1 = 0$ on ∂D^*. By differentiation with respect to k, it follows that w satisfies the same homogeneous boundary conditions. Therefore, from Green's theorem applied to u_1 and w we find that

$$2k_0 \int_{D^*} |u_1|^2 dx = \int_{D^*} \{\bar{w} \triangle u_1 - u_1 \triangle \bar{w}\}dx = 0,$$

whence $u_1 = 0$ first in D^* and then by analyticity everywhere outside $D_1 \cup D_2$. This implies that u^i satisfies the radiation and this is a contradiction. $\qquad \square$

For the same reasons as indicated above in connection with Theorem 1, obviously, Schiffer's approach to proving Theorem 3 cannot be extended to other boundary conditions whereas the extension of Jones' method is immediately evident.

According to Colton and Sleeman [3], a sound-soft scatterer is uniquely determined by the far field pattern for a finite number of incident plane waves provided a priori information on the size of the obstacle is available as formulated in the following Theorem 4. Its proof adopts Schiffer's idea and makes use of the following strong monotonicity property (see [11], Theorem 4.7): the n–th eigenvalue for a ball B containing the domains D_1 and D_2 is always smaller than the n–th eigenvalue for the subdomain $D^* \subset B$ where the eigenvalues are arranged according to increasing magnitude and taken with their respective multiplicity. For the details we refer to [2] or [3].

Theorem 4. *Let D_1 and D_2 be two scatterers which are contained in a ball of radius R, let*

$$N := \sum_{t_{nl} < kR} 2n + 1$$

where t_{nl}, $l = 0, 1, \ldots$, denote the positive zeros of the spherical Bessel functions j_n, $n = 0, 1, 2, \ldots$, and assume that the far field patterns coincide for $N + 1$ incident plane waves with distinct directions and one fixed wave number. Then $D_1 = D_2$.

Since the smallest positive zero of the spherical Bessel functions is given by π, as a simple corollary of Theorem 4 we have that a sound-soft scatterer is uniquely determined by the far field pattern for one incident plane wave provided the wave number satisfies $kR < \pi$.

As an interesting open problem, we wish to point out that it is not known if one incoming plane wave for one single direction and one single wave number completely determines a sound-soft scatterer (without any additional a priori information). And of course, it is also desirable to obtain results analogous to Theorem 4 for sound-hard obstacles and the impedance boundary condition.

As already mentioned above, Isakov [6] has proved the analogue of Theorem 1 for the scattering of plane waves from a penetrable obstacle, that is, for the inverse *transmission* problem. The variant indicated in the second proof of Theorem 1 can also be extended to the inverse transmission problem and seems to be slightly simpler than Isakovs original proof despite the fact that, due to the occurrence of the transmitted wave, some technical difficulties arise (for details see [9]). Following Isakov's ideas, Hettlich [5] was able to prove the analogue of Theorem 1 for the case of the *conductive* boundary condition. We expect, that Hettlich's result can also be obtained via the approach used in [9] for the transmission problem. However, this has not yet been carried out explicitly. It is another open problem to see whether Jones' approach works for the inverse transmission and conductive boundary value problems in order to prove analogues to Theorem 3.

Finally, we wish to conclude with a few remarks on related questions in *electromagnetic* scattering. Here, again Schiffer's approach for the same reasons as above cannot be used. However, for the case of the perfect conductor analogues to Theorems 1 and 3 can be established via the second variant of our proofs (see [2]). And it should also be possible to treat the impedance, the transmission and the conductive boundary condition as in [9]. However, again this has not yet been worked out in detail.

References

1. Colton, D., and Kress, R.: *Integral Equation Methods in Scattering Theory.* Wiley-Interscience Publication, New York 1983.
2. Colton, D., and Kress, R.: *Inverse Acoustic and Electromagnetic Scattering Theory.* Springer-Verlag, Berlin Heidelberg New York 1992.
3. Colton, D., and Sleeman, B.D.: Uniqueness theorems for the inverse problem of acoustic scattering. IMA J. Appl. Math. **31** (1983) 253–259.
4. Gieseke, B.: Zum Dirichletschen Prinzip für selbstadjungierte elliptische Differentialoperatoren. Math. Z. **68** (1964) 54–62.
5. Hettlich, F.: Ein inverses Streuproblem bei konduktiven Randbedingungen zur Helmholtzgleichung. Dissertation, Erlangen 1992.
6. Isakov, V.: On uniqueness in the inverse transmission scattering problem. Comm. Part. Diff. Equa. **15**(1990) 1565–1587.
7. Jones, D.S.: Note on a uniqueness theorem of Schiffer. Applicable Analysis **19** (1985) 181–188.
8. Kersten, H.: Grenz- und Sprungrelationen für Potentiale mit quadratsummierbarer Dichte. Resultate d. Math. **3** (1980) 17–24.
9. Kirsch, A. and Kress, R.: Uniqueness in inverse obstacle scattering. To appear in Inverse Problems.
10. Lax, P.D., and Phillips, R.S.: *Scattering Theory.* Academic Press, New York 1967.
11. Leis, R.: *Initial Boundary Value Problems in Mathematical Physics.* John Wiley, New York 1986.
12. Weck, N.: Klassische Lösungen sind auch schwache Lösungen. Arch. Math. **20** (1969) 628–637.

The Inverse Scattering Problem for a Homogeneous Bi-isotropic Slab Using Transient Data

Gerhard Kristensson and Sten Rikte

Department of Electromagnetic Theory, Lund Institute of Technology, P.O. Box 118, S-221 00 Lund, Sweden

Abstract *Transient wave propagation in a finite bi-isotropic slab is treated. The incident field impinges normally on the slab, which can be inhomogeneous wrt depth. Dispersion and bi-isotropy are modeled by time convolutions in the constitutive relations. Outside the slab the medium is assumed to be homogeneous, non-dispersive and isotropic, and such that there is no phase velocity mismatch at the boundaries of the slab. Two alternative methods of solution to the propagation problem are given—the imbedding method and the Green function approach. The second method is used to solve the inverse problem and the first to generate synthetic data. The inverse scattering problem is to reconstruct the four susceptibility kernels of the medium using a set of finite time trace of reflection and transmission data.*

1 Basic equations

The wave propagation problem in bi-isotropic media has received extensive attention during recent years. The fixed frequency problem has been analyzed by many authors, see [3, 10] for a review of recent work. Transient problems have also been addressed, see [8, 9]. The present analysis of the transient problem is a generalization of previous work for the isotropic slab, see Ref. [7], and applies the generalization of the wave splitting technique to the Maxwell equations [12].

Consider a dispersive bi-isotropic slab with boundaries, $z = 0$ and $z = d$. The slab can be inhomogeneous in the z-direction. Outside the slab, the medium is assumed to be non-dispersive with constant permittivity $\epsilon_0\epsilon$ and permeability $\mu_0\mu$. The phase velocity c and the wave impedance η of the surrounding medium is then given by

$$c = \frac{1}{\sqrt{\epsilon_0\epsilon\mu_0\mu}} = \frac{c_0}{\sqrt{\epsilon\mu}}, \qquad \eta = \sqrt{\frac{\mu_0\mu}{\epsilon_0\epsilon}} = \eta_0\sqrt{\frac{\mu}{\epsilon}}$$

respectively[1]. The slab is assumed to have no mismatch at the boundaries, $z = 0$ and $z = d$. The appropriate constitutive relations used in this paper are, see

---

[1] The subscript zero indicates the corresponding vacuum values.

Ref. [6, 9][2]:

$$\begin{cases} c\eta D(\boldsymbol{r},t) = \boldsymbol{E}(\boldsymbol{r},t) + ((G+F) * \boldsymbol{E})(\boldsymbol{r},t) + \eta((K+L) * \boldsymbol{H})(\boldsymbol{r},t) \\ c\boldsymbol{B}(\boldsymbol{r},t) = ((-K+L) * \boldsymbol{E})(\boldsymbol{r},t) + \eta \boldsymbol{H}(\boldsymbol{r},t) + \eta((G-F) * \boldsymbol{H})(\boldsymbol{r},t) \end{cases} \quad (1)$$

where time convolution is denoted by $*$, i.e.

$$(G * \boldsymbol{E})(\boldsymbol{r},t) = \int_{-\infty}^{t} G(z, t-t') \boldsymbol{E}(\boldsymbol{r}, t') \, dt'$$

The kernels G and F model the ordinary dispersive effects, while K and L model the bi-isotropy of the medium. If $L = 0$, the medium is reciprocal [6]. Causality implies that the functions G, F, K and L are identically zero for $t < 0$. Furthermore, all kernels are assumed to be continuously differentiable functions of t for $t > 0$ at each point $\boldsymbol{r}$ in the interior of the slab. Outside the slab, G, F, K and L are all identical to zero.

The slab is excited by a (known) general plane wave, with electric field $\boldsymbol{E}_i$, propagating in the positive z-direction. Moreover, the electric field $\boldsymbol{E}$ is everywhere assumed to be of the form

$$\boldsymbol{E}(\boldsymbol{r},t) = \boldsymbol{e}_x E_x(z,t) + \boldsymbol{e}_y E_y(z,t)$$

and similarly for the other electromagnetic vector fields[3]. Thus, by introducing the reflected, $\boldsymbol{E}_r$, and the transmitted, $\boldsymbol{E}_t$, electric fields, respectively, the total electric field outside the slab can be written as

$$\boldsymbol{E}(z,t) = \begin{cases} \boldsymbol{E}_i(z,t) + \boldsymbol{E}_r(z,t), & z \leq 0 \\ \boldsymbol{E}_t(z,t), & z \geq d \end{cases}$$

The source free Maxwell equations are the basic equations that model the dynamics of the electromagnetic fields.

$$\nabla \times \boldsymbol{E}(\boldsymbol{r},t) = -\frac{\partial \boldsymbol{B}(\boldsymbol{r},t)}{\partial t}$$

$$\nabla \times \boldsymbol{H}(\boldsymbol{r},t) = \frac{\partial \boldsymbol{D}(\boldsymbol{r},t)}{\partial t}$$

It is convenient to use a matrix notation and, therefore, vectors in the x-y-plane are identified with their column vector representations, i.e.

$$\boldsymbol{E}(\boldsymbol{r},t) = \begin{pmatrix} E_x(z,t) \\ E_y(z,t) \end{pmatrix}$$

The Maxwell equations can then be written

$$\frac{\partial}{\partial z} \begin{pmatrix} \boldsymbol{E} \\ \eta \boldsymbol{J} \boldsymbol{H} \end{pmatrix} = \frac{1}{c} \frac{\partial}{\partial t} \begin{pmatrix} c\boldsymbol{J}\boldsymbol{B} \\ c\eta \boldsymbol{D} \end{pmatrix}$$

---

[2] Note the change in notation between this paper and these references.
[3] $\boldsymbol{e}_x, \boldsymbol{e}_x, \boldsymbol{e}_z$ is the usual basis in $\mathbb{R}^3$.

where **J** is defined by

$$\mathbf{J} = \begin{pmatrix} 0 & -1 \\ 1 & 0 \end{pmatrix}$$

The constitutive relations (1) are now used to eliminate the $\mathbf{D}$ and the $\mathbf{B}$-fields in this equation. The result is

$$c\frac{\partial}{\partial z}\begin{pmatrix} \mathbf{E} \\ \eta\mathbf{J}\mathbf{H} \end{pmatrix} = \begin{pmatrix} (-\mathbf{K}+\mathbf{L})* & \mathbf{I}+(\mathbf{G}-\mathbf{F})* \\ \mathbf{I}+(\mathbf{G}+\mathbf{F})* & -(\mathbf{K}+\mathbf{L})* \end{pmatrix}\frac{\partial}{\partial t}\begin{pmatrix} \mathbf{E} \\ \eta\mathbf{J}\mathbf{H} \end{pmatrix} \tag{2}$$

where **I** is the 2×2 identity matrix and where

$$\mathbf{G} = G\mathbf{I}, \quad \mathbf{K} = K\mathbf{J}, \quad \mathbf{F} = F\mathbf{I}, \quad \mathbf{L} = L\mathbf{J}$$

2 Wave splitting

The wave splitting is a change of the dependent variables such that the new variables represent the general left- and right-going waves in a medium without dispersion, e.g., outside the slab. The idea of wave splitting has been used in several scattering problems, e.g., [1, 2, 11, 12]. Here, the same splitting as in an earlier paper, see [9], is used, i.e., the linear map

$$\begin{pmatrix} \mathbf{E}^+(z,t) \\ \mathbf{E}^-(z,t) \end{pmatrix} = \mathbf{P}\begin{pmatrix} \mathbf{E}(z,t) \\ \eta\mathbf{J}\mathbf{H}(z,t) \end{pmatrix} \tag{3}$$

where the 4×4 matrix **P** is defined as

$$\mathbf{P} = \frac{1}{2}\begin{pmatrix} \mathbf{I} & -\mathbf{I} \\ \mathbf{I} & \mathbf{I} \end{pmatrix}$$

is adopted. Note that $\mathbf{E}^\pm(z,t)$ are continuous at the boundaries of the slab, i.e., on the planes $z = 0$ and $z = d$, and hence they are continuous vector fields.

Outside the slab, $\mathbf{E}^\pm(z,t)$ are the electric fields of the general right- and left-going waves, respectively. Inside the slab, no interpretation in right- and left-going waves is possible, but the definition (3) is, nevertheless, well defined everywhere, and

$$\begin{cases} \mathbf{E}(z,t) = \mathbf{E}^+(z,t) + \mathbf{E}^-(z,t) \\ \mathbf{H}(z,t) = \dfrac{1}{\eta}\mathbf{J}\left(\mathbf{E}^+(z,t) - \mathbf{E}^-(z,t)\right) \end{cases}$$

3 Dynamics

The fields satisfy the system of equations in (2). The plus and minus fields $E^+(z,t)$ and $E^-(z,t)$, defined in Section 2, satisfy a similar system of equations. These equations are equivalent to (2) and straightforward to derive using (2) and (3). The result is

$$c\frac{\partial}{\partial z}\begin{pmatrix} E^+ \\ E^- \end{pmatrix} = \begin{pmatrix} -I & 0 \\ 0 & I \end{pmatrix}\frac{\partial}{\partial t}\begin{pmatrix} E^+ \\ E^- \end{pmatrix} + \begin{pmatrix} -G-K & -F+L \\ F+L & G-K \end{pmatrix} * \frac{\partial}{\partial t}\begin{pmatrix} E^+ \\ E^- \end{pmatrix}$$

The first term gives the dynamics for the free space contribution, and the second the perturbation due to the dispersion in the slab.

Propagation of singularities gives information about the rotation and the attenuation of the wave front as it propagates through the medium. Assume there is a finite jump discontinuity in the plus field E^+ at $z = 0$. This finite jump discontinuity propagates through the medium and the finite jump discontinuity at depth z is related to its value at $z = 0$ as

$$[E^+(z, t+z/c)] = Q(0, z)\,[E^+(0, t)]$$

where the 2×2 matrix $Q(0, z)$ quantifies the rotation and the attenuation of the discontinuity. The square bracket $[E^+(z, t+z/c)]$ denotes the finite jump discontinuity of the field at the point $(z, t+z/c)$, i.e.

$$[E^+(z, t+z/c)] = E^+(z, t+z/c+0) - E^+(z, t+z/c-0)$$

The field E^- shows no finite jump discontinuity of this kind and therefore the total field E has exactly the same finite jump discontinuity as the field E^+ above. Notice also, that the finite jump discontinuity $[E^+(0, t)]$ is the same on either side of the interface of the slab.

The matrix $Q(0, z)$ can be explicitly solved and expressed in the following general notation, see Ref. [9]:

$$Q(z_1, z_2) = \exp\left(a(z_1, z_2)\right)\begin{pmatrix} \cos\phi(z_1, z_2) & -\sin\phi(z_1, z_2) \\ \sin\phi(z_1, z_2) & \cos\phi(z_1, z_2) \end{pmatrix} \qquad (4)$$

where the angle of rotation of the wave front $\phi(z_1, z_2)$ is

$$\phi(z_1, z_2) = -\frac{1}{c}\int_{z_1}^{z_2} K(z', 0)\,dz'$$

and the attenuation factor $a(z_1, z_2)$ is

$$a(z_1, z_2) = -\frac{1}{c}\int_{z_1}^{z_2} G(z', 0)\,dz' \qquad (5)$$

Notice that this result holds for an inhomogeneous slab with arbitrary susceptibility kernels $G(z, t)$, $F(z, t)$, $K(z, t)$ and $L(z, t)$.

4 The scattering operators

By use of linearity, causality and time-invariance, the scattering operators that relate the incident field $E^+(0,t)$ to the reflected field $E^-(0,t)$ and the transmitted field $E^+(d, t + d/c)$ must be of the form

$$\begin{cases} E^-(0,t) = (\mathbf{R}_{\mathrm{ph}}(\cdot) * E^+(0,\cdot))(t) \\ E^+(d,t+d/c) = \mathbf{Q}(0,\mathbf{d})\left\{E^+(0,t) + (\mathbf{T}_{\mathrm{ph}}(\cdot) * E^+(0,\cdot))(t)\right\} \end{cases} \tag{6}$$

where $\mathbf{R}_{\mathrm{ph}}(t)$ and $\mathbf{T}_{\mathrm{ph}}(t)$ are the physical reflection and transmission kernels, respectively, and $\mathbf{Q}(0,\mathbf{d})$ defined in (4).

By the axial symmetry of the problem, it is obvious that the form of the scattering operators is

$$\mathbf{R}_{\mathrm{ph}}(t) = \begin{pmatrix} R_1^{\mathrm{ph}}(t) & -R_2^{\mathrm{ph}}(t) \\ R_2^{\mathrm{ph}}(t) & R_1^{\mathrm{ph}}(t) \end{pmatrix}, \qquad \mathbf{T}_{\mathrm{ph}}(t) = \begin{pmatrix} T_1^{\mathrm{ph}}(t) & -T_2^{\mathrm{ph}}(t) \\ T_2^{\mathrm{ph}}(t) & T_1^{\mathrm{ph}}(t) \end{pmatrix} \tag{7}$$

It is easy to see that all 2×2 matrices of this form commute.

5 The Imbedding equations

The wave splitting naturally leads to the study of a subsection $[z, d]$ of the slab $[0, d]$. For each subsection the reflection and transmission imbedding kernels are defined by

$$\begin{cases} E^-(z,t) = (\mathbf{R}(z,\cdot) * E^+(z,\cdot))(t) \\ E^+(d,t+(d-z)/c) = \mathbf{Q}(z,\mathbf{d})\left\{E^+(z,t) + (\mathbf{T}(z,\cdot) * E^+(z,\cdot))(t)\right\} \end{cases}$$

where

$$\mathbf{R}(z,t) = \begin{pmatrix} R_1(z,t) & -R_2(z,t) \\ R_2(z,t) & R_1(z,t) \end{pmatrix}, \qquad \mathbf{T}(z,t) = \begin{pmatrix} T_1(z,t) & -T_2(z,t) \\ T_2(z,t) & T_1(z,t) \end{pmatrix}$$

due to axial symmetry. By continuity of the split vector fields, the following boundary values are obtained:

$$\begin{cases} \mathbf{R}(0,t) = \mathbf{R}_{\mathrm{ph}}(t) \\ \mathbf{T}(0,t) = \mathbf{T}_{\mathrm{ph}}(t) \end{cases}, \qquad \begin{cases} \mathbf{R}(d,t) = 0 \\ \mathbf{T}(d,t) = 0 \end{cases}$$

The subsection problem generates a one parameter family of reflection and transmission kernels, which satisfy integro-differential equations. These imbedding equations are for the reflection kernel

$$c\partial_z \mathbf{R} - 2\partial_t \mathbf{R} = \partial_t \left\{\mathbf{F} + \mathbf{L} + 2\mathbf{G} * \mathbf{R} + (\mathbf{F} - \mathbf{L}) * \mathbf{R} * \mathbf{R}\right\} \tag{8}$$

$$\mathbf{R}(d,t) = 0$$

$$\mathbf{R}(z,+0) = -(\mathbf{F}(z,+0) + \mathbf{L}(z,+0))/2$$

$$[\mathbf{R}(z, 2(d-z)/c)] = \frac{\exp\left(2\mathbf{a}(z,\mathbf{d})\right)}{2}(\mathbf{F}(d,+0) + \mathbf{L}(d,+0))$$

and for the transmission kernel the result is

$$c\partial_z \mathbf{T} = \partial_t \{\mathbf{G} + \mathbf{K}\} * \mathbf{T} + \partial_t \{\mathbf{G} + \mathbf{K} + (\mathbf{F} - \mathbf{L}) * (\mathbf{R} + \mathbf{R} * \mathbf{T})\} \qquad (9)$$
$$\mathbf{T}(\mathbf{d}, \mathbf{t}) = 0$$

$$2c\mathbf{T}(\mathbf{z}, +0) = \int_z^d \{\mathbf{F}^2 - \mathbf{L}^2 - 2\partial_t(\mathbf{G} + \mathbf{K})\} (\mathbf{z}', +0)\, \mathrm{d}\mathbf{z}'$$

The attenuation factor $a(z, d)$ is defined in (5). The domain of definition for these equations is $\{(z, t) : t > 0,\ 0 < z < d \text{ and } t \neq 2(d - z)/c\}$.

If the medium is reciprocal, i.e., $L(z, t) = 0$, equation (8) shows that the cross-polarized reflection kernel $R_2(z, t) = 0$. The reason for this is that the imbedding equation (8) then has only diagonal entries and no off-diagonal terms (this assumes unique solubility of the equation). The reflection kernel then simplifies to $\mathbf{R}(\mathbf{z}, \mathbf{t}) = \mathbf{R}_1(\mathbf{z}, \mathbf{t})\mathbf{I}$. As a consequence of this, the cross-polarized part of the reflected electric field from the finite or semi-infinite reciprocal bi-isotropic slab will be zero even if the slab is inhomogeneous. This result holds for any susceptibility kernels G, F and K and is therefore a general result. If also $F(z, t) = 0$, then $\mathbf{R} = 0$, irrespective of the susceptibility kernel $G(z, t)$. This reflects the ill-posedness of the inverse scattering problem, since infinitely many dispersive profiles give the same reflection kernel $\mathbf{R}(\mathbf{z}, \mathbf{t})$, i.e., the same reflected field regardless of the incident field. Transmission data can be used to resolve this non-uniqueness.

6 The Green Functions

In the previous section the scattering problem was analyzed using imbedding arguments. In this section an independent approach—the Green function equations— is applied. The Green functions give the internal fields due to a delta function excitation. They are defined by

$$\begin{cases} \mathbf{E}^+(z, t + z/c) = \mathbf{Q}(0, z)\mathbf{E}^+(0, t) + (\mathbf{G}^+(\mathbf{z}, \cdot) * \mathbf{Q}(0, z)\mathbf{E}^+(0, \cdot)) (t) \\ \mathbf{E}^-(z, t + z/c) = (\mathbf{G}^-(\mathbf{z}, \cdot) * \mathbf{Q}(0, z)\mathbf{E}^+(0, \cdot)) (t) \end{cases}$$

where due to axial symmetry

$$\mathbf{G}^\pm(\mathbf{z}, \mathbf{t}) = \begin{pmatrix} G_1^\pm(z, t) & -G_2^\pm(z, t) \\ G_2^\pm(z, t) & G_1^\pm(z, t) \end{pmatrix}$$

By continuity of the split vector fields, the Green functions are related to the physical reflection and transmission kernels, i.e., the boundary conditions are

$$\begin{cases} G^+(0, t) = 0 \\ G^-(0, t) = R_{\mathrm{ph}}(t) \end{cases}, \qquad \begin{cases} G^+(d, t) = T_{\mathrm{ph}}(t) \\ G^-(d, t) = 0 \end{cases}$$

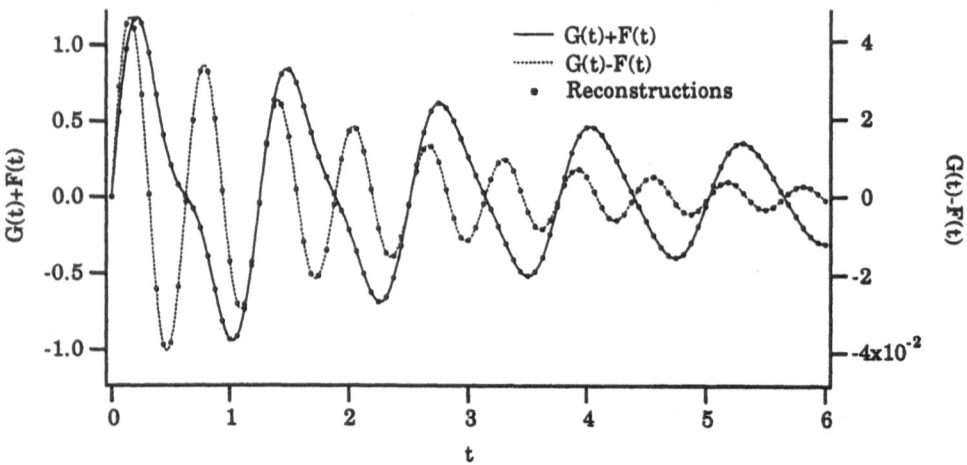

Figure 1: The susceptibility kernels $G(t) + F(t)$ and $G(t) - F(t)$ for the reciprocal Lorentz medium in Example 1 and their reconstructions. The time scale is given in units of d/c, and the vertical axis in units of c/d.

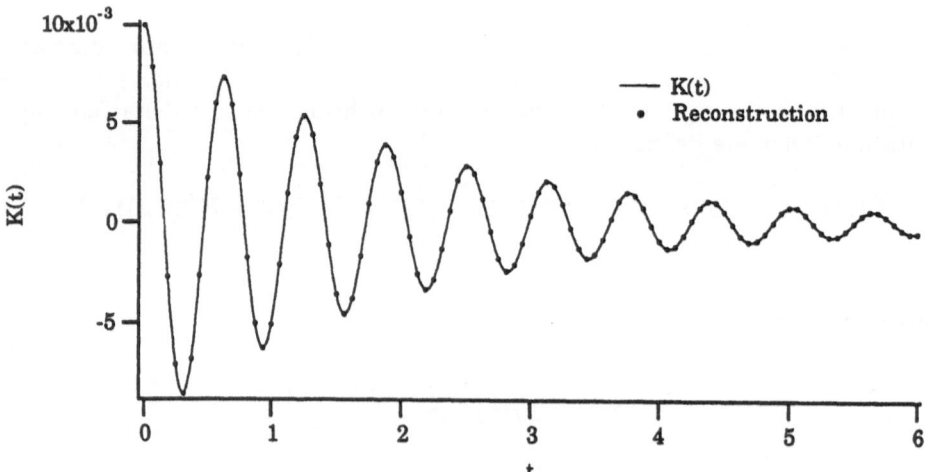

Figure 2: The susceptibility kernel $K(t)$ for the reciprocal Lorentz medium ($L(t) = 0$) in Example 1 and its reconstruction. The time scale is given in units of d/c, and the vertical axis in units of c/d.

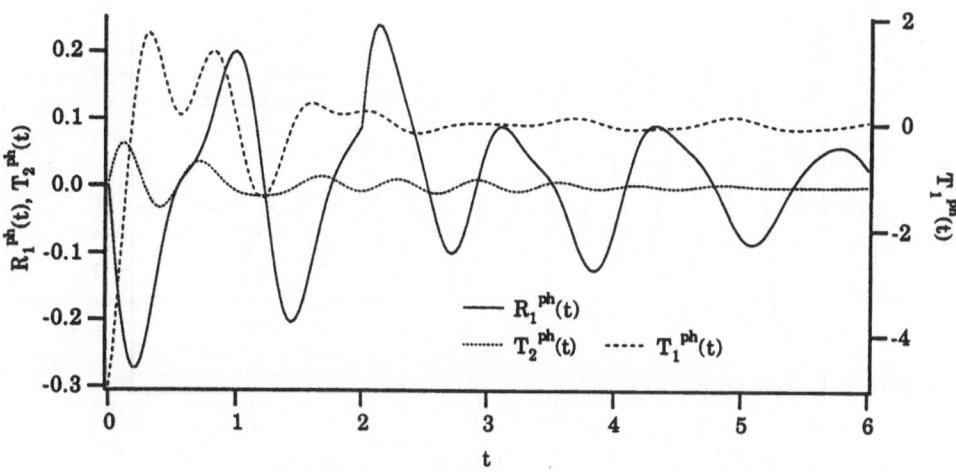

Figure 3: The reflection kernel $R_1^{ph}(t)$ and the transmission kernels $T_1^{ph}(t)$ and $T_2^{ph}(t)$ for the reciprocal Lorentz medium $(R_2^{ph}(t) = 0)$ in Example 1. The time scale is given in units of d/c, and the vertical axis in units of c/d.

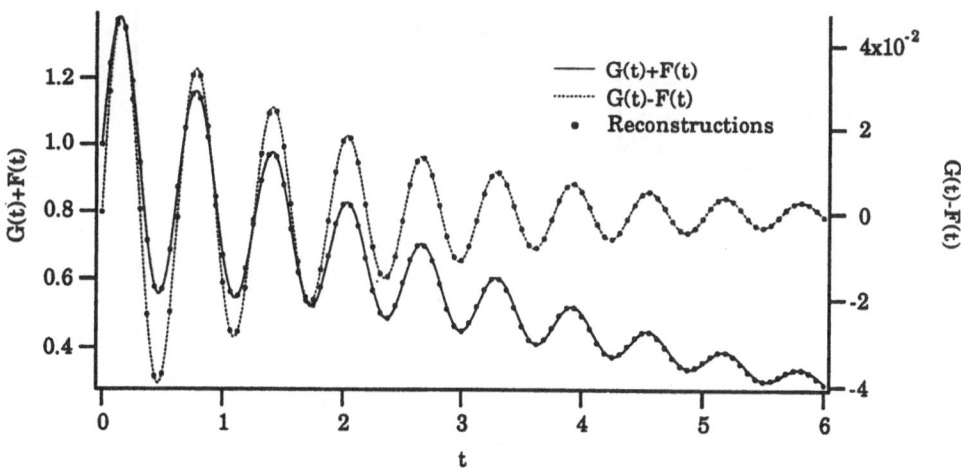

Figure 4: The susceptibility kernels $G(t) + F(t)$ and $G(t) - F(t)$ for the Debye-Lorentz medium in Example 2 and their reconstructions. The time scale is given in units of d/c, and the vertical axis in units of c/d.

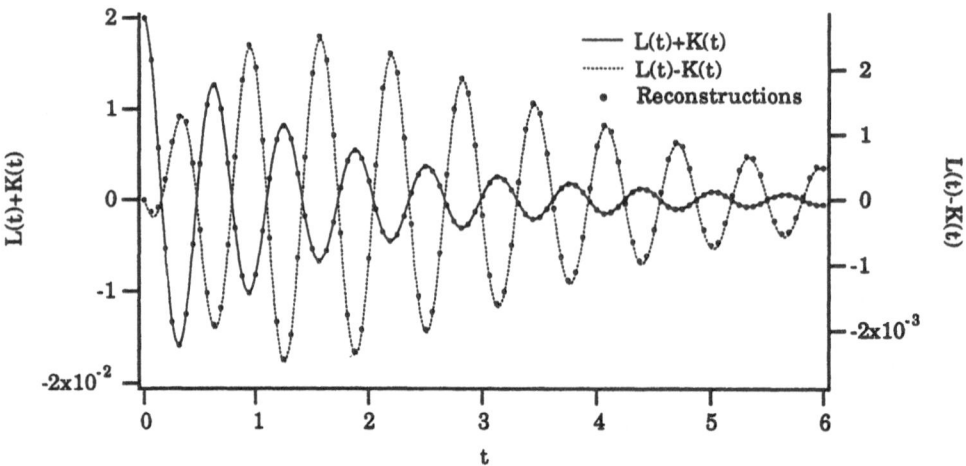

Figure 5: The susceptibility kernels $L(t) + K(t)$ and $L(t) - K(t)$ for the Debye-Lorentz medium in Example 2 and their reconstructions. The time scale is given in units of d/c, and the vertical axis in units of c/d.

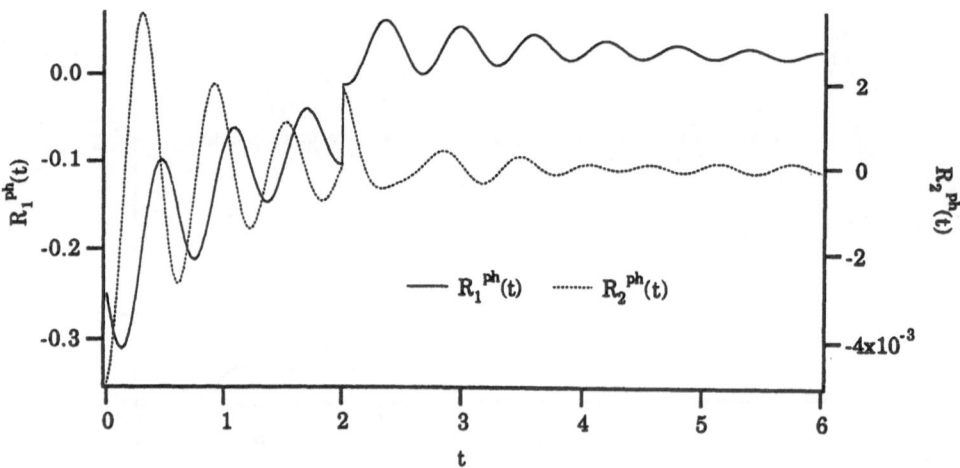

Figure 6: The reflection kernels $R_1^{ph}(t)$ and $R_2^{ph}(t)$ for the Debye-Lorentz medium in Example 2. The time scale is given in units of d/c, and the vertical axis in units of c/d. Notice the finite jump discontinuity in $R_1^{ph}(t)$ and $R_2^{ph}(t)$ at one round trip $t = 2$.

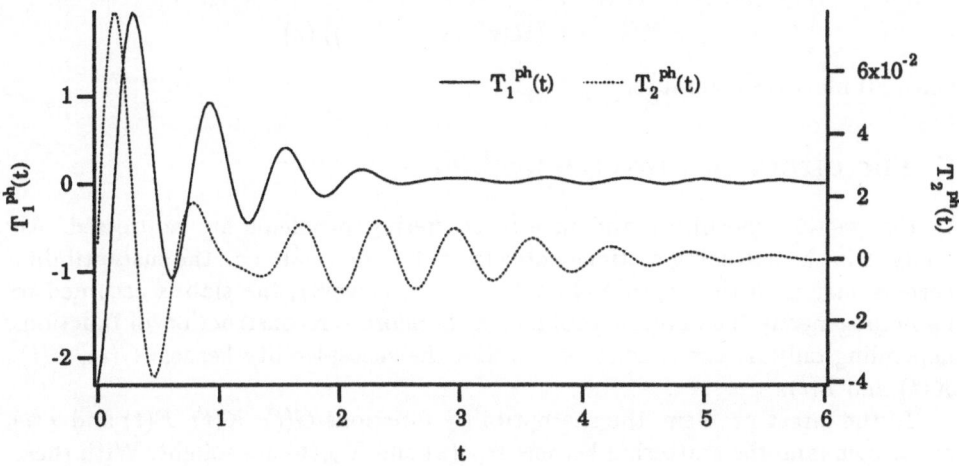

Figure 7:   The transmission kernels $T_1^{ph}(t)$ and $T_2^{ph}(t)$ for the Debye-Lorentz medium in Example 2. The time scale is given in units of d/c, and the vertical axis in units of c/d.

Analogous to the imbedding equations for the scattering kernels of the subsection problem, the Green functions satisfy a system of integro-differential equations. These equations, linear in $\mathbf{G}^{\pm}$, are for the matrix function $\mathbf{G}^{+}$

$$c\partial_z \mathbf{G}^+ = -\partial_t \{\mathbf{G} + \mathbf{K}\} * \mathbf{G}^+ - \partial_t \{\mathbf{G} + \mathbf{K} + (\mathbf{F} - \mathbf{L}) * \mathbf{G}^-\} \qquad (10)$$

$$\mathbf{G}^+(0, t) = 0$$

$$2c\mathbf{G}^+(z, +0) = \int_0^z \{\mathbf{F}^2 - \mathbf{L}^2 - 2\partial_t(\mathbf{G} + \mathbf{K})\}(z', +0)\, dz'$$

and for the matrix function $\mathbf{G}^{-}$

$$c\partial_z \mathbf{G}^- - 2\partial_t \mathbf{G}^- = 2\mathbf{G}(z, 0)\mathbf{G}^- + \partial_t \{\mathbf{F} + \mathbf{L}\} \qquad (11)$$
$$+ \partial_t \{(\mathbf{F} + \mathbf{L}) * \mathbf{G}^+\} + \partial_t \{\mathbf{G} - \mathbf{K}\} * \mathbf{G}^-$$

$$\mathbf{G}^-(d, t) = 0$$

$$\mathbf{G}^-(z, +0) = -(\mathbf{F}(z, +0) + \mathbf{L}(z, +0))/2$$

$$[\mathbf{G}^-(z, 2(d - z)/c)] = \frac{\exp(2a(z, d))}{2}(\mathbf{F}(d, +0) + \mathbf{L}(d, +0))$$

As in the imbedding formulation, the domain of definition in both these cases is $\{(z, t) : t > 0,\, 0 < z < d \text{ and } t \neq 2(d - z)/c\}$.

The explicit relation between the imbedding kernels $\mathbf{R}$ and $\mathbf{T}$ and Green functions is

$$\begin{cases} \mathbf{G}^+(d,t) = \mathbf{G}^+(z,t) + \mathbf{T}(z,t) + \left(\mathbf{T}(z,\cdot) * \mathbf{G}^+(z,\cdot)\right)(t) \\ \mathbf{G}^-(z,t) = \mathbf{R}(z,t) + \left(\mathbf{R}(z,\cdot) * \mathbf{G}^+(z,\cdot)\right)(t) \end{cases}$$

since all matrices commute.

7 The direct and inverse problems

In this section the direct and inverse scattering problems are addressed. All analysis in the preceding sections holds for a stratified slab, i.e., the susceptibility kernels vary with the depth z. In this section, however, the slab is assumed to be homogeneous. The inverse problem is therefore a reconstruction of functions depending only on the time t, i.e., to find the susceptibility kernels $G(t)$, $F(t)$, $K(t)$ and $L(t)$.

In the direct problem, the susceptibility functions $G(t)$, $K(t)$, $F(t)$ and $L(t)$ are known, and the scattering kernels $\mathbf{R}_{ph}(t)$ and $\mathbf{T}_{ph}(t)$ are sought. With these kernels the scattered fields can be calculated for any excitation, see (6). The internal fields can be calculated with the Green functions $\mathbf{G}^\pm(z,t)$. These direct problems have been addressed in two previous papers, see Refs. [8, 9]. Even though this paper emphasizes the inverse problem the solution of the direct scattering problem is used to reconstruct the dispersive profile.

In the inverse problem the kernels $\mathbf{R}_{ph}(t)$ and $\mathbf{T}_{ph}(t)$ and the matrix $\mathbf{Q}(0,d)$ are assumed to be known. These scattering kernels are obtained from the scattered fields by deconvolving (6). The unknowns in the inverse problem are four susceptibility kernels $G(t)$, $K(t)$, $F(t)$ and $L(t)$. Karlsson [4, 5] has shown that the forward algorithm can be used to solve the inverse problem for an isotropic dispersive medium. This method is a constructive method and does not use optimization. An outline of a generalization of this method is given below. To avoid bias in the solution of the inverse scattering problem, the imbedding equations are used to generate synthetic scattering data $\mathbf{R}_{ph}(t)$ and $\mathbf{T}_{ph}(t)$. The inverse scattering problem is then solved using the Green function formulations.

The method of Karlsson can be extended and used to solve the inverse scattering problem for a dispersive bi-isotropic medium. The outline of this extended Karlsson's method is:

Time step 0: Since $\mathbf{Q}(0,d)$ is assumed to be known through the transmission data, the initial values $G(0)$ and $K(0)$ are known, and the other initial values easily follow from

$$\begin{cases} \mathbf{F}(0) + \mathbf{L}(0) = -2\mathbf{R}_{ph}(0) \\ \mathbf{G}'(0) + \mathbf{K}'(0) = \frac{1}{2}(\mathbf{F}(0)^2 - \mathbf{L}(0)^2) - \frac{c}{d}\mathbf{T}_{ph}(0) \\ \mathbf{F}'(0) + \mathbf{L}'(0) = -2(\mathbf{R}'_{ph}(0) + \mathbf{G}(0)\mathbf{R}_{ph}(0)) \end{cases}$$

These initial values are easily found from the results above, see (8) and (9) or (10) and (11).

Time step J: Suppose that $G'(j)$, $K'(j)$, $F'(j)$ and $L'(j)$[4] are known for time $j < J$. Analysis of the discretized version of the Green functions equations[5], (10) and (11), for the direct scattering problem using the trapezoidal rule implies that

$$\begin{cases} \mathbf{T}_{ph}(\mathbf{J}) = \mathbf{A}^+(\mathbf{J})G'(\mathbf{J}) + \mathbf{B}^+(\mathbf{J})K'(\mathbf{J}) + \mathbf{C}^+(\mathbf{J})F'(\mathbf{J}) + \mathbf{D}^+(\mathbf{J})L'(\mathbf{J}) + \mathbf{E}^+(\mathbf{J}) \\ \mathbf{R}_{ph}(\mathbf{J}) = \mathbf{A}^-G'(\mathbf{J}) + \mathbf{B}^-K'(\mathbf{J}) + \mathbf{C}^-F'(\mathbf{J}) + \mathbf{D}^-L'(\mathbf{J}) + \mathbf{E}^-(\mathbf{J}) \end{cases}$$

The coefficient matrices $\mathbf{A}^-$, $\mathbf{B}^-$, $\mathbf{C}^-$ and $\mathbf{D}^-$ are constant matrices and their explicit values are easily found. The matrices $\mathbf{A}^+$, $\mathbf{B}^+$, $\mathbf{C}^+$, $\mathbf{D}^+$, $\mathbf{E}^+$ and $\mathbf{E}^-$ depend only on $G'(j)$, $K'(j)$, $F'(j)$ and $L'(j)$ for time $j < J$, and vary, as indicated, with J. All matrices in this expression have the form as in (7) and therefore they all commute.

Thus, by running the forward program five times (four times in the reciprocal case) for different sets of $G'(J)$, $K'(J)$, $F'(J)$, $L'(J)$ at time J, a linear system of equations in the unknown coefficient matrices is obtained. After calculating these matrices, a linear system of equations in $G'(J)$, $K'(J)$, $F'(J)$ and $L'(J)$ is achieved. This system of equations can be solved since $\mathbf{T}_{ph}(\mathbf{J})$ and $\mathbf{R}_{ph}(\mathbf{J})$ are known. Finally, the Green functions $\mathbf{G}^\pm$ at all grid points for $j = J$ are computed by running the forward program once again. The procedure is then repeated for the next time step $J + 1$.

Two explicit examples in the next section illustrate the numerical performance of this algorithm.

8 Numerical results

In the following explicit examples, all frequencies (i.e., all numerical values) are given in units of c/d and time t in units of d/c.

Example 1

In this first example, a reciprocal multi-frequency Lorentz medium is presented. The susceptibility kernels are

$$\begin{cases} G(t) = 0.5e^{-0.2t}\sin 5t + 0.25e^{-0.5t}\sin 10t \\ F(t) = 0.5e^{-0.2t}\sin 5t + 0.2e^{-0.5t}\sin 10t \\ K(t) = 0.01e^{-0.5t}\cos 10t \\ L(t) = 0 \end{cases}$$

Since the medium is reciprocal ($L(t) = 0$), there is no cross coupling in the reflected field ($R_2^{ph}(t) = 0$). The susceptibility kernels are shown in Figures 1 and 2. Notice that the chirality of the medium, modeled by $K(t)$, is two orders

---

[4] $G'(j) = \frac{d}{dt}G(j\Delta t)$ etc.
[5] The same conclusion can be made from (8) and (9) if the imbedding equations are used to solve the inverse scattering problem.

of magnitude smaller than the kernel $G(t) + F(t)$. This reduction of magnitude in the chirality kernel is believed to model realistic data. The reflection and transmission data for this medium are shown in Figure 3. The scattering data are continuous functions of t, since $F(0) = L(0) = 0$. The reconstructions of the susceptibility kernels are shown in Figures 1 and 2. In these reconstructions 64 data points per round trip ($t = 2$) are used.

Example 2

The second example is a non-reciprocal multi-frequency Debye-Lorentz medium. These susceptibility kernels are

$$
\begin{cases}
G(t) = 0.5e^{-0.2t} + 0.25e^{-0.5t}\sin 10t \\
F(t) = 0.5e^{-0.2t} + 0.2e^{-0.5t}\sin 10t \\
K(t) = 0.01e^{-0.5t}\cos 10t \\
L(t) = 0.01e^{-t}\cos 10t
\end{cases}
$$

The susceptibility kernels are depicted in Figures 4 and 5. The reflection and transmission data for this medium are shown in Figures 6 and 7. As in Example 1, the chirality of the medium is two orders of magnitude smaller than the kernel $G(t) + F(t)$. Notice that the reflection kernels $R_1^{\mathrm{ph}}(t)$ and $R_2^{\mathrm{ph}}(t)$ have finite jump discontinuities at one round trip ($t = 2$), since $F(0), L(0) \neq 0$. The reconstructions of the susceptibility kernels are shown in Figures 4 and 5. In these reconstructions 64 data points per round trip ($t = 2$) are used.

Both these numerical experiments show that excellent reconstructions of the susceptibility kernels $G(t)$, $F(t)$, $K(t)$ and $L(t)$ are obtained.

References

1. Corones, J. P., Davison, M. E., Krueger, R. J.: Direct and inverse scattering in the time domain via invariant imbedding equations. J. Acoust. Soc. Am. **74** (5) (1983) 1535–1541.
2. Corones, J. P., Davison, M. E., Krueger, R. J.: Wave splittings, invariant imbedding and inverse scattering. In: Devaney, A. J. (ed.): *Inverse Optics*, pages 102–106, SPIE Bellingham, WA, 1983. Proc. SPIE 413.
3. Engheta, N., Jaggard, D. L.: Electromagnetic chirality and its applications. IEEE Antennas and Propagation Society Newsletter, pages 6–12, October 1988.
4. Karlsson, A.: Inverse scattering for viscoelastic media using transmission data. Inverse Problems **3** (1987) 691–709.
5. Karlsson, A.: Direct and inverse electromagnetic scattering from a dispersive medium. Technical Report TRITA-TET 89-2, Department of Electromagnetic Theory, S-100 44 Stockholm, Sweden, 1989.
6. Karlsson, A., Kristensson, G.: Constitutive relations, dissipation and reciprocity for the Maxwell equations in the time domain. J. Electro. Waves Applic. **6** (5/6) (1992) 537–551.

7. Kristensson, G., Krueger, R. J.: Direct and inverse scattering in the time domain for a dissipative wave equation. part 3: Scattering operators in the presence of a phase velocity mismatch. J. Math. Phys. **28** (2) (1987) 360–370.

8. Kristensson, G., Rikte, S.: Scattering of transient electromagnetic waves in reciprocal bi-isotropic media. Technical Report LUTEDX/(TEAT-7015)/1–17/(1991), Lund Institute of Technology, Department of Electromagnetic Theory, P.O. Box 118, S-211 00 Lund, Sweden, 1991. J. Electro. Waves Applic. (in press).

9. Kristensson, G., Rikte, S.: Transient wave propagation in reciprocal bi-isotropic media at oblique incidence. Technical Report LUTEDX/(TEAT-7019)/1–25/(1992), Lund Institute of Technology, Department of Electromagnetic Theory, P.O. Box 118, S-211 00 Lund, Sweden, 1992.Accepted for publication in J. Math. Phys.

10. Lakhtakia, A.: Recent contributions to classical electromagnetic theory of chiral media: what next? Speculations in Science and Technology **14**(1) (1991) 2–17.

11. Weston, V. H.: Invariant imbedding for the wave equation in three dimensions and the applications to the direct and inverse problems. Inverse Problems **6** (1990) 1075–1105.

12. Weston, V. H.: Time-domain wave-splitting of Maxwell's equations. Technical Report LUTEDX/(TEAT-7016)/1–25/(1991), Lund Institute of Technology, Department of Electromagnetic Theory, P.O. Box 118, S-211 00 Lund, Sweden, 1991.

On the Inverse Scattering Problem for Rational Reflection Coefficients

P. B. Kurasov

Department of Mathematical and Computational Physics, Institute for Physics, St. Petersburg University, St.Petersburg, 198904 Russia and Manne Siegbahn Institute, Freskativägen 24, 10405 Stockholm, Sweden

Abstract *The inverse scattering problem on the half-axis for long range potentials is studied. It is shown that the solution of the inverse problem contains arbitrary real parameters iven if no bound states are present. Connections with the inverse problem on the whole axis are discussed.*

1 Introduction

The present paper is devoted to the inverse scattering problem on the half line [1],[2]. New exact solutions of this inverse problem are constructed. Such exact solutions play an important role in the theory of nonlinear equations. Nonuniqueness of the solution of the inverse problem in the presence of bound states allows to construct soliton solutions of nonlinear equations. Several examples of the ambiguity potentials without any bound states for the case of the whole line scattering [3] were discovered during the last years [4, 5, 6, 7, 8, 9]. The most general description of such potentials is given in [2]. This class of ambiguities is connected with a slow decrease of the potentials at infinity (like const./x^2). But such potentials do not produce any ambiguities for half line scattering. In order to obtain ambiguities in that case the class of admissible potentials must be extended. In the present paper we shall investigate the solution of the inverse problem on the half line for potentials with the following behaviour at infinity

$$V(x) \sim \frac{\sum_{j=1}^{M} A_j \sin k_j x}{x}.$$

The corresponding scattering matrix does not satisfy the Levinson condition. Similar scattering matrices were obtained for delta-functional potentials and selfadjoint perturbations constructed with the help of the extension-restriction procedure [10, 11, 12]. We shall restrict our consideration to the case of rational reflection coefficients only in order to avoid additional unessential difficulties.

The direct scattering problem for oscillating potentials is discussed in the second part of the paper. The third part is devoted to the solution of the inverse problem for rational reflection coefficients corresponding to the short range potentials. We recall simple analytical formulas for the solution, obtained from the Gelfand-Levitan-Marchenko equation. In the fourth part we discuss the solution of the inverse problem for the rational reflection coefficients violating the Levinson theorem. It is shown that the resulting oscillating potential is not defined uniquely by the reflection coefficient even if no bound state is present. The family of ambiguity potentials contains a finite number of real parameters.

2 The direct scattering problem

We shall discuss the scattering problem for the Schroedinger operator

$$\mathcal{A} = -\frac{d^2}{dx^2} + V(x) \tag{1}$$

on the half axis $[0, \infty)$ with Dirichlet boundary condition at the zero point: $Dom(\mathcal{A}) = \{u \in W_2^2(\mathbb{R}_+), u(0) = 0\}$. The unperturbed operator is

$$\mathcal{A}_0 = -\frac{d^2}{dx^2}.$$

Usually the scattering problem is investigated in the case of real potentials satisfying the condition

$$\int_0^\infty |V(x)| dx < \infty. \tag{2}$$

Under this condition the Jost solutions exist for all real nonzero values of the spectral parameter k and the asymptotics of the regular solution

$$\varphi(k, x) : -\frac{d^2}{dx^2}\varphi(k, x) + V(x)\varphi(k, x) = k^2\varphi(k, x), \varphi(k, 0) = 0$$

for large x is a combination of plane waves

$$\varphi(k, x) \sim A_+(k)e^{ikx} + A_-(k)e^{-ikx} + o(\frac{1}{\sqrt{x}}), \quad k \in [0, \infty). \tag{3}$$

The scattering operator coincides with the reflection coefficient $S(k) = -A_-(k)/A_+(k)$. Under the condition of the finitness of the first moment:

$$\int_0^\infty x|V(x)| dx < \infty. \tag{4}$$

the Jost solution $f(k, x), -f''(k, x) + V(x)f(k, x) = k^2 f(k, x), f(k, x) = e^{ikx} + o(1/\sqrt{x}), x \to \infty$ and the Jost function $F(k) = f(k, 0)$ are analytic functions of the spectral parameter k in the upper half plane, continuous up to the real axis. The scattering matrix $S(k)$ coincides with the ratio of two Jost functions $S(k) = F(-k)/F(k)$.

The Jost function for potentials violating the condition (4) can have singularities on the real axis. This phenomenon was studied for the first time in connection with the zero energy bound state, when potential decreases at infinity like $1/|x|^2$ and the Jost function has singularity at point zero. Condition (2) is sufficient but not necessary condition for the wave operators to be complete.

The usual scattering operator exists also in the case when the Jost solution is well defined for almost all k on the real axis. An additional condition on the difference between the regular and free solutions at infinity, $x \to \infty$ is necessary [13] . Hence the class of admissible potentials can be extended. For example, the wave operators are complete for potentials with the following asymptotic behaviour at infinity

$$V(x) = \frac{\sum_{j=1}^{M} A_j \sin(k_j x)}{x} + V_0(x); \tag{5}$$

$$|V_0(x)| \leq C(1+x^2)^{-1/2-\epsilon}$$

for some $C, \epsilon > 1/4$. This class of potentials was investigated in connection with the phenomenon of bound states imbedded into the continuous spectrum [14, 15, 16] . The Jost solution for a potential from the class (5) exists for $k \neq 0, k \neq k_j/2$. The reflection coefficient is defined for almost every real k.

3 Inverse scattering problem for short-range potentials

The inverse problem is the problem to restore the potential $V(x)$ from the known reflection coefficient. For the class of potentials defined by the conditions (2, 4) this problem can be solved by Gelfand-Levitan-Marchenko procedure [1, 2].

An important class of analytically solvable inverse problems is formed by the rational Jost functions. All rational Jost functions can be presented in the following form

$$F(k) = \prod_{j=1}^{M} \frac{k + \imath a_j}{k + \imath b_j} \tag{6}$$

with $\Re b_j > 0$. The sets $\{a_j\}, \{b_j\}$ are symmetric over the real axis. The constants a_j with negative real part can be only real and correspond to bound states. Such Jost functions define scattering matrices of the form

$$S(k) = \prod_{j=1}^{M} \frac{k - \imath a_j}{k - \imath b_j} \frac{k + \imath b_j}{k + \imath a_j} \tag{7}$$

corresponding to potential exponentially decreasing at infinity. Simple analytical formulas for the potential and regular solution can be obtained by solution of the Gelfand-Levitan-Marchenko equation [1, 2]

$$V(x) = -2\frac{d^2}{dx^2} \ln \det W(x),$$

$$\varphi(k, x) = \frac{\det \begin{vmatrix} W(x) & f(x) \\ \beta(k, x) & \dfrac{\sin kx}{k} \end{vmatrix}}{\det W(x)}, \tag{8}$$

where

$$W_{nm}(x) = \frac{e^{-a_n x}}{2b_m}\left(\frac{e^{b_m x}}{a_n - b_m} - \frac{e^{-b_m x}}{a_n + b_m}\right),$$

$$f_n(x) = e^{-a_n x}$$

$$\beta_n(k, x) = \int_0^x \frac{\sin \imath b_n t}{\imath b_n}\frac{\sin kt}{k}dt.$$

4 Inverse problem for oscillating potentials

The scattering matrix for potentials satisfying condition (5) is a generalized function. We shall consider the case when this function is equivalent to some rational function of the form

$$S(k) = \prod_{j=1}^{M}\frac{k - \imath a_j}{k + \imath a_j} \tag{9}$$

$\Re a_j > 0$. As the phase shift on the real axis

$$\delta(\infty) - \delta(-\infty) = \frac{1}{2\imath}\int_{-\infty}^{+\infty}\frac{S'(k)}{S(k)}dk = \pi M$$

is positive the Levinson theorem can not be fulfilled for such reflection coefficients. Then the inverse problem does not have a solution in the class of short range potentials even if we assume that some bound states are present.

The Gelfand-Levitan method is based on the representation of the scattering matrix as a ratio of the Jost functions analytical in the upper half plane $\Pi_+ = \{k, \Im k > 0\}$ continuous in the closed upper half plane $\bar{\Pi}_+ = \{k, \Im k \geq 0\}$ with the unit limit at infinity. Similar representation can be introduced in the investigated case also

$$S(k) = \frac{F_0(-k)}{F_0(k)}, \quad F_0(k) = \prod_{j=1}^{M}\frac{k + \imath a_j}{k + \imath b_j}, \quad \Re b_j = 0. \tag{10}$$

The formally introduced Jost function $F_0(k)$ has singularities on the real axis. The singularities $k = -\imath b_j$ are situated symmetrically with respect to the origin and do not give any contribution to the scattering matrix. The Jost function $F_0(k)$ contains $[M/2]$ arbitrary real parameters. To solve the inverse problem for this class of the reflection coefficients the limit of the Gelfand-Levitan procedure can be used. In the first step we approximate the Jost functions $F_0(k)$ by the functions with singularities at the lower half plane, which are Jost functions for some short-range potentials

$$F_\epsilon = \prod_{j=1}^{M}\frac{k + \imath a_j}{k + \imath(b_j + \epsilon)}, \epsilon > 0. \tag{11}$$

The potential $V_\epsilon(x)$ and the regular solution $\varphi_\epsilon(x)$ corresponding to the Jost function $F_\epsilon(k)$ can be calculated with the help of the formulas (8). Then the pointwise limits of the potential and regular solution when $\epsilon \to 0$ are considered. The potentials V_ϵ exponentially decrease at infinity, but the limit potential V_0 is from the class (5).

The following theorem was formulated in [12] as a conjecture and it was proved there for $M = 1, 2$ only.

Theorem 1. *The limit of the regular solution $\varphi_0(k, x) = \lim_{\epsilon \to 0} \varphi_\epsilon(k, x), x \in [0, \infty)$ is a regular solution for the Schroedinger equation for every $k \in [0, \infty), k \neq 0, \imath b_j$ with the limit potential $V_0(x) = \lim_{\epsilon \to 0} V_\epsilon(x), x \in [0, \infty)$. The reflection coefficient for the limit potential is given by the formula*

$$S(k) = \prod_{j=1}^{M} \frac{k - \imath a_j}{k + \imath a_j}.$$

Proof. The Theorem 1 can be proved by an iteration procedure because every scattering matrix $S(k)$ with the symmetrically situated zeroes and poles can be presented as a product of elementary unimodular functions containing not more than two factors.

The formulas (8) can be generalized for the case when the background operator has the form

$$\mathcal{A}_0 = -\frac{d^2}{dx^2} + V_0(x) \tag{12}$$

with the potential V_0 from the class (5). The generalized formulas allow to calculate the potential $V(x)$ and the regular solution corresponding to the reflection coefficient $S(k) = \prod_{j=1}^{M}(k - \imath a_j)/(k + \imath a_j)S_0(k)$, where $S_0(k)$ is the reflection coefficient for the potential $V_0(x)$. The standard Gelfand-Levitan-Marchenko procedure can not be used directly and one needs to consider the approximation procedure similar to one discussed earlier. To prove the Theorem 1 one needs to consider this approximation procedure for the case $M = 1, 2$ only.

Lemma 2. *Let the background operator be of the form (12) and let the logarithmic derivative of the corresponding Jost function be bounded at the point $t = 0$. Then for $M = 1$ the limit of the regular solution $\varphi_{\epsilon=0}(k, x)$ is a regular solution for the Schroedinger equation for every $k \in [0, \infty), k \neq 0, \imath b_j$ with the limit potential $V_{\epsilon=0}(x)$. The reflection coefficient for the limit potential is given by the formula*

$$S(k) = \frac{k - \imath a_1}{k + \imath a_1} S_0(k).$$

Proof. The Jost function and approximate Jost functions are introduced as follows:

$$F(k) = \frac{k + \imath a_1}{k} F_0(k) \Rightarrow F_\epsilon(k) = \frac{k + \imath a_1}{k + \imath \epsilon} F_0(k) \tag{13}$$

The potential and the regular solution, corresponding to the approximate Jost function are [1, 2]:

$$V_\epsilon(x) - V_0(x) =$$

$$= -2\frac{(\epsilon^2 - a^2)(f^2(\imath a_1)\varphi'^2_x(\imath\epsilon) - f'^2_x(\imath a_1)\varphi^2(\imath\epsilon)) - (\epsilon^2 - a^2)^2 f^2(\imath a_1)\varphi^2(\imath\epsilon)}{(f(\imath a_1)\varphi'_x(\imath\epsilon) - f'_x(\imath a_1)\varphi(\imath\epsilon))^2}$$

$$(14)$$

$$\varphi_\epsilon(k,x) = \varphi(k) + \frac{a_1^2 - \epsilon^2}{k^2 + \epsilon^2}\frac{\mathbf{W}[\varphi(\imath\epsilon), \varphi(k)]}{\mathbf{W}[\varphi(\imath\epsilon), f(\imath a_1)]}f(\imath a_1) \tag{15}$$

where the following notations for the Jost and regular solutions corresponding to the nonperturbed operator $\mathcal{A}_0$ were used $f(k) \equiv f_0(k,x), \varphi(k) \equiv \varphi_0(k,x)$. The limits of the potential and regular solution for $\epsilon \to 0$ are

$$V_{\epsilon=0}(x) - V_0(x) = 2a^2 \times \tag{16}$$

$$\frac{\left(\frac{F_0'(0)}{F_0(0)}f_x'(0) - f''_{xk}(0)\right)^2 f^2(\imath a_1) + \left(\frac{F_0'(0)}{F_0(0)}f(0) - f'_k(0)\right)^2 \left(a^2 f^2(\imath a_1) - f'^2_x(\imath a_1)\right)}{\left(f(\imath a_1)(f''_{xk}(0) - \frac{F_0'(0)}{F_0(0)}f_x'(0)) + f'_x(\imath a_1)\left(\frac{F_0'(0)}{F_0(0)}f(0) - f'_k(0)\right)\right)^2}$$

$$\varphi_{\epsilon=0}(k,x) = \varphi(k) + \frac{a_1^2}{k^2}f(\imath a_1)\times \tag{17}$$

$$\frac{\frac{F_0'((0)}{F_0(0)}\mathbf{W}[f(0), F_0(k)f(-k) - F_0(-k)f(k)] - \mathbf{W}[f'_k(0), F_0(k)f(-k) - F_0(-k)f(k)]}{\left(\frac{F_0'(0)}{F_0(0)}\mathbf{W}[f(0), f(\imath a_1)] - \mathbf{W}[f'_k(0), f(\imath a_1)]\right)(F_0(k)f'_{0x}(-k,0) - F_0(-k)f'_{0x}(k,0))}$$

By direct calculation one can prove that the limit of the regular solution is a regular solution for the limit potential. The asymptotics of the potential for large x is

$$V_{\epsilon=0}(x) - V_0(x) \sim \frac{2a^2}{\left(1 + a(\imath\frac{F_0'(0)}{F_0(0)} + x)\right)^2} \sim \frac{2}{x^2}$$

We note that the Jost function $F_0(k)$ has the following property: $F_0(-\bar{k}) = \overline{F_0(k)}$. Hence the logarithmic derivative of the Jost function at point zero is purely imaginary $\imath F_0'(0)/F_0(0) \in \mathbb{R}$ and the calculated potential is real. The asymptotics of the regular solution for $x \to \infty$ is

$$\varphi_{\epsilon=0}(k,x) \sim \frac{1}{F_0(k)f'_{0x}(-k,0) - F_0(-k)f'_{0x}(k,0)}\frac{1}{k}$$

$$\{e^{-\imath kx}(k + \imath a_1)F_0(k) - e^{\imath kx}(k - \imath a_1)F_0(-k)\} \tag{18}$$

and the reflection coefficient is

$$S(k) = \frac{k - \imath a_1}{k + \imath a_1}S_0(k), \quad k \neq 0, \pm \imath b_j \tag{19}$$

The Jost function and the Jost solution have singularities on the real axis at the points $k = \pm \imath b_j, 0$. The case of the multiple singularities can be studied separately. This complets the proof of Lemma 1 .   □

We note that the condition of the finiteness of the logarithmic derivative of the Jost function at the point zero is fulfilled automatically if the potential V_0 was constructed by the iteration procedure.

Lemma 3. *Let the background operator be of the form (12). Then for $M = 2$ the limit of the regular solution $\varphi_{\epsilon=0}(k, x)$ is a regular solution for the Schroedinger equation for every $k \in [0, \infty), k \neq 0, \imath b_j$ with the limit potential $V_{\epsilon=0}(x)$. The reflection coefficient for the limit potential is given by the formula*

$$S(k) = \frac{k - \imath a_1}{k + \imath a_1} \frac{k - \imath a_2}{k + \imath a_2} S_0(k).$$

Proof. To avoid complicated formulas we shall discuss here the proof of this Lemma for the case of $V_0 = 0$ only. The original scattering matrix is given by the expression

$$S(k) = \frac{k - \imath a_1}{k + \imath a_1} \frac{k - \imath a_2}{k + \imath a_2}, \quad a_1 = \bar{a}_2 \quad or \quad a_1, a_2 \in \mathbb{R}.$$

There is arbitrariness in the definition of the Jost function in this case. The family of the possible Jost functions depends on the real parameter b_0

$$F_{b_0}(k) = \frac{k + \imath a_1}{k - b_0} \frac{k + \imath a_2}{k + b_0} \tag{20}$$

Approximate Jost functions are introduced as follows

$$F_{b_0, \epsilon}(k) = \frac{k + \imath a_1}{k - b_0 + \imath \epsilon} \frac{k + \imath a_2}{k + b_0 + \imath \epsilon}, \epsilon > 0 \tag{21}$$

The limit potential when $\epsilon \to 0$ is

$$V_{a_1, a_2, b_0}(x) = 16 b_0^2 \frac{1 - (b_0 x + B) \sin 2(b_0 x + \delta(b_0)) - \cos 2(b_0 x + \delta(b_0))}{(2 b_0 x + 2B - \sin 2(b_0 x + \delta(b_0)))^2} \tag{22}$$

where we used the following notations

$$B = B(b_0, a_1, a_2) = \frac{b_0(a_1 + a_2)(a_1 a_2 + b_0^2)}{(a_1^2 + b_0^2)(a_2^2 + b_0^2)}$$

$$e^{2\imath \delta(b_0)} = S(b_0) \tag{23}$$

The corresponding regular solution is

$$\varphi(k, x) = \frac{\sin kx}{k} + \tag{24}$$

$$+ \frac{1}{2\imath k} \frac{e^{\imath kx} \left(-w(b_0, -\imath k, a_2) + w(b_0, -\imath k, a_1)\right) + e^{-\imath kx} \left(w(b_0, \imath k, a_2) - w(b_0, \imath k, a_1)\right)}{w(b_0, a_1, a_2)}$$

where the function w is
$w(b_0, a_1, a_2) =$

$$= \frac{a_2 - a_1}{(a_1^2 - b^2)(a_2^2 - b^2)} (2 b_0 x + 2B(b_0, a_1, a_2) - \sin 2(b_0 x + \delta(b_0, a_1, a_2))) \tag{25}$$

The asymptotics of the solution for $x \to \infty$ is:

$$\varphi(k,x) \sim \frac{e^{ikx}}{2ik} \frac{(k-ia_1)(k-ia_2)}{k^2-b_0^2} - \frac{e^{-ikx}}{2ik} \frac{(k+ia_1)(k+ia_2)}{k^2-b_0^2} \tag{26}$$

for $k \neq \pm b_0$ and the scattering matrix can be easily calculated:

$$S(k) = \frac{k-ia_1}{k+ia_1} \frac{k-ia_2}{k+ia_2} \tag{27}$$

The calculated Jost function and Jost solution have singularities at the points $k = \pm b_0$. This proof can be generalized for the case $V_0 \neq 0$. Then the additional condition $b_0 \neq ib_j$ is necessary. We have now finished the proof of Lemma 2 and Theorem 1. $\qquad\square$

The proof of Theorem 1 shows, that in the case $M = 1$ the unique potential was calculated. Arbitrariness of the solution of the inverse problem in the second case is connected with the arbitrariness of the definition of the Jost function.

Acknowledgements The author wishes to thank professors C.Chadan, N.Elander, B.Pavlov, B.Karlson, R.Newton and J.Sjostrand for the interest to the work and fruitful discussions. He is grateful for professor L.Päivärinta for the invitation to a very stimulating conference at Saariselka. He is indebted to Manne Siegbahn Institute for Atomic Physics and Stockholm University for the financial support and hospitality.

References

1. Faddeev, L.D., Uspekhi Mat. Nauk **14**, 57 (1959); English translation: J.Math.Phys.**4** (1963) 73.
2. Chadan K., Sabatier P.C. "Inverse problems in quantum scattering theory", 2nd edition, Springer-Verlag, Berlin, 1989.
3. Deift, P., Trubowitz, E., Comm. Pure and Appl. Math. **XXXII**, 121 (1979).
4. Abraham, P.B., De Facio, B., Moses, H.E., Phys. Rev. Let. **46**, 1657 (1981).
5. Brownstein, K.R., Phys. Rev. D **25**, 2704 (1982).
6. Moses, H.E., Phys. Rev. A **27**, 2220 (1983).
7. Aktosun, T., Newton, R.G., Inverse Problems **1**, 291 (1985).
8. Degasperis, A., Sabatier, P.C., Inverse Problems **3**, 73 (1987).
9. Aktosun, T., Inverse Problems **4**, 347 (1988).
10. Albeverio S., Gesztesy F., Hoegh-Krohn R., Holden H. "Solvable models in quantum mechanics", Springer-Verlag, Berlin, 1988.
11. Pavlov, B.S., Russian Math. Surveys **42:6**, 127 (1987).
12. Kurasov P.B., Let.Math.Phys. **25**, 287 (1992).
13. Reed M., Simon B. "Methods of Modern Mathematical Physics" III "Scattering Theory", Academic Press, New York, 1979, p.155-168.
14. Matveev V.B., Skriganov M.M. Dokl. Acad. Nauk SSSR **202** (1972), p.775-758.
15. Skriganov M.M. Zap. Nauch. Sem. LOMI **38** (1973), p.149.
16. Simon B. Comm. Pure Applied Math., vol **XXII**, 531-538 (1967).

Some Geometric Aspects of Multidimensional Inverse Spectral Problems

Ya. Kurylev

St. Petersburg Branch, Steklov Mathematics Institute, St. Petersburg, Russia and Purdue University

1 Introduction

The paper concerns reconstruction procedures and uniqueness results for inverse boundary problems for second-order elliptic PDO. We are mostly interested in the analysis of the operators on Riemannian manifolds though some results on operators in a domain in $\mathbb{R}^m$ are also included.

The method used is the BC-method put forward by M. Belishev (for the pioneering work see [1]). In its present state the method was elaborated for the operators in $\mathbb{R}^m$ in [2, 3], and applied to an inverse problem of a Riemannian manifold reconstruction (both topology and Riemannian metrics) in [4]. In the paper we modify considerations of [4] (concerning originally the Laplace-Beltrami operator with Neumann boundary data) for the case of more general operators on a Riemannian manifold and derive from here some consequences for the operators in $\mathbb{R}^m$. In the presentation, we skip almost all technical details (referring to the paper [4]) and pay attention to some geometrical ideas underlying the results.

2 Geometry of rays and Schrödinger operator on a Riemann manifold

Let Ω be a smooth (C^∞), compact Riemannian manifold with the border $\Gamma :=\partial\Omega$, $\Gamma \neq \phi$, $dim\Omega = m \geq 2$. For a Schrödinger operator $A = A(\Omega, q, \sigma)$:

$$Au = -\Delta u + qu, \quad \mathcal{D}(A) := \{u \in H^2(\Omega) : \partial_\nu u - \sigma u|_\Gamma = 0\} \qquad (1)$$

where Δ and ∂_ν are relatively the Laplace-Beltrami operator and outward normal derivative and smooth q and σ are real valued, we define the boundary spectral data (BSD) of A as follows:

BSD of the operator $A(\Omega, q, \sigma)$ is the triple

$$(\Gamma, \{\lambda_k\}_{k=1}^\infty, \{\psi_k|_\Gamma\}_{k=1}^\infty), \qquad (2)$$

where λ_k, $k = 1, 2, \ldots$, are the eigenvalues of A (counting multiplicity) and $\psi_k|_\Gamma$, $k = 1, 2, \ldots$, are the traces on Γ of the corresponding orthonormalized eigenfunctions.

Theorem A *Let the system (Ω, q) belong to the class $\mathcal{N}$. Then BSD (2) determine Ω (both topology and Riemannian metrics), q and σ uniquely.*

The class $\mathcal{N}$ here is the set of all systems (Ω, q) controllable from Γ. This means that for the initial-boundary value problem

$$\begin{cases} \partial_t^2 w - \Delta w + qw = 0 & \text{in } Q^T = \Omega \times (0, T) \\ \partial_\nu w - \sigma w|_{\Sigma^T} = f \in L^2(\Sigma^T), & \Sigma^T = \Gamma \times (0, T) \\ w|_{t=0} = \partial_t w|_{t=0} = 0 \end{cases} \qquad (3)$$

we have the following property: The set of solutions of (3) for $f \in L^2(\Sigma^T)$ (denoted by $u^f(x, t)$) at the moment $t = T$ is dense in $L^2(\Omega^T)$, where

$$L^2(\Omega^T) = \{a \in L^2(\Omega) : supp\, a \in \Omega^T\} \qquad (3)$$

$$\Omega^T = \{x \in \Omega : \tau(x) \leq T\} \qquad (4)$$

and

$$\tau(x) = dist(x, \Gamma), \qquad (6)$$

(for any $T \geq 0$). The controllability is related to the Holmgren-John uniqueness theorem for the Cauchy problem for the wave equation (3) with Cauchy data on the time-like cylinder Σ^{2T}. It is yet proved only for the piecewise real-analytic manifolds Ω and potentials q [5]. However, the recent result by Hörmander [6] raises hopes in its validity in more general case.

The proof of the Theorem A is, in fact, a constructive one. We give the procedure of Ω, q, σ recovering. To describe it let us introduce some notions.

Let $x(\gamma, \xi)$ be the normal geodesics starting from the point $\gamma \in \Gamma, \xi$ being the arclength along the geodesics. There exists a critical value $\tau^*(\gamma)$ such that

$$\tau(x(\gamma, \xi)) = \xi \text{ for } \xi \leq \tau^*(\gamma), \ \tau(x(\gamma, \xi)) < \xi \text{ for } \xi > \tau^*(\gamma). \qquad (7)$$

The set ω of the corresponding points in Ω:

$$\omega = \bigcup_{\gamma \in \Gamma} x(\gamma, \tau^*(\gamma)) \qquad (8)$$

is the cut locus of Ω (with respect to Γ). Introducing on $\Omega \setminus \omega$ the semigeodesic coordinates (s.g.c) $i(x) = (\gamma, \tau)$, where γ is the starting point of the unique shortest geodesics coming to x from Γ and τ is defined by (6), we denote the range by Θ:

$$\Theta := i(\Omega \setminus \omega), \qquad (9)$$

Θ being the pattern of Ω.

The basic result obtained in [2]-[4], that may be generalized for the case of the Schrödinger operator on Ω, is the following.

Let $f \in C_0^\infty(\Sigma^T)$ and $(\Omega, q) \in \mathcal{N}$. Then

$$(\partial_t C^T \mathcal{P}_\perp^{T,\xi} f)(\gamma, T - \xi - 0) = -\kappa(\gamma, \xi) u^f(\gamma, \xi; T) H(\tau^*(\gamma) - \xi), \quad 0 \le \xi \le T \quad (10)$$

Here C^T and $P_\perp^{T,\xi}$ are operators that may be directly constructed in terms of the BSD (2). Thus

$$C^T f = \sum_{k=1}^{\infty} (f, s_k^T)_{L^2(\Sigma^T)} s_k^T, \quad (11)$$

$$s_k^T(\gamma, t) = \frac{\sin \sqrt{\lambda_k}(T - t)}{\sqrt{\lambda_k}} \psi_k(\gamma), \ \gamma \in \Gamma, \ s_k^T \in L^2(\Sigma^T), \quad (12)$$

$\mathcal{P}^{T,\xi}$ is a projector onto the subspace of functions f, such that $\mathrm{supp} f \in \Gamma \times [T - \xi, T]$, in the auxiliary Hilbert space defined as the completion of $L^2(\Sigma^T)$ in the norm

$$|||f|||^2 = (C^T f, f)_{L^2(\Sigma^T)}, \quad (13)$$

and $\mathcal{P}_\perp^{T,\xi} = \mathcal{E} - \mathcal{P}^{T,\xi}$, $\mathcal{E}$ being the identity operator in the auxiliary space. At last,

$$\kappa(\gamma, \xi) = \left[\frac{g(\gamma, \xi)}{g(\gamma, 0)} \right]^{1/4}, \quad (14)$$

g being the volume element in s.g.c., and $H(t)$ is the Heaviside function. Moreover, taking in (10) instead of f the elements q_k^T defined by

$$C^T q_k^T = s_k^T, \ k = 1, 2, \ldots, \quad (15)$$

we have

$$(\partial_t C^T \mathcal{P}_\perp^{T,\xi} q_k^T)(\gamma, T - \xi - 0) = \kappa(\gamma, \xi) \psi_k(\gamma, \xi) H(\tau^*(\gamma) - \xi). \quad (16)$$

Remark 1. For a direct way of defining q_k^T (without solving the equation (15)) see [4].

Using (16) and taking into account that $\psi_k(x)$ are the eigenfunctions of the operator (1), we have that $\Psi_k(\gamma, \xi) = \kappa(\gamma, \xi) \psi_k(\gamma, \xi)$ satisfy the equations

$$\{\kappa(-\Delta + q)\kappa^{-1}\}\Psi_k)\gamma, \xi) = \lambda_k \Psi_k(\gamma, \xi) \quad (16')$$

The relation (16') together with some density properties of eigenfunctions indicate the possibility of finding $\tau^*(\gamma)$, i.e. the pattern Θ, and the metric tensor g_{ij}, $i, j = 1, \ldots, m$, in s.g.c;

$$ds^2 = d\tau^2 + g_{\alpha\beta}(\gamma, \tau) d\gamma^\alpha d\gamma^\beta, \ \alpha, \beta = 1, \ldots, m - 1.$$

Determining κ (see (14)) we find $\psi_k(\gamma, \xi)$, $q(\gamma, \xi)$ and inner products $\langle d\psi_k, d\psi_r \rangle$, $k, r = 1, 2, \ldots$, for $(\gamma, \xi) \in \Theta$. Continuing them onto the $cl\Theta$ (in $\Gamma \times \mathbb{R}_+$) we glue $cl\Theta$ by means of the identification

$$(\gamma_1, \tau_1) \sim (\gamma_2, \tau_2) \Leftrightarrow \psi_k(\gamma_1, \tau_1) = \psi_k(\gamma_2, \tau_2), \ k = 1, 2, \ldots,$$

and provide the resulting manifold Θ^* (on $\Theta^* \backslash \Theta$) with local coordinates made of continued eigenfunctions $\psi_{k(r)}$, $r = 1, \ldots, m$, and the Riemannian metric tensor

$$\hat{g}^{rs} = \langle d\psi_{k(r)}, d\psi_{k(s)} \rangle_{r,s=1}^m.$$

Then the constructed Riemannian manifold turns out to be isometric to Ω. At last, using the boundary data (1), we find $\sigma(\gamma), \gamma \in \Gamma$.

3 Other types of the operators on manifolds and operators in $\mathbb{R}^n$

Using the theorem A and the observation that the weighted transformation $\beta : u \to \beta u$, where $\beta \in C^\infty(\Omega)$, $\beta \geq \beta_0 > 0$, converts an operator $A = A(\Omega, \mu, q, \sigma)$ of the form

$$Au = -\delta\mu du + qu, \quad \mathcal{D}(A) = \{u \in H^2(\Omega) : \partial_v u - \sigma u|_\Gamma = 0\} \quad (17)$$

into the operator $\tilde{A} = A(\tilde{\Omega}, \tilde{\mu}, \tilde{q}, \tilde{\sigma})$ of the same form (17), we can obtain a result concerning the operators A of the form (17). (We denote by d and δ the differentiation and co-differentiation on Ω, $\mu(x)$ is a positive function, $\mu \geq \mu_0 > 0$, $\mu \in C^\infty(\Omega)$. At last, $\tilde{\Omega}$ is the same differentiable manifold as Ω but with the other Riemannian structure).

Theorem B *Let $A = A(\Omega, \mu, q, \sigma)$ be an operator of the form (17) and (Ω, μ, q) $\in \mathcal{N}$. Then the BSD (2) determine A, i.e. Ω, μ, q, σ, uniquely to within the group of weighted transformations*

$$\beta : u \to \beta u, \quad A \to \beta A \beta^{-1} \quad (18)$$

In its turn, the Theorem B may be used for deriving some results on inverse boundary problems in a smooth domain $\Omega \subset \mathbb{R}^m$, $m \geq 2$.

Let $A = A(a^{ij})$ be an anistropic conductivity operator

$$Au = -\partial_i a^{ij} \partial_j u; \quad \mathcal{D}(A) = \{u \in H^2(\Omega), \ a^{ij} n_i \partial_j u - \sigma u|_{\Gamma=0}\} \quad (19)$$

where $a^{ij}(x) = a^{ji} \in C^\infty(\overline{\Omega})$, $a^{ij}\xi_i\xi_j \geq a_0|\xi|^2$.

Theorem C *Let A be an operator of the form (19), and $(\Omega, a^{ij}) \in \mathcal{N}$. Then the BSD (2) determine a^{ij}, $i, j = 1, \ldots, m$, and σ uniquely to within the group of unimodular diffeomorphisms $y = y(x)$ of Ω that are identical on Γ:*

$$y : \Omega \to \Omega, \ y|_\Gamma = id|_\Gamma, \ y^*(dx^1 \wedge \ldots dx^m) = dx^1 \wedge \ldots \wedge dx^m, \quad (20)$$

Remark 2. It is interesting to compare the Theorem C with the uniqueness results obtained in terms of a given Dirichlet-to-Neumann map. It is shown in [7]–[9] that D-to-N map determines a^{ij}, $i, j = 1, \ldots, m$, uniquely to within the group of diffeomorphisms $z = z(x)$ of Ω that are identical on Γ.

$$z : \Omega \to \Omega, \ z|_\Gamma = id|_\Gamma \quad (21)$$

At last, we describe one result concerning isotropic operators in $\Omega \subset \mathbb{R}^m$, $m \geq 2$. So let $A = A(\epsilon, \mu, q, \sigma)$ be an operator in $L^2(\Omega, \epsilon^{-1})$ of the form

$$Au := -\epsilon \, div(\mu \nabla u) + qu, \quad \mathcal{D}(A) = \{u \in H^2(\Omega) : \partial_n u - \sigma u|_\Gamma = 0\}, \quad (22)$$

where $\epsilon, \mu \in C^\infty(\overline{\Omega})$, $\epsilon, \mu \geq \epsilon_0, \mu_0 > 0$, and ∂_n is outward normal derivative to Γ in Euclidean metric.

Theorem D *Let A be an operator of the form (22) and $(\Omega, \epsilon, \mu, q) \in \mathcal{N}$. Then BSD (2) determine A, i.e. ϵ, μ, q, σ, uniquely to within the group of weighted transformations (18).*

Corollary *Let A be an operator of the form (22); $(\Omega, \epsilon, \mu, q) \in \mathcal{N}$ and either ϵ, or μ or q are known. Then BSD (2) determines A, i.e. ϵ, μ, q, σ uniquely.*

In the conclusion we note that quite analogous results may be obtained in the case when, instead of BSD, we possess the response operator R^T defined by (3) as follows:

$$R^T : f = w^f|_{\Sigma^T} \tag{23}$$

for $T > 2T^*$, T^* being the geodesic radius of Ω, $T^* = \max_{\gamma \in \Gamma} \tau^*(\gamma)$ (for a nonstationary variant of the BC method see [2, 3], for applications to anisotropic problems see [10]).

It gives pleasure to the author to thank M. Belishev and L. D. Faddeev for their interest in work and fruitful discussions. The author is particularly grateful to A. Katchalov, Yu. Burago and H. Karcher whose attention and suggestions were inestimable in writing the paper.

References

1. Belishev, M. I.: On an approach to multidimensional inverse problems for the wave equation. Dokl. Akad. Nauk SSSR (in Russian) **297** (3) (1987) 524–527.
2. Belishev, M. I.: Boundary control and wave field continuation. Preprint LOMI P-I-90 (in Russian); Leningrad, 1990.
3. Belishev, M. I., Kurylev, Ya. V.: Boundary control, wave field continuation and inverse problems for the wave equation. Comput. Math. Appl. **22** (1991) 27–52.
4. Belishev, M. I., Kurylev, Ya. V.: To the reconstruction of a Riemannian manifold via its spectral data (BC-method). Comm. PDE **17** (1992) 767–804.
5. Kurylev, Ya. V.: To the Holmgren-John uniqueness theorem for the wave equation with piecewise analytic coefficients. Zap. Nauchn. Semin. POMI (in Russian) **203** (1992) 113–136.
6. Hörmander, L.: A uniqueness theorem for second-order hyperbolic differential equations. Comm. PDE **17** (1992) 699–714.
7. Lee, J. M., Uhlmann, G.: Determining anisotropic real-analytic conductivity by boundary measurements. Comm. Pure Appl. Math. **42** (1989) 1097–1112.
8. Sylvester, J.: An anisotropic inverse boundary value problem. Comm. Pure Appl. Math. **43** (1990) 201–233.
9. Sylvester, J., Uhlmann, G.: Inverse problems in anisotropic media. Contemp. Math. **122** (1991) 105–117.
10. Belishev, M., Katchalov, A.: Quasiphotons in the dynamic inverse problem for controlled Riemannian manifolds, Preprint POMI, P-12-91 (in Russian), St. Petersburg, 1991.

Three Dimensional Time Harmonic Inverse Electromagnetic Scattering*

Pierluigi Maponi¹, Luciano Misici¹, Francesco Zirilli²

¹ Dipartimento di Matematica e Fisica, Università di Camerino - 62032 Camerino (Italy)
² Dipartimento di Matematica "G. Castelnuovo", Università di Roma "La Sapienza" - 00185 Roma (Italy)

Abstract *A numerical method for the three dimensional inverse electromagnetic time harmonic scattering is presented. We consider an obstacle D that is a bounded, simply connected domain with smooth boundary ∂D contained in the three dimensional euclidean space $\mathbb{R}^3$. The far field patterns of the vector Helmholtz equation generated by a known electromagnetic wave incident on the obstacle D are used as data. From these data the boundary of the obstacle ∂D is reconstructed. The reconstruction procedure proposed here generalizes the "Herglotz function method" introduced by Colton and Monk [1] in the acoustic problem and is effective in the so called resonance region.*

1 Introduction

Let $\mathbb{R}^3$ be the three dimensional real euclidean space, $\underline{x} = (x, y, z)^T \in \mathbb{R}^3$ be a generic vector, where the superscript T means transpose, $(\cdot, \cdot)$ will denote the euclidean scalar product and $\| \cdot \|$ the euclidean norm. In the following we will use complex vectors without changing the notations.

Let $D \subset \mathbb{R}^3$ be a bounded simply connected domain with smooth boundary ∂D; without loss of generality we assume that D contains the origin. The region $\mathbb{R}^3 \setminus D$ is filled with an homogeneous isotropic medium that does not contains charges.

Let $\underline{E}^i(\underline{x})$ be the part depending on $\underline{x}$ of the electric field associated to a linearly polarized time harmonic incoming wave propagating in a homogeneous isotropic medium, that is:

$$\underline{E}^i(\underline{x}) = \underline{w}e^{ik(\underline{x},\underline{\alpha})} \tag{1.1}$$

where $\underline{w}, \underline{\alpha} \in \mathbb{R}^3$ with $\|\underline{\alpha}\| = 1$ are given and $k > 0$ is the wave number, moreover we assume that $(\underline{w}, \underline{\alpha}) = 0$ so that

$$\operatorname{div} \underline{E}^i(\underline{x}) = ik(\underline{w}, \underline{\alpha})e^{ik(\underline{x},\underline{\alpha})} = 0 \tag{1.2}$$

* The research reported in this paper has been made possible through the support and sponsorship of the Italian Government through the Ministero per l'Università e per la Ricerca Scientifica under contract MURST40%,1990

where $\underline{E}^i(\underline{x}) = (E_x^i(\underline{x}), E_y^i(\underline{x}), E_z^i(\underline{x}))^T$ and $\operatorname{div} \underline{E}^i(\underline{x}) = \frac{\partial E_x^i(\underline{x})}{\partial x} + \frac{\partial E_y^i(\underline{x})}{\partial y} + \frac{\partial E_z^i(\underline{x})}{\partial z}$.
We note that $\underline{w}$ is the polarization vector and $\underline{\alpha}$ is the propagation direction of the incomig electric field. We note that the magnetic field $\underline{H}^i(\underline{x})$ associated to this incoming wave is given by

$$\underline{H}^i(\underline{x}) = \frac{1}{ik} \operatorname{curl} \underline{E}^i(\underline{x}) \tag{1.3}$$

where $\operatorname{curl} \underline{E}^i(\underline{x}) = \left(\frac{\partial E_z^i(\underline{x})}{\partial y} - \frac{\partial E_y^i(\underline{x})}{\partial z}, \frac{\partial E_x^i(\underline{x})}{\partial z} - \frac{\partial E_z^i(\underline{x})}{\partial x}, \frac{\partial E_y^i(\underline{x})}{\partial x} - \frac{\partial E_x^i(\underline{x})}{\partial y} \right)^T$.

Let us denote with $\underline{E}^s(\underline{x}) = (E_x^s, E_y^s, E_z^s)^T$ the part depending on $\underline{x}$ of the time harmonic electric field scattered by the obstacle D when hit by the incoming wave $\underline{E}^i(\underline{x})$ and with

$$\underline{E}(\underline{x}) = \underline{E}^i(\underline{x}) + \underline{E}^s(\underline{x}) \tag{1.4}$$

the corresponding total electric field. It is easy to see [2] that, in the previous hypotheses the Maxwell's equations can be reduced to the vector Helmholtz equation with the divergence free condition, that is:

$$\Delta \underline{E}^s(\underline{x}) + k^2 \underline{E}^s(\underline{x}) = \underline{0} \quad \text{in } \mathbb{R}^3 \setminus D \tag{1.5}$$

$$\operatorname{div} \underline{E}^s(\underline{x}) = 0 \quad \text{in } \mathbb{R}^3 \setminus D \tag{1.6}$$

where $\Delta = \frac{\partial^2}{\partial x^2} + \frac{\partial^2}{\partial y^2} + \frac{\partial^2}{\partial z^2}$ and $\Delta \underline{E}^s = (\Delta E_x^s, \Delta E_y^s, \Delta E_z^s)^T$.

The partial differential equations (1.5),(1.6) must be equipped with boundary conditions. We assume the Silver-Müller radiation condition at infinity:

$$\operatorname{curl} \underline{E}^s(\underline{x}) \times \hat{\underline{x}} - ik \underline{E}^s(\underline{x}) = o\left(\frac{1}{\|\underline{x}\|} \right) \quad , \quad \|\underline{x}\| \to \infty \tag{1.7}$$

where $\hat{\underline{x}} = \frac{\underline{x}}{\|\underline{x}\|}$, $\underline{x} \neq 0$ and $\times$ is the vector product. Moreover let $\underline{\nu}(\underline{x})$ be the exterior unit normal to ∂D at the point $\underline{x} \in \partial D$ we assume the following boundary condition:

$$\chi_1 \underline{\nu} \times \frac{1}{ik} \operatorname{curl} \underline{E}(\underline{x}) + \chi_2 \, \underline{\nu} \times (\underline{\nu} \times \underline{E}(\underline{x})) = \underline{0} \quad , \quad \underline{x} \in \partial D \tag{1.8}$$

where $\chi_1, \chi_2 \in \mathbb{C}$ are given complex constants. We note that $\chi_1 = 0$, $\chi_2 \neq 0$ gives the boundary condition for a perfectly conducting obstacle, $\chi_1 \neq 0, \chi_2 = 0$ gives the boundary condition for a perfectly insulating obstacle and that $\chi_1 \neq 0, \chi_2 \neq 0$ gives a mixed boundary condition corresponding to an obstacle characterized by an electromagnetic impedance.

In [2] it is shown that $\underline{E}^s(\underline{x})$, solution of the boundary value problem (1.5), (1.6), (1.7), (1.8) has the following expansion:

$$\underline{E}^s(\underline{x}) = \frac{e^{ik\|\underline{x}\|}}{\|\underline{x}\|} \underline{E}_0(\hat{\underline{x}}, k, \underline{\alpha}, \underline{w}) + O\left(\frac{1}{\|\underline{x}\|^2} \right) \quad , \quad \|\underline{x}\| \to \infty \tag{1.9}$$

where $\underline{E}_0(\hat{\underline{x}}, k, \underline{\alpha}, \underline{w})$ is the (electric) far field pattern generated by the incoming wave (1.1) that hits the obstacle D.

In this paper we introduce a numerical method to solve the following inverse problem: from the knowledge of the nature of the obstacle (that is the constants χ_1, χ_2) and of the (electric) far fields patterns generated by several (known) incoming waves we want to recover the boundary of the obstacle ∂D.

The method suggested here generalizes the "Herglotz function" method introduced in acoustics in [1] and used to study the problem considered in this paper for a perfectly conducting obstacle in [3]. This method is very effective in the resonance region, that is when

$$kL \cong 1 \qquad (1.10)$$

where L is a characteristic length of the obstacle D.

To be more precise let λ_n $n = 1, 2, \ldots$ be the eigenvalues of the "vector" Laplace operator restricted to the divergence free vector fields in the interior of D with the homogeneous boundary condition (1.8). Moreover let $B = \{\underline{x} \in \mathbb{R}^3 \mid \|\underline{x}\| < 1\}$ and ∂B be the boundary of B. We will consider the following inverse problem:

Problem 1.1 *Let $\Omega_1 \subseteq \partial B$, $\Omega_2 \subset \partial B \times \mathbb{R}^3$, $\Omega_3 \subset \mathbb{R}$ be three given sets such that $-\lambda_i \notin \Omega_3$ $i = 1, 2, \ldots$ and let $\underline{E}_0(\hat{\underline{x}}, k, \underline{\alpha}, \underline{w}_\alpha)$ be the electric far field defined in (1.9). ¿From the knowledge of $\underline{E}_0(\hat{\underline{x}}, k, \underline{\alpha}, \underline{w}_\alpha)$ for $\hat{\underline{x}} \in \Omega_1$, $(\underline{\alpha}, \underline{w}_\alpha) \in \Omega_2$, $k^2 \in \Omega_3$ determine the boundary of the obstacle ∂D.*

We note that the condition $-\lambda_i \notin \Omega_3$ $i = 1, 2, \ldots$ is a non resonant condition. In section 2 we derive some mathematical relations that relate the far fields patterns to the unknown boundary ∂D. In section 3 some numerical experience obtained exploiting the relations derived in section 2 is shown.

2 Some relations used to solve the inverse problem

For $\underline{x}, \underline{y} \in \mathbb{R}^3$ let

$$\Phi(k\|\underline{x} - \underline{y}\|) = \frac{e^{ik\|\underline{x} - \underline{y}\|}}{4\pi\|\underline{x} - \underline{y}\|} \qquad (2.1)$$

be the Green's function of the (scalar) Helmholtz operator with the Sommerfeld radiation condition at infinity.

Let $\underline{g}(\hat{\underline{x}})$ be a square integrable vector valued complex function defined on the surface of the unit sphere ∂B, such that:

$$(\hat{\underline{x}}, \underline{g}(\hat{\underline{x}})) = 0 \quad, \ \forall \hat{\underline{x}} \in \partial B \qquad (2.2)$$

and let

$$\underline{E}_1(\underline{y}) = \int_{\partial B} \underline{g}(\hat{\underline{x}}) e^{ik(\hat{\underline{x}}, \underline{y})} d\lambda(\hat{\underline{x}}) \qquad (2.3)$$

where $d\lambda(\hat{\underline{x}})$ is the surface measure on ∂B. It is easy to see that $\underline{E}_1(\underline{y})$ is a divergence free vector field that satisfies the vector Helmholtz equation for any $\underline{y} \in \mathbb{R}^3$. Let $\underline{v} \in \mathbb{R}^3$ be a given vector we define the vector function

$$\underline{M}(\underline{y}) = -4\pi\left\{\underline{v}\overline{\Phi(k\|\underline{x}-\underline{y}\|)} + \frac{1}{k^2}\nabla_{\underline{x}}(\underline{v}, \nabla_{\underline{x}}\overline{\Phi(k\|\underline{x}-\underline{y}\|)})\right\}\bigg|_{\underline{x}=\underline{0}} \tag{2.4}$$

where $\nabla_{\underline{x}}$ is the gradient operator with respect to $\underline{x}$, and $\overline{\Phi}$ is the complex conjugate of Φ.

Definition 2.1 *The domain D is said to be a generalized Herglotz domain if the unique solution of the boundary value problem:*

$$(\Delta + k^2)\underline{E}_1(\underline{y}) = \underline{0} \qquad\qquad in\ D \tag{2.5}$$

$$\mathrm{div}\underline{E}_1(\underline{y}) = 0 \qquad\qquad in\ D \tag{2.6}$$

$$\overline{\chi_1}\left(\mathrm{curl}\,\underline{E}_1(\underline{y}) \times \underline{\nu}\right) + ik\overline{\chi_2}\left((\underline{E}_1(\underline{y}) \times \underline{\nu}) \times \underline{\nu}\right) =$$
$$\overline{\chi_1}\left(\mathrm{curl}\,\underline{M}(\underline{y}) \times \underline{\nu}\right) + ik\overline{\chi_2}\left((\underline{M}(\underline{y}) \times \underline{\nu}) \times \underline{\nu}\right) \qquad on\ \partial D \tag{2.7}$$

is given by (2.3) for a suitable choice of $g(\hat{\underline{x}})$. The function $g(\hat{\underline{x}})$ that corresponds to the solution of (2.5), (2.6), (2.7) is said to be the generalized Herglotz kernel associated to the domain D.

We note that $g(\hat{\underline{x}})$ depends on $\underline{v}$ and that the class of the generalized Herglotz domains is not empty since the sphere with centre in $\underline{x} = \underline{0}$ belongs to it for every $\underline{v}$. Moreover we remark that the boundary value problem (2.5), (2.6), (2.7) has a unique solution due to the non-resonant assumption made on k^2.

Let us restrict our attention to the class of the generalized Herglotz domains and let $g(\hat{\underline{x}})$ be the generalized Herglotz kernel of D, it is easy to see [1], [3] that:

$$\int_{\partial B} (\overline{g(\hat{\underline{x}})}, \underline{E}_0(\hat{\underline{x}}, k, \underline{\alpha}, \underline{w}_{\underline{\alpha}}))d\lambda(\hat{\underline{x}}) = (\underline{w}_{\underline{\alpha}}, \underline{v}) \quad \forall \underline{\alpha} \in \partial B, \underline{w}_{\underline{\alpha}} \in \mathbb{R}^3 \tag{2.8}$$

The inverse Problem 1.1 proposed in section 1 will be solved in three steps:

i) from the knowledge of the far fields $\underline{E}_0(\hat{\underline{x}}, k, \underline{\alpha}, \underline{w}_{\underline{\alpha}})$ for several $\hat{\underline{x}}$, $\underline{\alpha}$ and $\underline{w}_{\underline{\alpha}}$ determine, using (2.8), an approximation of the generalized Herglotz kernel $g(\hat{\underline{x}})$ of the domain D.

ii) from $g(\hat{\underline{x}})$, using (2.3), determine $\underline{E}_1(\underline{y})$

iii) from $\underline{E}_1(\underline{y})$, using (2.7), determine ∂D .

The numerical implementation of the steps i)–iii) of the method suggested to solve Problem 1.1 is analogous to the implementation described in [3] for the perfectly conducting obstacle and will be omitted.

3 Numerical experience

In this section we describe some numerical results obtained using the reconstruction method presented in section 2.

Let us consider the spherical coordinates:

$$x = r \sin \theta \cos \phi$$
$$y = r \sin \theta \sin \phi \tag{3.1}$$
$$z = r \cos \theta$$

where $r \geq 0, 0 \leq \theta \leq \pi, 0 \leq \phi < 2\pi$. The surfaces ∂D considered are the following ones:

Reverse Platelet	$r = \dfrac{5}{4} + \dfrac{1}{4} \cos 4\theta$	(3.2)
Long Cylinder	$(x^2 + y^2)^3 + \left(\dfrac{2}{3} z\right)^6 = 1$	(3.3)
Pseudo Apollo	$r = \dfrac{3}{5} \left(\dfrac{17}{4} + 2 \cos 3\theta\right)^{1/2}$	(3.4)
Ellipsoid	$r = \left(\dfrac{\sin\theta}{h(\theta)}\right)^2 + \left(\dfrac{2}{3} \cos\theta\right)^2$	(3.5)
Corrugated Cylinder	$r = \left(\left(\dfrac{2\sin\theta}{3H(\theta)}\right)^6 + \cos^6\theta\right)^{-1/2}$	(3.6)

where

$$h(\theta) = \left(\left(\frac{3}{4}\cos\phi\right)^2 + \sin^2\phi\right)^{-1/2} \tag{3.7}$$

$$H(\theta) = (R_h + A_h \cos 4\phi + B_h \cos 8\phi + C_h \cos 16\phi)^2 \tag{3.8}$$

and

$$A_h = \frac{0.3}{1.34} f_h \ , \ B_h = \frac{0.05}{1.34} f_h \ , \ C_h = \frac{0.01}{1.34} f_h \ , \ R_h = 1 - (A_h + B_h + C_h) \tag{3.9}$$

The parameter f_h is called corrugation parameter and is chosen equal to 0.2 in our numerical experience.

The obstacles D corresponding to (3.2), (3.3), (3.4) are bodies symmetric with respect to the z-axis, the obstacles corresponding to (3.3) and (3.5) are convex bodies. The obstacle corresponding to (3.6) is nonconvex and nonsymmetric with respect to the z-axis. The characteristic length L of all the obstacles considered here can be chosen equal one.

The data of our numerical experience are obtained solving numerically the boundary value problem (1.5), (1.6), (1.7), (1.8) using the T-matrix approach [4]. Let $\hat{\underline{x}}, k, \underline{\alpha}, \underline{w}_\alpha$ be given and $\underline{\tilde{E}}_0$ be the computed value of $\underline{E}_0(\hat{\underline{x}}, k, \underline{\alpha}, \underline{w}_\alpha)$.

Let $\hat{r}(\underline{x})$, $\hat{\theta}(\underline{x})$, $\hat{\phi}(\underline{x})$ be the orthonormal basis of $\mathbb{R}^3$ associated to the spherical coordinates system at the point $\underline{x}$, in order to simulate experimental errors, we substitute $\underline{E}_0 = \tilde{E}_{0,\theta}\hat{\theta} + \tilde{E}_{0,\phi}\hat{\phi}$ with

$$(1 + \varepsilon\zeta_1)\tilde{E}_{0,\theta}\hat{\theta} + (1 + \varepsilon\zeta_2)\tilde{E}_{0,\phi}\hat{\phi} \qquad (3.10)$$

where ε is a real parameter and ζ_i, $i = 1,2$ are independent random complex numbers with real and imaginary part uniformly distributed in $[-1,1]$. We remark that the $\hat{r}$ component of a far field $\underline{E}_0$ is always zero. Moreover we choose:

$$\Omega_1 = \left\{ \hat{\underline{x}}(\theta_i, \phi_j) \mid \theta_i = \frac{\pi}{10}i, \quad i = 1, ..., 9 \, ; \quad \phi_j = \frac{\pi}{9}j, \quad j = 0, ..., 8 \right\} \cup \{0,0\} \cup \{\pi, 0\}$$

$$\Omega_2 = \left\{ (\underline{\alpha}_{ij}, \underline{\omega}_{ijk}) \Big| \underline{\alpha}_{ij} = \hat{\underline{x}}(\theta_i, \phi_j), \underline{\omega}_{ijk} = (-\sin\gamma_k \sin\phi_j + \cos\gamma_k \cos\theta_i \cos\phi_j, \right.$$

$$\sin\gamma_k \cos\phi_j + \cos\gamma_k \cos\theta_i \sin\phi_j, -\cos\gamma_k \sin\theta_i)^T : \theta_i = (1 + \frac{17}{4}i)\frac{\pi}{36},$$

$$\left. i = 0, ..., 8; \phi_j = \frac{2\pi}{7}j \, , \, j = 0, ..., 6 \, ; \quad \gamma_k = \frac{\pi}{2}k \, , \quad k = 0, 1 \right\}$$

$$\Omega_3 = \{ 9 \}$$

Let $r = f(\theta, \phi)$ be the representation in spherical coordinates of ∂D (see (3.2), (3.3),...,(3.6)) and let $f_c(\theta, \phi)$ be the corresponding value computed using the reconstruction procedure of section 2. Let $(\theta_i, \phi_j) = (i\pi/36, j\pi/18)$, $i = 1, 2, ..., 35$, $j = 0, 1, ..., 35$ we use as performance index of our reconstruction procedure a relative L^2 error that is:

$$E_{L^2} =$$

$$\frac{\left((f(0,0) - f_c(0,0))^2 + (f(\pi,0) - f_c(\pi,0))^2 + \sum_{i=1}^{35}\sum_{j=0}^{35}(f(\theta_i, \phi_j) - f_c(\theta_i, \phi_j))^2 \right)^{\frac{1}{2}}}{\left((f(0,0))^2 + (f(\pi,0))^2 + \sum_{i=1}^{35}\sum_{j=0}^{35}(f(\theta_i, \phi_j))^2 \right)^{\frac{1}{2}}}$$

$$(3.14)$$

The results obtained with the reconstruction procedure of section 2 are shown in Tables 3.1, 3.2 and in Figures 3.1, 3.2, 3.3.

All the reconstructions presented are satisfactory in particular Table 3.2 shows that our method is robust when applied to noisy data.

Original 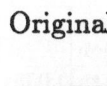   Reconstruction n. 6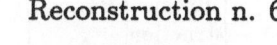

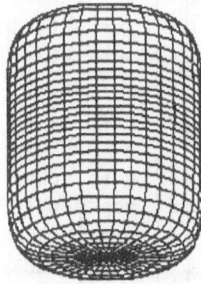   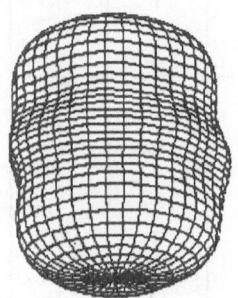

Fig. 3.1 (view angle $\phi = -45^o$, $\theta = 70^o$)

Original   Reconstruction n. 7

Fig. 3.2 (view angle $\phi = -45^o$, $\theta = 70^o$)

Original   Reconstruction n. 11

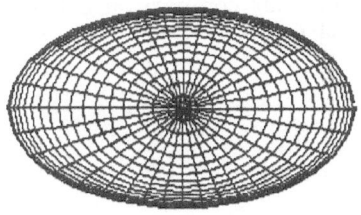   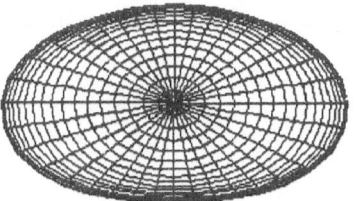

Fig. 3.3 (view angle $\phi = 0^o$, $\theta = 0^o$)

Table 1. Numerical results. $\epsilon = 0.05$, $L_{max} = L_g = 7$. (*) Reconstruction shown in Figure 3.1 or 3.2 or 3.3

Object	Recon-struction	Boundary Condition χ_1	χ_2	v	L_ρ	Penali-zation	E_{L^2}
Reverse Platelet	1	1	$1+i$	$1, 0, 1$	4	yes	0.0351
	2	1	0	$1, 0, 1$	4	yes	0.0521
	3	0	1	$0, 0, 1$	4	no	0.0742
Long Cylinder	4	1	$1+i$	$1, 0, 1$	4	yes	0.0597
	5	1	0	$1, 0, 1$	4	yes	0.1280
	6(*)	0	1	$0, 0, 1$	6	no	0.0441
Pseudo Apollo	7(*)	1	$1+i$	$1, 0, 1$	4	yes	0.0262
	8	1	0	$1, 0, 1$	4	yes	0.0740
	9	0	1	$1, 0, 1$	4	yes	0.0722
Ellipsoid	10	1	$1+i$	$0, 0, 1$	4	no	0.0333
	11(*)	1	0	$0, 0, 1$	4	no	0.0147
	12	0	1	$0, 0, 1$	4	no	0.0403
Corrugated Cylinder	13	1	$1+i$	$0, 0, 1$	4	no	0.0763
	14	1	0	$0, 0, 1$	4	no	0.0554
	15	0	1	$0, 0, 1$	4	no	0.0555

Table 2. Reconstructions with noisy data. $L_{max} = L_g = 7$, $L_\rho = 4$, $\chi_1 = 1$, $\chi_2 = 1 + i$, $v = (0, 0, 1)$

Object	Reconstruction	ϵ	E_{L^2}
Ellipsoid	16	0.01	0.0290
	17	0.05	0.0333
	18	0.10	0.0399
	19	0.20	0.0586
	20	0.30	0.0910
	21	0.40	0.1271
	22	0.50	0.1517

The parameters L_{max}, L_g, L_ρ and the penalization term are defined in [3] and their value is reported here to make possible the comparison between the results presented in [3] and the results presented here.

Determination of a Radially Symmetric Speed of Sound from Transmission Eigenvalues

Joyce R. McLaughlin

Mathematical Sciences Department, Rensselaer Polytechnic Institute, Troy, New York 12180-3590

1 Introduction

For this talk we consider the inverse acoustic scattering problem for an inhomogeneous medium of compact support in a ball Ω_b of radius b in R^3. The local speed of sound $c(\mathbf{r})$ has the property $c(\mathbf{r}) \equiv c_0 > 0$ for $r = |\mathbf{r}| > r_0$ where c_0 is constant and $b > r_0$. Denoting $n(\mathbf{r}) = [c_0/c(\mathbf{r})]^2$, ω-frequency of the incident wave, α the direction of the incident wave, $k^2 = \omega^2/c_0^2$, we have, as in [CM] the following equations satisfied by the velocity potential $u(\mathbf{r})$:

$$\triangle u + k^2 n(\mathbf{r})u = 0$$

$$u(\mathbf{r}) = e^{ik(\mathbf{r}\cdot\alpha)} + u^s(\mathbf{r}) \tag{1}$$

$$\lim_{r\to\infty} r\left(\frac{\partial u^s}{\partial r} - iku^s\right) = 0.$$

The last condition (1) is the Sommerfeld radiation condition. We can consider solutions of (1) for complex k and as such one has the expansion for the scattered field

$$u^s(\mathbf{r}) = \frac{e^{ikr}}{r}F(\hat{r}, k, \alpha) + O\left(\frac{1}{r^2}\right), \qquad \text{Im } k \geq 0.$$

where $\hat{r}$ are the spherical coordinates for the boundary of the unit sphere $\partial\Omega$ in R^3. The function $F(\hat{r}, k, \alpha)$ is the scattering amplitude.

We assume that the sound speed $c(\mathbf{r}) = c(r)$ is spherically symmetric, that $n(r) = [c_0/c(r)]^2 \in H^2(\Omega_b)$, and begin with the problem of reconstructing $c(r)$ from the scattering amplitude, F, which is a function of five variables. Our approach is to extract functionals of F which will determine $c(r)$. Accordingly, first fix α. Then, the functionals, which are nonlinear functionals of $c(r)$, will be values of k for which the integral average of F over $\partial\Omega$ is zero. The functionals can thus be determined from the scattering amplitude; they will be shown to be eigenvalues for the interior transmission problem. This last property allows us to solve the inverse scattering problem as an inverse spectral problem.

The paper is organized as follows. In Section 2 we show that the zeros of an integral average of F determine spectra of the interior transmission problem

References

1. Colton, D., Monk, P.: The numerical solution of the three dimensional inverse scattering problem for time harmonic acoustic waves. SIAM J. Sci. Stat. Comput. **8** (1987) 278–291
2. Colton, D., Kress, R.: *Integral equation methods in scattering theory.* J.Wiley & Sons Publ., New York 1983
3. Maponi, P., Misici, L., Zirilli, F.: An inverse problem for the three dimensional vector Helmholtz equation for a perfectly conducting obstacle. Computers Math. Applic. **22** (1991) 137–146
4. Waterman, P. C.:New formulation of acoustic scattering. J.Acoust. Soc. of America **45** (1988) 1417–1429

where the corresponding eigenfunctions are spherically symmetric. In Section 3, asymptotic forms for this spectra are given and uniqueness theorems are presented. It is shown that all or part of the sound speed is uniquely determined by the spectra depending on the 'size' of $c(r)$. In Section 4, a reconstruction procedure is presented for the case where the 'size' is large enough so that $c(r)$ is uniquely determined. In Section 5, numerical results are presented.

2 Determining the spectra

Consider the homogeneous interior transmission problem, see [CM], where we seek values of k^2, called transmission eigenvalues, for which

$$\begin{aligned}
\Delta v + k^2 n(r)v &= 0, & \mathbf{r} \text{ in } \Omega_b, \\
\Delta w + k^2 w &= 0, & \mathbf{r} \text{ in } \Omega_b, \\
v(\mathbf{r}) - w(\mathbf{r}) &= 0, & \mathbf{r} \text{ in } \partial\Omega_b, \\
\frac{\partial}{\partial r}(v(\mathbf{r}) - w(\mathbf{r})) &= 0, & \mathbf{r} \text{ in } \partial\Omega_b,
\end{aligned} \tag{2}$$

has a nontrivial solution pair $\{v, w\}$. We confine ourselves to those transmission eigenvalues of (2) for which the corresponding pair $\{v, w\}$ is spherically symmetric. These eigenvalues can be considered as the eigenvalues of an associated one-dimensional Sturm-Liouville problem which is not self adjoint.

The reduction is made as in [CM], [MP], [MPS] as follows. Since n is radially symmetric, and $\{v, w\}$ are radially symmetric, $v = a_{00}y(r, k)/r$ and $w = b_{00}j_0(kr) = b_{00}\sin kr/kr$. We can then rewrite problem (2) for the radially symmetric solutions as

$$y'' + k^2 n(r)y = 0, \qquad 0 \le r \le b,$$

$$\lim_{r \to 0} \left(\frac{1}{r}y(r) - 1\right) = 0, \tag{3}$$

$$\det \begin{pmatrix} \dfrac{1}{r}y(r) & -\dfrac{\sin kr}{kr} \\ \dfrac{d}{dr}\left(\dfrac{1}{r}y(r)\right) & -\dfrac{d}{dr}\left(\dfrac{\sin kr}{kr}\right) \end{pmatrix} = 0 \qquad \text{for } r = b.$$

The eigenvalues of this eigenvalue problem are the nonlinear functionals of $n(r)$ (and hence $c(r)$) that we will use as data for the inverse problem.

To characterize the same set of transmission eigenvalues in terms of $F(\hat{r}, k, \alpha)$, suppose that k^2 is not an eigenvalue of (3). Then there exists a solution $\{a_{00}y(r, k)/r, b_{00}j_0(kr)\}$ of the nonhomogeneous interior transmission problem [CM]:

$$\begin{aligned}
\Delta v + k^2 n(r)v &= 0 & \mathbf{r} \in \Omega_b, \\
\Delta w + k^2 w &= 0, & \mathbf{r} \in \Omega_b,
\end{aligned}$$

$$v(\mathbf{r}) - w(\mathbf{r}) = \frac{e^{ikr}}{r} \qquad \text{for } \mathbf{r} \in \partial\Omega_b,$$

$$\frac{\partial}{\partial r}(v(\mathbf{r}) - w(\mathbf{r})) = \frac{d}{dr}\frac{e^{ikr}}{r} \quad \text{for } \mathbf{r} \in \partial\Omega_b,$$

with

$$b_{00} = \frac{\det\begin{pmatrix} \frac{1}{r}y(r,k) & \frac{e^{ikr}}{r} \\ \frac{d}{dr}\left(\frac{1}{r}y(r,k)\right) & \frac{d}{dr}\left(\frac{e^{ikr}}{r}\right) \end{pmatrix}\Big|_{r=b}}{\det\begin{pmatrix} \frac{1}{r}y(r) & -j_0(kr) \\ \frac{d}{dr}\left(\frac{1}{r}y(r)\right) & -\frac{d}{dr}j_0(kr) \end{pmatrix}\Big|_{r=b}}$$

Using the identity

$$\int_{\partial\Omega} e^{ik(\hat{t},\mathbf{r})}d\sigma(\hat{t}) = 4\pi j_0(kr) = \frac{4\pi}{b_{00}}w$$

then we have that w is a Herglotz wave function [CM], [HW] and applying the representation (2.19) from [CM] we obtain

$$\frac{1}{b_{00}} = \frac{1}{4\pi}\int_{\partial\Omega} F(\hat{r},k,\alpha)d\sigma(\hat{r}) \tag{4}$$

for Im $k \geq 0$, k^2 not an eigenvalue of (3). Since both sides of the equation (5) are continuous for all k, Im $k \geq 0$, since complex eigenvalues of (3) occur in complex conjugate pairs, and since the eigenvalues of (3) occur when $\frac{1}{b_{00}} = 0$, we conclude that the eigenvalues of (3) are exactly determined by the zeros of the integral average of F on the right hand side of (5).

3 Uniqueness of the sound speed

The uniqueness result depends on the asymptotics of the real eigenvalues of (3) and on the size of the inhomogeneity which is measured by $a = [(1/b)\int_0^b \sqrt{n(r)}dr]^{-1} = [(1/b)\int_0^b (c_0/c(r))dr]^{-1}$. Roughly speaking if the size, a, is large enough, $a \geq 3$, then the transmission eigenvalues with radially symmetric eigenfunctions uniquely determine $n(r)$, and hence $c(r)$. If the size a satisfies $0 < a < 3$, $a \neq 1$, then $n(r)$, and hence $c(r)$, are uniquely determined only on a subinterval.

We begin with a transformation of the eigenvalue problem (3). We let $B = \int_0^b \sqrt{n(r)}dr$, $\lambda = B^2 k^2$, $a = \frac{b}{B}$. Making the change of variables $x = (1/B)\times \int_0^r \sqrt{n(t)}dt$, $\tilde{y}(x) = y(r(x),k)$, $\tilde{n}(x) = n(r(x))$, and taking into account that $n'(b) = 0$, $n(b) = 1$, we obtain the Sturm-Liouville problem in impedance form [CM], [MP], [MPS], [RS]

$$[\tilde{n}^{1/2}(x)\tilde{y}_x(x)]_x + \lambda\tilde{n}^{1/2}(x)\tilde{y} = 0,$$

$$\tilde{y}(0) = 0, \tag{5}$$

$$\tilde{y}(1) \cos \sqrt{\lambda} a - \tilde{y}'(1) \frac{\sin \sqrt{\lambda} a}{\sqrt{\lambda}} = 0.$$

Existence of an infinite number of real eigenvalues for this problem was shown in [CM]. Letting $q(x) = [\tilde{n}^{1/4}]_{xx}/\tilde{n}^{1/4}$ and $\|q\| = [\int_0^1 q^2]^{1/2}$, the following detailed result for the asymptotics of the real eigenvalues was obtained in [MP].

Theorem 1. *Suppose real $q(x) \in L^2(0,1)$, $a \neq 1$ and let $A = e^{\|q\|}[1 + \|q\|] + 1$. Then there is a positive integer $i_0 >$ max $[A, 25A|a - 1|/\pi^2]$ such that for each $i > i_0$,*

i) there are at least i eigenvalues, λ, of (5) satisfying $|\lambda| < (i + \frac{1}{2})^2 \pi^2/(a - 1)^2$;

ii) there is an isolated eigenvalue μ_i of (4) satisfying $\mu_i > (i_0 \pi)^2/(a - 1)^2$ and

$$\left| \mu_i(q) - \frac{\pi^2 i^2}{(a-1)^2} + \frac{1}{a-1} \int_0^1 q(t)dt - \frac{1}{(a-1)} \int_0^1 q(t) \cos(\frac{2\pi i t}{a-1})dt \right| < \frac{C}{i}$$

where

$$C = 0.2 + \frac{0.16}{|a-1|}(1 + A \times |1 - a|) + \|q\| \left(3.6A + \frac{(6.1)A}{|a-1|} + \frac{0.65}{|a-1|} + \frac{(1.1)A}{|a-1|^2} \right).$$

Note that the asymptotic forms for the μ_i suggest the uniqueness theorem. That is we can determine $|a - 1|$ and $(1/(a - 1)) \int_0^1 q(t)dt$ by

$$\lim_{i \to \infty} \frac{\mu_i(q)}{\pi^2 i^2} = \frac{1}{(a-1)^2}, \qquad \lim_{i \to \infty} \left[\mu_i(q) - \frac{\pi^2 i^2}{(a-1)^2} \right] = \frac{-1}{(a-1)} \int_0^1 q(t)\, dt.$$

The next term in the asymptotic form, $(1/(a - 1) \int_0^1 q(t) \cos(2\pi i t/(a - 1))dt$ suggests that we may have enough trigonometric moments to uniquely determine $q(x)$ if $a \geq 3$ but not if $0 < a < 1$ or $1 < a < 3$. Finally, if q is uniquely determined, then we obtain $\tilde{n}$ by solving the initial value problem $(\tilde{n}^{1/4})_{xx} - q(\tilde{n})^{1/4} = 0$, $0 \leq x \leq 1$, $\tilde{n}(1) = 1$, $\tilde{n}_x(1) = 0$. We state the uniqueness theorem and refer the reader to [MP] for the (rigorous) proof.

Theorem 2. *Suppose there is a common sequence of eigenvalues $\lambda_j, j = 1, 2, \ldots$ of (5) for $\tilde{n}_1, \tilde{n}_2 \in H^2(0,1)$ with $\tilde{n}_1(1) = \tilde{n}_2(1) = 1$, $\tilde{n}_1'(1) = \tilde{n}_2'(1) = 0$, and a positive integer i_0 satisfying:*

1) for every $i > i_0, |\lambda_j| < (i + \frac{1}{2})^2 \pi^2/(a - 1)^2$ for $j = 1, 2, \ldots, i$;

2) λ_j is real for $|\lambda_j| > (i_0 + \frac{1}{2})^2 \pi^2/(a - 1)^2$.

Then

1) if $a \geq 3$, a is uniquely determined and $\tilde{n}_1 \equiv \tilde{n}_2$;

2) *if* $1 < a < 3$ *then* a *is uniquely determined;*
   and if $\tilde{n}_1 \equiv \tilde{n}_2$ *for* $(a-1)/2 \leq x \leq 1$,
   then $\tilde{n}_1 \equiv \tilde{n}_2$ *for* $0 \leq x < (a-1)/2$;

3) *if* $0 < a < 1$ *then* a *is uniquely determined;*
   and if $\tilde{n}_1 \equiv \tilde{n}_2$ *for* $(1-a)/2 \leq x \leq 1$,
   then $\tilde{n}_1 \equiv \tilde{n}_2$ *for* $0 \leq x < (1-a)/2$.

We have stated the uniqueness theorem for $\tilde{n}(x) = n(r(x))$. However the goal is to reconstruct $n(r)$ from the transmission eigenvalues. Our method then will be to be given transmission eigenvalues $\{k_j^2\}_{j=1}^{\infty}$ so that k_j^2 is real for $j > i_0$, some positive integer i_0. We then will determine $B > b$ or $B < b$ from

$$|B - b| = \lim \frac{j\pi}{k_j}$$

and hence determine the corresponding $\{\lambda_j\}_{j=1}^{\infty}$. We thus find the uniqueness theorem for $n(r)$ as a direct corollary of Theorem 2 as follows, see [MP].

Theorem 3. *Let* $n_i(r) \in H^2(0, b)$ *with* $n_i(r) \equiv 1$ *for* $r \geq r_0 > 0$, $b > r_0$, $i = 1, 2$. *Let* $B_i = \int_0^b (n_i(r))^{1/2} dr$, *and suppose simultaneously* $B_i > b$ *or* $B_i < b, i = 1, 2$. *Suppose there is a common sequence of eigenvalues* k_j^2 *of (3) for* n_1 *and* n_2 *and a positive integer* m_0 *satisfying:*

   i) *for every* $m > m_0$, $|k_j^2| < (m + \frac{1}{2})^2 \pi^2/(B_i - b)^2$, $\quad j = 1, \ldots, m, m \geq m_0, i = 1, 2$;

   ii) *for* $j > m_0$, k_j^2 *are real and satisfy*

$$|k_j^2| > \frac{(m_0 + \frac{1}{2})^2 \pi^2}{(B_i - b)^2}, \qquad i = 1, 2;$$

Then $B_1 = B_2$ *and denote the common value* $B_1 = B_2 = B$. *If also*

   i) $3B \leq b$,

   or

   ii) $B < b < 3B$ *and* $n_1(r) = n_2(r)$ *for* r *satisfying* $0 \leq \int_r^b [n_i(r)]^{1/2} dr \leq (3B - b)/2$,

   or

   iii) $0 \leq b < B$ *and* $n_1(r) = n_2(r)$ *for* r *satisfying* $0 \leq \int_r^b [n_i(r)]^{1/2} dr \leq (B + b)/2$,

then $n_1 \equiv n_2$.

4 The method for reconstruction of c(r)

In this section we present the reconstruction procedure for $n(r)$ for the case $3B \leq b$ (equivalently $3 \leq a$). Our method is to:

1. find $B = b + \lim_{j \to \infty} (j\pi/k_j)$, $a = b/B$ and $\lambda_j = (k_j)^2 B^2$, $j = 1, 2, \ldots$,

2. reconstruct $\tilde{n}(x)$ from the $\{\lambda_j\}_{j=1}^{\infty}$;

3. construct $n(r) = \tilde{n}(x(r))$ by solving

$$\frac{dx}{dr} = \sqrt{\tilde{n}(x)}, \qquad x(0) = 0,$$

   for $x(r)$;

4. calculate $c(r) = c_0/\sqrt{n(r)}$.

It remains then to explain only the second of the above steps, that is the reconstruction of $\tilde{n}(x)$.

As in [RS] and [MPS] and similar to the procedure in [GL], we note that solutions of

$$(\tilde{n}^{1/2}\tilde{y}_x)_x + \lambda\tilde{n}^{1/2}\tilde{y} = 0$$

$$\tilde{y}(0) = 0 \tag{6}$$

$$\tilde{y}'(0) = [\tilde{n}(0)]^{-1/4}$$

can be represented by

$$y = \frac{1}{\tilde{n}^{1/4}}\frac{\sin\sqrt{\lambda}x}{\sqrt{\lambda}} + \int_0^x \frac{K(x,t)}{\tilde{n}^{1/4}(x)}\frac{\sin\sqrt{\lambda}t}{\sqrt{\lambda}}dt. \tag{7}$$

If we let $M(x,t) = (1 + \int_t^x K(x,s)ds)/\tilde{n}^{1/4}(x)$, then $M(x,t)$ satisfies the conditions

$$\tilde{n}^{1/2}(x)M_{tt} - (\tilde{n}^{1/2}(x)M_x)_x = 0, \qquad \text{for } 0 \leq t \leq x \leq 1,$$

$$M_t(x,0) = 0 \qquad\qquad\qquad \text{for } 0 \leq x \leq 1, \tag{8}$$

$$M(x,x) = \frac{1}{\tilde{n}^{1/4}(x)} \qquad\qquad \text{for } 0 \leq x \leq 1.$$

The second condition in (8) ensures that $M(x,t)$ can be extended to an even function in t, so that the system (8) can be rewritten as

$$\tilde{n}^{1/2}(x)M_{tt} - (\tilde{n}^{1/2}M_x)_x = 0, \qquad 0 \leq |t| \leq x \leq 1, \tag{9}$$

$$M(x,x) = M(x,-x) = \frac{1}{\tilde{n}^{1/4}(x)}, \qquad 0 \leq x \leq 1.$$

The possibility of a reconstruction arises from the fact that y satisfies a boundary condition at $x = 1$ for each eigenvalue. That is for each $\lambda = \lambda_j, j = 1, 2, \ldots$ it can be shown that

$$\cos \sqrt{\lambda} a \int_0^1 M_t(1, t) \sin \sqrt{\lambda} t \, dt + \sin \sqrt{\lambda} a \int_0^1 M_x(1, t) \cos \sqrt{\lambda} t \, dt = \sin \sqrt{\lambda}(1 - a).$$

(10)

From this the values of $M_t(1, t)$ and $M_x(1, t)$ are recovered. Once this is done then $\tilde{n}(x)$ will be found from the resultant overposed problem for $M(x, t)$. Before describing this last step, we will present the procedure for finding $M_t(1, t)$ and $M_x(1, t)$.

One first notes that the asymptotic forms for λ_j can be re-expressed as

$$\left\{ \lambda_j - \frac{\pi^2 j^2}{(a-1)^2} + \frac{2M_x(1, 1)}{(a-1)} \right\}_{j=1}^\infty \in \ell_2$$

so that $M_x(1, 1)$ can be determined from the real eigenvalues. Then $M_x(1, t)$ and $M_t(1, t)$ are expressed as

$$M_t(1, t) = -t M_x(1, 1) + \sum_{i=1}^\infty x_{2i} \sin i\pi t,$$

$$M_x(1, t) = M_x(1, 1) + \sum_{i=1}^\infty x_{2i-1} \cos \left(i - \frac{1}{2} \right) \pi t.$$

These series representations are inserted in the equation (10) for $\lambda = \lambda_j$ and the infinite set of linear equations is to be solved for the coefficients $\{x_i\}_{i=1}^\infty \in \ell_1^2$ [we note $\{x_i\} \in \ell_1^2$ implies $\sum_{i=1}^\infty (i x_i)^2 < \infty$.] One quickly sees that when $a > 3$ then the infinite coefficient matrix will likely not be invertible unless the number of equations is reduced. Roughly speaking, for each positive integer s, one wants only one equation from an eigenvalue with $\sqrt{\lambda_{j(s)}}$ near $s\pi/2$, $s = 1, 2, \ldots$. Denoting $j(s) = [(a-1)/2s]$ for positive integers s, the set of linear equations becomes

$$\sum_{i=1}^\infty x_{2i} \left(\cos \sqrt{\lambda_{j(s)}} a \right) \int_0^1 \sin i\pi t \sin \sqrt{\lambda_{j(s)}} t \, dt$$

$$+ \sum x_{2i-1} \left(\sin \sqrt{\lambda_{j(s)}} a \right) \int_0^1 \cos \left(i - \frac{1}{2} \right) \pi t \cos \sqrt{\lambda_{j(s)}} t \, dt$$

$$= \sin \left(\sqrt{\lambda_{j(s)}}(1 - a) \right) + M_x(1, 1) \frac{\cos \sqrt{\lambda_{j(s)}} a \sin(\sqrt{\lambda_{j(s)}})}{\lambda_{j(s)}}$$

$$- M_x(1, 1) \frac{\cos \sqrt{\lambda_{j(s)}}(1 - a)}{\sqrt{\lambda_{j(s)}}},$$

$s = 1, \ldots, \infty$. The following theorem shows the solvability of this system and hence the recovery of $M_t(1,t)$ and $M_x(1,t)$, for large enough a.

Theorem 4. *Denote α_0 as the minimal positive solution of the equation*

$$(1 - \cos \pi\alpha + \sin \pi\alpha)^2 + (2 - \cos \pi\alpha + \sin \pi\alpha)^2 \cdot (\tan \pi\alpha)^2 = \frac{1}{2}$$

and let $a_0 = (1 + \alpha_0)/\alpha_0 \sim 9.4$. If $a > a_0$ then the system (11) is a Fredholm type system in ℓ_1^2, namely it is equivalent to a system:

$$(I + N + \mathcal{K})x = y$$

where I is the identity operator, N is a bounded operator with $\|N\|_{\ell_1^2} < 1$, $\mathcal{K}$ is finite-dimensional and $y \in \ell_1^2$.

See [MPS] for the proof. Here we note only that the theorem states that up to a finite dimensional subspace in H^1, the functions $M_t(1,t)$ and $M_x(1,t)$ can be reconstructed from the spectrum of the boundary value problem (4) in the case $a > a_0$.

The final step for the reconstruction of $\tilde{n}(x)$ then comes from finding the pair $\tilde{n}(x), M(x,t)$ which satisfy (9) with given $M_t(1,t)$ and $M_x(1,t)$, $0 \le |t| \le 1$. The method is iterative. Choosing $\tilde{n}_0(x)$ one solves

$$\tilde{n}_0^{1/2} M_{0,tt}(x,t) - (\tilde{n}_0^{1/2} M_{0,x})_x = 0 \qquad 0 \le |t| \le x \le 1,$$
$$M_{0,t}(1,t) = M_t(1,t), \qquad\qquad -1 \le t \le 1,$$
$$M_{0,x}(1,t) = M_x(1,t), \qquad\qquad -1 \le t \le 1,$$

for $M_0(x,t)$ and then obtains the next value $\tilde{n}_1$, using the characteristic condition, $\tilde{n}_1 = 1/[M_0(x,x;\tilde{n}_0)]^4$. Continuing,

$$\tilde{n}_i = \frac{1}{[M_{i-1}(x,x;\tilde{n}_{i-1})]^4}, \qquad i = 1,2,3,\ldots$$

and $\tilde{n}(x)$ is obtained in the limit as the fixed point.

$$[\tilde{n}(x)] = \frac{1}{[M(x,x;\tilde{n}(x))]^4}.$$

For details of the convergence this iteration scheme, see [RS].

5 Numerical results

Here we show the results of numerical implementation of the procedure. If we are given m real eigenvalues $(k_j^2)_{j=1}^m$, we first estimate B, then a, then $M_x(1,1)$. Letting $\lambda_j = (k_j)^2 B^2$ we then select the subset $\{\lambda_{j(s)}\}_{s=1}^N$ and solve for the coefficients $\{\lambda_i\}_{i=1}^N$ to determine an approximation to $M_x(1,t), M_t(1,t)$. Using these approximations, we invoke the interaction scheme described in the previous section, obtaining an approximation to $\tilde{n}(x)$. Finally we follow steps three and four at the beginning of Section III to obtain the approximate $c(r)$.

The following figure displays the result for one choice of $c(r)$. Here $b = 1$ and we have chosen $m = 10$ and $m = 30$ which results in $N = 8$ and $N = 24$ respectively.

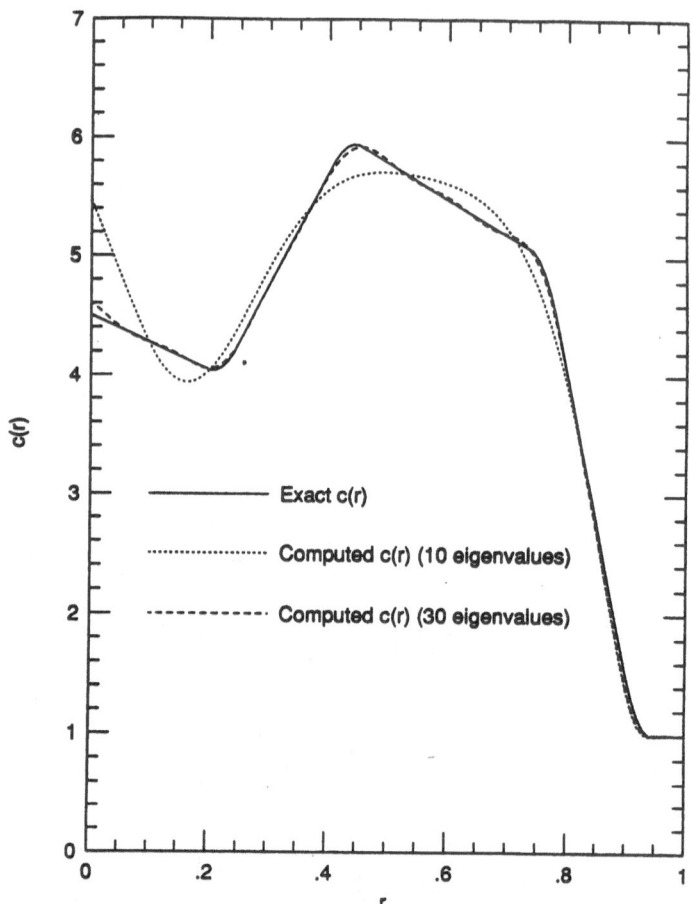

Acknowledgement The author's research was partially supported by ONR Grant N00014-91J-1166.

References

[CM] Colton, D., Monk, P.: The inverse scattering problem for time-harmonics acoustic waves in an inhomogeneous medium. Q. J. Mech. and Math. **41** (1988) 97–125.

[GL]   Gel'fand, I. M., Levitan, B. M.: On the determination of a differential equation from its spectral function. Amer. Math. Soc. Trans. **1** (1951) 253–304.

[HW] Hartman, P., Wilcox, C.: On solutions of the Helmholtz equation in exterior domains. Math. Zeit. **75** (1961) 228–255.

[MP]   McLaughlin, J. R., Polyakov, P.L: On the uniqueness of a spherically symmetric speed of sound from transmission eigenvalues. To appear in J. Diff. Eq.

[MPS] McLaughlin, J. R., Polyakov, P. L., Sacks, P. E.: Reconstruction of a spherically symmetric speed of sound. Submitted.

[RS]   Rundell, W., Sacks, P. E.: The Reconstruction of Sturm-Liouville Operators. Inverse Problems **8** (1992) 457–482.

A Finite Difference Method for the Inverse Scattering Problem at Fixed Frequency

F. Natterer, F. Wübbeling

Fachbereich Mathematik, Universität Münster, Germany

Abstract *We consider the problem of computing numerically the potential f in the Helmholtz equation $\Delta u + k^2(1 - f)u = 0$ from plane wave irradiation at a fixed frequency k. We discretize the differential equation in the plane by a five point difference star on a grid with stepsize h. It turns out that the resulting bilinear system can be solved recursively for f and u. We study the stability of this recursion. We find that the method enjoyes some stability properties provided hk is chosen properly. The complexity of the method is $O(h^{-4})$.*

1 Introduction

Consider the scattering problem: Find a function u in $\mathbb{R}^n$ such that

$$
\begin{aligned}
\Delta u + k^2(1 - f)u &= 0 \quad \text{in } \mathbb{R}^n , \\
u(x) &= e^{ikx \cdot \theta} + w(x) \quad , \theta \in S^{n-1} ,
\end{aligned}
\tag{1.1}
$$

where w satisfies the Sommerfeld radiation condition, and f is a function of compact support. The inverse problem calls for the computation of f from knowing u outside the support of f for all directions θ and a single value of k. See [1], [8] for the general background. The uniqueness of the inverse problem has recently been settled for $n = 3$ by Sylvester and Uhlmann [10], Nachmann [7] and Ramm [9]. As for numerical methods, [5] seems to be the state of the art.

In this note we study the numerical solution of the inverse problem for $n = 2$ by a finite difference method. The method is based on a 5 point discretization of (1.1) which takes into account the high frequency behaviour of u. Following a suggestion of Curtis and Morrow [2] and Grünbaum [3] the resulting bilinear system for u and f is solved recursively for f and u. Our aim is to analyse the stability of this process. This has been completely ignored in [2], while numerical results in [3] indicate that the recursion is awfully unstable. Ideally we would like to have stability if the stepsize h is tied to the irradiating frequency k by the Nyquist condition $hk = \pi$, in which case it would be possible to resolve details of size $2h = \frac{2\pi}{k}$, the optimal value in the sense of the sampling theorem, see e.g. [4].

As in [3], the recursion proceeds diagonal by diagonal in the finite difference mesh. On each diagonal we do a solution step, in which we solve a linear system for f on the current diagonal, followed by a propagation step, in which we extend the field to the next diagonal. We investigate the stability of these two steps separately.

In the propagation step we solve the initial value problem for the elliptic equation (1.1) by a finite difference method. It belongs to the folklore of numerical analysis that such a method is necessarily unstable. However, a closer look reveals that this instability is exclusively a high frequency phenomenon. If the frequency is restricted to a certain interval $[k_{\min}, k_{\max}]$ - either by choosing an apropriate step size h or by band-pass filtering - then the propagation step turns out to be stable. We call $[k_{\min}, k_{\max}]$ the stability interval.

If the physical and numerical directions of propagation coincide, the stability interval is $[-k, k]$. This is completely satisfactory, since we do not expect to be able to resolve frequencies beyond the irradiating frequency k. If the said two directions are different, then the stability interval is shifted and does not contain a neighbourhood of zero when they make a right angle. By a suitable organization of the computational process and by sacrifying a certain amount of resolution (i.e. by chosing hk larger than its optimal value π) we can avoid this extreme case.

In the solution step we have to solve a linear system whose matrix is of the Van der Monde type. Such system are notoriously unstable. However, by exploiting the analyticity of the field u in the direction variable θ we find that the system is not as unstable as one might expect. Its solution can be reduced to stable cases of analytic continuation, provided that hk is sufficiently large. Thus, again by sacrifying a certain amount of resolution, we can make the solution process stable. Of course this does not imply the stability of the combined process.

The outline of the paper is as follows. In section 2 we describe the finite difference approximation and the recursive solution process. In section 3 we prove the stability of the initial value problem for the Helmholtz equation for frequencies in the stability interval, establishing in this way the stability of the propagation step. In section 4 we study - in a rather heuristic way - the stability of the solution step.

2 The finite difference method

In the following we assume f to be zero outside the unit square $D = \{x \in \mathbb{R}^2 : 0 \le x_1, x_2 \le 1\}$, and we assume u to be known outside D for a finite number of direction θ_j, $j = 1, \ldots, p$. We put

$$u = e^{ikx \cdot \theta}(1 + v)$$

where v satisfies the differential equation

$$\Delta v + 2ik\theta \cdot \nabla v - k^2(1 + v)f = 0 \ . \tag{2.2}$$

For v small and k large we have approximately

$$2ik\theta \cdot \nabla v - k^2 f = 0$$

or

$$v(x, \theta) = \frac{k}{2i} \int\limits_0^\infty f(x - t\theta)dt \; .$$

In particular, for $x \in D + \mathbb{R}_+\theta$, we have

$$v(x, \theta) = \frac{k}{2i} \, Rf(\theta^\perp, x \cdot \theta^\perp) \qquad (2.3)$$

where R is the Radon transform

$$(Rf)(\theta, s) = \int\limits_{x \cdot \theta = s} f(x)dx$$

and $\theta^\perp \cdot \theta = 0$.

Doing the discretization on Δu would force us to use a stepsize h much smaller than $\frac{1}{2}$ of the wavelength $2\pi/k$ of the irradiating waves $e^{ikx \cdot \theta}$. This term is no longer present in v, as can be seen from (2.3). Thus for the discretization of v the stepsize $h = \pi/k$ corresponding to the aimed - at resolution suffices.

A natural discretization of (2.2) is

$$-4v^j_{\ell,m} + (1 + i\varepsilon\theta_1)v^j_{\ell+1,m} + (1 - i\varepsilon\theta_1)v^j_{\ell-1,m} + (1 + i\varepsilon\theta_2)v^j_{\ell,m+1}$$
$$+ (1 - i\varepsilon\theta_2)v^j_{\ell,m-1} - (1 + v^j_{\ell,m})\varepsilon^2 f_{\ell,m} = 0 \qquad (2.4)$$

where $\varepsilon = hk$. $v^j_{\ell,m}$ stands for the discrete values of $v(x, \theta_j)$ at $x = h(\ell, m) = x_{\ell,m}$. With the values of $v^j_{\ell,m}$ for $x_{\ell,m} \neq D$ known we can solve (2.4) simultaneously for $v^j_{\ell,m}, f_{\ell,m}$ ($x_{\ell,m} \in D$) in the following way.

Assume $v^j_{\ell,m}$ to be already known for $\ell + m \leq n$ and $f_{\ell,m}$ for $\ell + m < n$. Renumber the gridpoints $x_{\ell,m}$ along the diagonal $\ell + m = n$ from left to right by x_ℓ, $\ell = 0, \ldots, n$, see Fig. 1 and the gridpoints on the diagonal $\ell + m = n + 1$, $\ell = 1, \ldots, n$ by $x_{n+1}, \ldots, x_{2n}$. Write down the difference equations (2.4) at the gridpoints $0, \ldots, n$. Denoting terms involving only quantitative which are already known (i.e. $v^j_{\ell,m}$ for $\ell + m \leq n$ and $f_{\ell n}$ for $\ell + m < n$) by b, these equations read

$$\begin{aligned}
(1 + v^j_0)\varepsilon^2 f_0 \quad & - w_j v^j_{n+1} && = b^j_0 \\
(1 + v^j_1)\varepsilon^2 f_1 \quad & - s_j v^j_{n+1} - w^j v^j_{n+2} && = b^j_1 \\
& \qquad \cdots \\
(1 + v^j_{n-1})\varepsilon^2 f_{n-1} \quad & - s^j v^j_{2n-1} - w^j v^j_{2n} = b^j_{n-1} \\
(1 + v^j_n)\varepsilon f_n \quad & - s^j v^j_{2n} = b^j_n
\end{aligned} \qquad (2.5)$$

where

$$s_j = 1 + i\varepsilon\theta^j_2 \quad , \quad w_j = 1 + i\varepsilon\theta^j_1 \; .$$

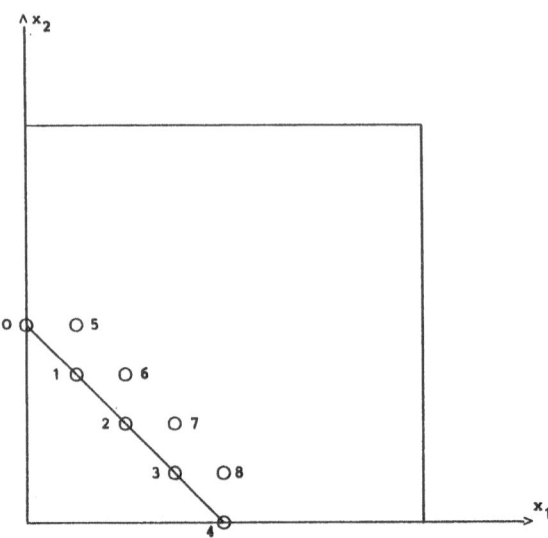

Fig. 1. Renumbering of gridpoints $(n = 4)$.

The unknown quantities $v_{n+1}^j, \ldots, v_{2n}^j$ in (2.5) are readily eliminated by forming suitable linear combination of successive equations, yielding for each direction θ_j a single equation for $f_0, \ldots, f_n$:

$$\epsilon^2 \sum_{\ell=0}^{n} (1 + v_\ell^j) \lambda_j^\ell f_\ell = \sum_{\ell=0}^{n} b_\ell^j \lambda_j^\ell, \tag{2.6}$$

$$\lambda_j = -\frac{w_j}{s_j}.$$

For $p > n$, (2.6) is an overdetermined linear system for $f_0, \ldots, f_n$. Once (2.6) is solved for $f_0, \ldots, f_n$ ("solution step"), the values of $v_{n+1}^j, \ldots, v_{2n}^j$ on the diagonal $\ell + n = n + 1$ are readily computed by successive substitution from (2.5) ("propagation step"). In the light of our stability analysis in §3, the order in which $v_{n+1}^j, \ldots, v_{2n}^j$ are computed does matter: For directions θ_j making an angle $\leq 45°$ with the x_1-axes we have to start with v_{n+1}^j, for the other directions with v_{2n}^j.

The number p of directions one needs to make (2.5) overdetermined is $O(h^{-1})$. This agrees with a well known result from tomography which states that for a resolution h the number of directions has to be $O(h^{-1})$. In each propagation step we need $O(pn) = O(h^{-1})$ operations, while $O(n^3) = O(h^{-3})$ operations are needed in each solution step. Thus the complexity of the whole process is $O(h^{-4})$. At a first glance this may look prohibitive. But one has to bear in mind that the standard algorithm in tomography requires $O(h^{-3})$ operations and is executed on special hardware within seconds.

3 Stability of the propagation step

We investigate the stability of the initial value problem for (2.2). We restrict ourselves to the case $f = 0$, i.e. we consider

$$\Delta v + 2ik\theta \cdot \nabla v = 0 ,$$
$$v(0, x_2) = v_0(x_2) , \quad \frac{\partial v}{\partial x_1}(0, x_2) = v_1(x_2) .$$

(3.7)

Let $\hat{v}(x_1, \xi)$ be the partial Fourier transform of v with respect to x_2, i.e.

$$v(x_1, x_2) = (2\pi)^{-1/2} \int_{\mathbb{R}^1} e^{ix_2\xi} \hat{v}(x_1, \xi) d\xi .$$

The initial value problem (3.7) is readily solved in terms of $\hat{v}$ by

$$\hat{v}(x_1, \xi) = e^{-ik\theta_1 x_1} \{\hat{v}_0(\xi) \cos x_1 \kappa + \frac{\hat{v}_1(\xi)}{\kappa} \sin x_1 \kappa \} ,$$

$$\kappa = \sqrt{k^2\theta_1^2 - 2k\theta_2\xi - \xi^2} .$$

Thus $\hat{v}$ depends on these initial values in a stable way for real $\kappa > 0$, i.e. for

$$h^2\theta_1^2 - 2k\theta_2\xi - \xi^2 > 0 ,$$

or, equivalently,

$$- k(1 + \theta_2) < \xi < k(1 - \theta_2) .$$

(3.8)

Hence the initial value problem (3.7) which describes wave propagation in the direction x_1 is stable in the frequency range (3.8). If θ points in the direction of x_1, i.e. $\theta = \begin{pmatrix} 1 \\ 0 \end{pmatrix}$, then the stability interval is $[-k, k]$. On the other hand, if θ is perpendicular to the x_1 direction, i.e. $\theta = \begin{pmatrix} 0 \\ 1 \end{pmatrix}$, then the stability interval is $[-2k, 0]$ and does not contain all small frequencies. If θ makes an angle of $45°$ with the x_1-axes, then the stability interval is $[-k(1 + \sqrt{\frac{1}{2}}); k(1 - \sqrt{\frac{1}{2}})]$.

A lengthy analysis reveals that the same holds for the finite difference approximation (2.4), with a stability interval practically identical to (3.8). In order to have a reasonable stability interval for each direction θ_j we do the propagation in x_1-direction for all direction θ_j which make an angle $\leq 45°$ with the x_1-axes and in the x_2-direction for the other direction θ_j. In the algorithm of §2 this can be achieved by computing $v_{n+1}^j, \ldots, v_{2n}^j$ from (2.5) starting with v_{n+1}^j in the former case and with v_{2n}^j in the latter case.

4 Stability of the solution step

In this section we show - in a rather heuristic manner - that the system (2.6) can be solved in a stable way, provided that ϵ is sufficiently large and that enough directions θ_j are available. In fact we shall assume that the data are available for $\theta = \begin{pmatrix} \cos\varphi \\ \sin\varphi \end{pmatrix}$, $0 \le \varphi < 2\pi$. Then, (2.6) reads

$$\epsilon^2 \sum_{\ell=0}^{n} (1 + v_\ell(\varphi))\lambda^\ell(\varphi)f_\ell = \sum_{\ell=0}^{n} b_\ell(\varphi)\lambda^\ell(\varphi) \tag{4.9}$$

where

$$\lambda(\varphi) = -\frac{1 + i\varepsilon\cos\varphi}{1 + i\varepsilon\sin\varphi}, \quad 0 \le \varphi < 2\pi.$$

$v_\ell(\varphi)$ is a discrete approximation to $v(x_\ell, \theta)$ which is an entire function of φ. Hence $v_\ell(\varphi)$ may be viewed, at least approximately, as an entire function of φ. The same applies to $b_\ell(\varphi)$ which is a linear combination of expressions such as $v_\ell(\varphi)$ with linear functions of $\sin\varphi$, $\cos\varphi$ as coefficients. Putting $w = \cos\varphi$, $\sqrt{1 - w^2} = \sin\varphi$ for $0 \le \varphi \le \pi$, $-\sqrt{1 - w^2} = \sin\varphi$ for $\pi \le \varphi \le 2\pi$ we may view $v_\ell(\varphi)$, $b_\ell(\varphi)$, $\lambda(\varphi)$ as analytic functions $v_\ell(w)$, $b_\ell(w)$, $\lambda(w)$ on the Riemannian surface of $\sqrt{1 - w^2}$ with the cut along $-1 \le w \le 1$, and $v_\ell(w)$, $b_\ell(w)$ are known along both shores of this cut. By the monodromy theorem we have on this Riemannian surface

$$\epsilon^2 \sum_{\ell=0}^{n}(1 + v_\ell(w))\lambda^\ell(w)f_\ell = \sum_{\ell=0}^{n} b_\ell(w)\lambda^\ell(w), \tag{4.10}$$

$$\lambda(w) = -\frac{1 + i\varepsilon w}{1 + i\varepsilon\sqrt{1 - w^2}}. \tag{4.11}$$

We want to replace the variable w by λ. Solving (4.11) for w we obtain after some algebra

$$w = w(\lambda) = \frac{1}{\varepsilon(1 + \lambda^2)}\{i(1 + \lambda) - \lambda\sqrt{(1 + \lambda)^2 + \varepsilon^2(1 + \lambda^2)}\}. \tag{4.12}$$

$w(\lambda)$ is an algebraic function of λ with branch points λ_0 and $\overline{\lambda}_0$ where

$$\lambda_0 = -\frac{1}{1 + \varepsilon^2}\left(1 - i\varepsilon\sqrt{\varepsilon^2 + 2}\right).$$

λ_0, $\overline{\lambda}_0$ lie on the unit circle. We construct the two-sheeted Riemannian surface Λ of $w(\lambda)$ by cutting the λ-plane along the vertical lines joining λ_0, $\overline{\lambda}_0$ with ∞. In the upper sheet we define the $\sqrt{}$ in (4.12) by its principle value near $\lambda = -1$, i.e. we put $w(-1) = \sqrt{\frac{1}{2}}$ in the upper sheet and $w(-1) = -\sqrt{\frac{1}{2}}$ in the lower sheet.

Besides the branch points λ_0 $\overline{\lambda}_0$, the function $w(\lambda)$ has singularities at $\lambda = \pm i$. We study these singularities.

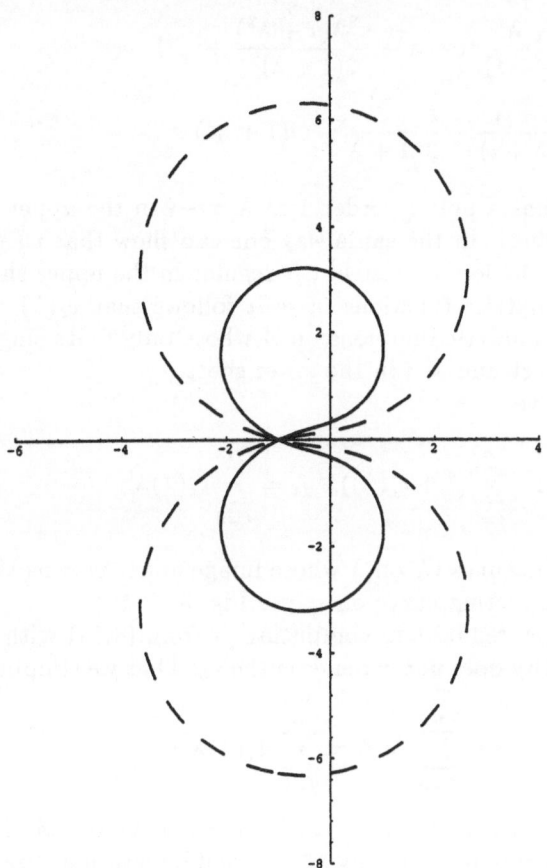

Fig. 2. The curve $\lambda = \lambda(\varphi)$, $0 \le \varphi < 2\pi$ for $\varepsilon = \pi$ (solid) and $\varepsilon = 2\pi$ (dashed).

For the principle value we have $\sqrt{z^2} = z$ if $Re\, z > 0$. Hence we have in the upper sheet of Λ for λ close to $\pm i$

$$\sqrt{(1+\lambda)^2 + \varepsilon^2(1+\lambda^2)} = \sqrt{(1+\lambda)^2(1 + \frac{\varepsilon^2(1+\lambda^2)}{(1+\lambda)^2})}$$

$$= (1+\lambda)\sqrt{1 + \frac{\varepsilon^2(1+\lambda^2)}{(1+\lambda)^2}}$$

$$= (1+\lambda)(1 + \frac{\varepsilon^2(1+\lambda^2)}{2(1+\lambda)^2} + \ldots)$$

where the dots stand for powers of $1 + \lambda^2$ higher than 1. Thus,

$$w(\lambda) = \frac{1}{\varepsilon(1+\lambda^2)}\{i(1+\lambda) - \lambda(1+\lambda)(1 + \frac{\varepsilon^2(1+\lambda^2)}{2(1+\lambda)^2} + \ldots)\}$$

$$= \frac{1+\lambda}{\varepsilon(1+\lambda^2)}\left\{i - \lambda - \frac{\varepsilon^2\lambda(1+\lambda^2)}{2(1+\lambda)^2} + \ldots\right\}$$

$$= -\frac{1+\lambda}{\varepsilon(\lambda+i)} - \frac{\varepsilon}{2}\frac{\lambda}{1+\lambda} + \mathcal{O}(1+\lambda^2)\,.$$

It follows that $w(\lambda)$ has a pole of order 1 at $\lambda = -i$ in the upper sheet but is regular in the lower sheet. In the same way one can show that $w(\lambda)$ has a pole of order 1 at $\lambda = i$ in the lower sheet but is regular in the upper sheet.

Since v_ℓ, b_ℓ are analytic functions of w it follows that $v_\ell(\lambda) := v_\ell(w(\lambda))$, $b_\ell(\lambda) := b_\ell(w(\lambda))$ are analytic functions on Λ whose only finite singularities are at $-i$ in the upper sheet and at i in the lower sheet.

From (4.10) we have

$$\varepsilon^2 \sum_{\ell=0}^{n}(1 + v_\ell(\lambda))\lambda^\ell f_\ell = \sum_{\ell=0}^{n} b_\ell(\lambda)\lambda^\ell\,. \tag{4.13}$$

v_ℓ, b_ℓ are known on the curve C_ε on λ whose image under $w = w(\lambda)$ is $[-1, +1]$. C_ε is a closed nonintersecting curve on Λ, see Fig. 3.

Our problem is now reduced to computing f_ℓ from (4.13) with $\lambda \in C_\varepsilon$. We believe that the stability does not depend on the v_ℓ. Thus we simplify (4.13) into

$$\varepsilon^2 \sum_{\ell=0}^{n} \lambda^2 f_\ell = \sum_{\ell=0}^{n} b_\ell(\lambda)\lambda^\ell\,. \tag{4.14}$$

Let C be the curve on Λ depicted in Fig. 3. In each of the sheets of Λ C is part of the unit circle. Multiplying (4.14) by λ^{-m-1} and integrating over C we obtain

$$f_m = \frac{1}{2\pi i\varepsilon^2} \int_C \sum_{\ell=0}^{n} b_\ell(\lambda)\lambda^{\ell-m-1}d\lambda\,. \tag{4.15}$$

Note that b_ℓ is not known on C. For the stable evaluation of the right hand side we use Cauchy's theorem to deform C into parts of C_ε on which $|\lambda| \geq 1$ for $\ell - m < 0$ and $|\lambda| \leq 1$ otherwise. We refer to fig. 3 where $a, b \ldots$ denote points on the upper sheet of Λ and $a', b' \ldots$ their counterparts on the lower sheet. For $\ell - m < 0$ we move the arc $a'\lambda_0 c$ of C upwards until it merges with the arc ac of C_ε. This is possible because there is no singularity between these two arcs - remember that i is a pole only in the lower sheet. In the same way we move the arc $a\overline{\lambda}_0\overline{c}'$ of C downwards to become the arc $a\overline{c}'$ of C_ε. For $\ell - m \geq 0$ the situation is less obvious. We deform the arc $a'\lambda_0 c$ into $a'\overline{b}'\lambda_0 bc$ and the arc $a\overline{\lambda}_0\overline{c}'$ into $ab\overline{\lambda}_0\overline{b}'\overline{c}'$. This is not quite what we want since $\overline{b}'\lambda_0 b$ and $b\overline{\lambda}_0\overline{b}$ do not belong to C_ε. This means that we do not known the values of b_ℓ on $\overline{b}'\lambda_0 b$ and $b\overline{\lambda}_0\overline{b}'$, and the same applies to the arc $c\overline{c}$ of C which we have not yet considered. In all these cases, b_ℓ can be computed from its values on C_ε by analytic continuation. From [6], Theorem III.2.2 we know that analytic continuation can be done in a stable way except in a neighbourhood of the singularities λ_0, $\overline{\lambda}_0$, $\pm i$. Thus the

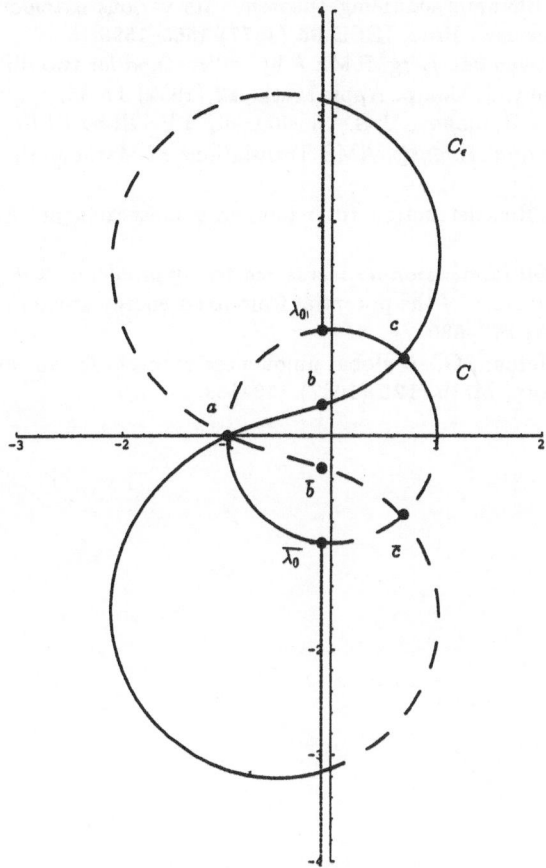

Fig. 3. The curves C_ϵ (for $\epsilon = \pi$) and C on the Riemanian surface Λ. Solid line: upper sheet. Dashed line: lower sheet. Dots indicate the cuts joining the two sheets.

arc $c\bar{c}$ is no problem at all, while the singularities at the endpoints of $b'\lambda_0 b$ and $b\bar{\lambda}_0\bar{b}'$ are doing little harm since the integrands are zero there.

The conclusion is that the solution step is stable to some degree. Of course this does not guarantee that the whole process is stable. Since we do not know how to carry the theoretical analysis farther the next step is to do numerical experiments.

References

1. Colton, D., Kress, R.: *Inverse Acoustic and Electromagnetic Scattering Theory.* Springer 1992.
2. Curtis, E.B., Morrow, J.A.: Determining the resistors in a network, SIAM J. Appl. Math. **50** (1990) 918–930.
3. Grünbaum, F.A.: Diffuse tomography: the isotopic case, Inverse Problems **8** (1992) 409–420.

4. Jerry, A.J.: The Shannon sampling theorem - its various extenious and applications: a tutorial review. Proc. IEEE **65** (1977) 1565–1596.
5. Kleinmann, R.E., van den Berg, P.M.: A hybrid method for two-dimensional problems in tomography. J. Comp. Appl. Math. **42** (1992) 17–35.
6. Lavrent'ev, M.M., Romanov, V.G., Shishatskii, S.P.: *Ill-posed Problems of Mathematical Physics and Analysis*. AMS Translations of Mathematical Monographs, Vol. **64** (1986).
7. Nachmann, A.I.: Reconstruction from boundary measurement. Ann. Math. **128** (1988) 531–676.
8. Ramm, A. G.: *Multidimensional inverse scattering problems*. Wiley 1992.
9. Ramm, A. G.: Recovery of the potential from fixed-energy scattering data. Inverse Problems **4** (1988) 877–886.
10. Sylvester, J., Uhlmann, G.: A global uniqueness theorem for an inverse boundary value problem. Ann. Math. **125** (1987) 153–169.

Present Status of the Generalized Marchenko Method for the Solution of the Inverse Scattering Problem in Three Dimensions

Roger G. Newton

Department of Physics, Indiana University,Bloomington, IN 47405, USA

Abstract *The present status of the generalized Marchenko method for the solution of the inverse scattering problem of the Schrödinger equation in three dimensions is summarized. Gaps and errors in previous publications are filled and corrected, and a new connection between the scattering amplitude and the number of bound-state eigenvalues is pointed out.*

The generalized Marchenko method for the solution of the inverse scattering problem in three dimensions seeks not only to reconstruct a potential that is known to underlie the given data (the full scattering amplitude as a function of angles and energy), but to determine if such a potential exists, to construct it if it does, and to prove that the constructed potential, in fact, corresponds to the data. It was given in the book [1]. At the time of its publication, however, there were still certain gaps in the method; specifically, some of the sufficient conditions for the existence of an underlying potential had not been proved to be necessary. Furthermore, one of the needed lemmas (Lemma 2.4.1) was not correctly stated and proved. These gaps were filled and the errors corrected in the two subsequent papers [2] and [3]. This paper presents the first complete summary of the method, which leads to a set of conditions that are both necessary and sufficient for a function to be admissible as a scattering amplitude for the Schrödinger equation in three dimensions, even in the presence of bound states (point eigenvalues). These conditions fall short of being a full characterization of admissible amplitudes in that they do not specify the class to which the underlying potential belongs if a scattering amplitude is given.

We note, to start with, that, while the existence of an underlying potential for a given scattering amplitude is not guaranteed (and is, indeed, very improbable, as a count of the variables indicates), its uniqueness is a priori assured by the fact that, as the energy tends to infinity with fixed "momentum transfer" (the difference between the initial and final wave vectors or momenta), the scattering amplitude approaches the Fourier transform of the potential. This implies that, in contrast to the one-dimensional case, the point eigenvalues and their needed characteristics must all be contained in the scattering amplitude and need to be extraced from it. A method for this extraction was described in [1].

We begin with the solution of the Schrödinger equation in $\mathbf{R}^3$ with the potential V,

$$(\Delta + k^2)\psi = V\psi, \quad x \in \mathbf{R}^3,$$

defined by the outgoing-wave boundary condition as $r \to \infty$,

$$\psi^+(k, \theta, x) = e^{ik\theta \cdot x} + (e^{ikr}/r)A(k, \hat{x}, \theta) + o(1/r), \quad k \in \mathbf{R}, \theta \in \mathbf{S}^2, \ \hat{x} := x/r, \ r := |x|,$$

which at the same time defines the scattering amplitude $A(k, \theta', \theta)$ as a function $\mathbf{R} \times \mathbf{S}^2 \times \mathbf{S}^2 \mapsto \mathbf{C}$. This function will be regarded as the integral kernel of a function $A(k)$ on $\mathbf{R}$ with values in the ring of bounded operators $L^2(\mathbf{S}^2) \mapsto L^2(\mathbf{S}^2)$, and we define the S matrix as an operator by

$$S(k) = \mathbf{1} - \frac{k}{2\pi i}A(k).$$

It is not hard to prove that, for $V \in \mathbf{R}$, S is unitary.

The class of potentials within which we shall work is the following:

$$\mathcal{V}_0 = \{V \mid V \in \mathbf{R}, \ \lim_{|x| \to \infty} V(x) = 0, \text{ and } \exists a, C, \epsilon > 0,$$

$$\text{such that } \forall x \in \mathbf{R}^3, \ |\nabla V(x)| < C(a + |x|)^{-4-\epsilon}\}.$$

Our results will be stated in terms of a larger class $\mathcal{W}$ that contains $\mathcal{V}_0$ and is defined in [1], but which is too cumbersome to specify here.

It is convenient to define the following functions in terms of the scattering amplitude:

$$G(\alpha, \theta, \theta') := \frac{i}{(2\pi)^2}\int_{-\infty}^{\infty} dk\, kA(k, -\theta, \theta')e^{-ik\alpha}, \quad \alpha \in \mathbf{R}, \ \theta, \theta' \in \mathbf{S}^2,$$

$$\mathcal{G}(\alpha, \theta; \beta, \theta') := G(\alpha + \beta; \theta, \theta'), \quad \alpha, \beta \in \mathbf{R}_+, \ \theta, \theta' \in \mathbf{S}^2.$$

This function defines an operator $L^2(\mathbf{R}_+ \times \mathbf{S}^2) \mapsto L^2(\mathbf{R}_+ \times \mathbf{S}^2)$ in the sense that

$$(\mathcal{G}f)(\alpha, \theta) = \int_0^{\infty} d\beta \int_{\mathbf{S}^2} d\theta'\, G(\alpha + \beta, \theta, \theta')f(\beta, \theta').$$

Similarly we define operators $\mathcal{G}^\#$, $\mathcal{H}$ and $\mathcal{H}^\#$ by the following integral kernels:

$$\mathcal{G}^\#(\alpha, \theta; \beta, \theta') := G(-\alpha - \beta, \theta\theta'),$$

$$\mathcal{H}(\alpha, \theta; \beta, \theta') := G(\alpha - \beta, \theta, \theta'),$$

$$\mathcal{H}^\#(\alpha, \theta; \beta, \theta') := G(-\alpha + \beta, \theta, \theta').$$

The operators $\mathcal{G}$ and $\mathcal{G}^\#$ are self-adjoint, and the unitarity of S implies that $\|\mathcal{G}\| \leq 1$ and $\|\mathcal{G}^\#\| \leq 1$, whereas $\mathcal{H}^\#$ is the adjoint of $\mathcal{H}$. We have the following first result, which is stated as Theorem 2.3.1 in [1] and proved there:

Lemma 1 *If $V \in \mathcal{W}$ and A is the corresponding scattering amplitude then the operators $\mathcal{G}$ and $\mathcal{G}^\#$ are bounded and self-adjoint as operators $L^2(\mathbf{R}_+ \times \mathbf{S}^2) \mapsto$*

$L^2(\mathbf{R}_+ \times \mathbf{S}^2)$, and $\|\mathcal{G}^2\|_2 < \infty$ and $\|\mathcal{G}^{\#2}\|_2 < \infty$. Here $\|\cdot\|_2$ is the Hilbert-Schmidt norm.

Note that the essential reason for the boundedness of the operators $\mathcal{G}$ and $\mathcal{G}^{\#}$ is that α and β are constrained to be non-negative. The operator $\mathcal{H}$, on the other hand is a Wiener-Hopf kernel and one of the main points of the Marchenko method is to avoid the use of integral equations with these kernels.

We next define a useful set of classes of scattering amplitudes:

Definition $S \in \mathcal{S}$ and $A \in \mathcal{A}$ if and only if $S(k) = 1 - \frac{k}{2\pi i} A(k)$ and the following six conditions are satisfied:

(i) the kernel $A(k, \theta, \theta')$ that defines the operator family $A(k)$, $k \in \mathbf{R}$, with values in the ring of bounded operators $L^2(\mathbf{S}^2) \mapsto L^2(\mathbf{S}^2)$, is a continuous, uniformly bounded, differentiable function $\mathbf{R} \times \mathbf{S}^2 \times \mathbf{S}^2 \mapsto \mathbf{C}$;

(ii) $QAQ = \tilde{A}$ (this is called reciprocity); here $Qf(\theta) := f(-\theta)$ and the tilde denotes the transpose; [this simply means that $A(k, -\theta', -\theta) = A(k, \theta, \theta')$]

(iii) $A(-k) = \overline{A(k)}$;

(iv) $S^{\dagger}S = SS^{\dagger} = \mathbf{1}$; unitarity; (the dagger denotes the adjoint)

(v) $\|S - \mathbf{1}\| \in L^2(\mathbf{R})$; $\|\cdot\|$ here is the operator norm;

(vi) the operators $\mathcal{G}$ and $\mathcal{G}^{\#}$ defined in terms of A are compact.

The subclasses $\mathcal{A}_n$ are defined by adding the following two requirements:

(vii) $\delta := \frac{1}{2} \arg \det S$, defined as a continuous function of k, is such that $\exists \delta_\infty$ and

$$\mathrm{ind}_L S := \delta(0) - \lim_{k \to \infty} [\delta(k) + k\delta_\infty] = \pi n;$$

(viii) $A(k, \theta, \theta)$ is, for each $\theta \in \mathbf{S}^2$, the boundary value of an analytic function meromorphic in $\mathbf{C}^+$ with simple poles at points $k = i\kappa_m$ on the positive imaginary axis (forward analyticity) which are all normal as defined in [1].

If S is admissible as an S matrix of the Schrödinger equation with a potential that is in the class $\mathcal{V}_0$, then $S \in \mathcal{S}$. We shall for simplicity always assume that $k = 0$ is not an exceptional point of the Lippmann-Schwinger equation; so there is assumed to be no bound or half-bound state at $k = 0$. The following is Theorem 1.5.16 in [1].

Theorem A   If $V \in \mathcal{W}$ and $H = V - \Delta$ has no bound or half-bound states, then $A \in \mathcal{A}_0$.

If

(i) $V \in \mathcal{W}$ and $(1 + |x|^{N_0})V \in L^1(\mathbf{R}^3)$,

(ii) H has M bound states of eigenvalues $-\kappa_m^2$ with degeneracy N_m, $m = 1, \ldots, M$,

(iii) these are all normal (as defined in [1]),

(iv) $k = 0$ is not an exceptional point,

then $A \in \mathcal{A}_n$, where $n = \sum_{m=1}^{b} N_m$ and the number N_0 in the hypothesis equals the largest of the numbers N_m.

Now the solution ψ^+ of the Schrödinger equation, as a function of the variable k, has an analytic continuation into the upper half of the complex plane (see [1]), where it is meromorphic, with simple poles on the positive imaginary axis at $i\kappa_m$ if $-\kappa_m^2$ is an eigenvalue. The residue of the pole at $i\kappa_m$ lies in the span of the finitely many functions $Y_{\kappa_m}^b(\theta)$, $b = 1, \ldots N_m$, which we call the *characters* of the N_m linearly independent eigenfunctions at $-\kappa_m^2$. These functions are determined by the asymptotic angle dependence of the eigenfunctions; for their precise definition, see [1]. (For central potentials, they are spherical harmonics.) Items *(iii)* in Theorem A and *(viii)* in the definition of $\mathcal{A}$ allow us to find the eigenvalues and the characters from the scattering amplitude; see [1]. Finally, as $|k| \to \infty$ in $\mathbf{C}^+$,

$$\zeta(k, \theta, x) := \psi^+(k, \theta, x)e^{-ik\theta \cdot x} \to 1.$$

The function ζ, of course, shares the analytic properties of ψ^+.

The crucial property of ψ^+ for the solution of the inverse problem is the well-known equation

$$\psi^+(-k, \theta, x) = \int_{\mathbf{S}^2} d\theta' \, S(-k, -\theta', \theta)\psi^+(k, \theta', x),$$

which is proved in [1]. In terms of ζ it reads

$$\zeta(-k, \theta, x) = \int_{\mathbf{S}^2} d\theta' \, S_x(-k, -\theta, \theta')\zeta(k, \theta', x),$$

where

$$S_x(k, \theta, \theta') := S(k, \theta, \theta')e^{ikx \cdot (\theta - \theta')}.$$

In vector and operator form, this equation may be written simply

$$\zeta(-k) = QS_x(-k)\zeta(k).$$

It thus gives rise to the following generalized Riemann-Hilbert problem:

Riemann-Hilbert Problem $H_\sigma^1(S_x)$: *Let $\Omega(k)$ be defined in terms of the given distribution kernel S_x with $A(k, \theta, \theta') \in \mathcal{A}$ by $\Omega(k) := QS_x(-k)$. Let κ_m be a given set of positive numbers and $\mathcal{H}_m$ a given set of finite-dimensional subspaces of $L^2(\mathbf{S}^2)$ in one-to-one correspondence with them, the sum of whose dimensions equals n_σ. (We denote the set of positive numbers κ_m together with the corresponding spaces $\mathcal{H}_m$ by σ.) Find a function f, $\mathbf{R} \times \mathbf{S}^2 \mapsto \mathbf{C}$, which is such that*

(i) $f - \hat{1} \in L^2(\mathbf{R} \times \mathbf{S}^2)$ $(\hat{1} := 1, \forall \theta \in \mathbf{S}^2, \forall k \in \mathbf{R})$;

(ii) $f(k)$, $\mathbf{R} \mapsto L^2(\mathbf{S}^2)$, *is the boundary value of an analytic function meromorphic in $\mathbf{C}^+$ such that $\lim_{|k| \to \infty} \|f(k) - \hat{1}\| = 0$; [Here $\| \cdot \|$ is the norm in $L^2(\mathbf{S}^2)$.]*

(iii) *in $\mathbf{C}^+$ this analytic function has a finite number of simple poles at the points $i\kappa_m$ and its residue at $i\kappa_m$ lies in the space $\mathcal{H}_m$;*

(iv) *on $\mathbf{R}$ f satisfies the equation*

$$f(-k) = \Omega(k)f(k). \tag{1}$$

One way of solving this problem is by Fourier transformation. Setting $g(\alpha) :=$
$G(\alpha)\hat{1}$, for $\alpha \in \mathbf{R}$, and defining

$$\eta(\alpha) := \frac{1}{2\pi} \int_{-\infty}^{\infty} dk \, e^{-ik\alpha}[\zeta(k) - \hat{1}],$$

we obtain the Fourier transform of equation (1):

$$\eta(\alpha) = Q\eta(-\alpha) + g(\alpha) + \int_{-\infty}^{\infty} d\beta \, G(\alpha + \beta)\eta(\beta).$$

Split the equation into two parts, for $\alpha > 0$ and $\alpha < 0$, and use the fact that
ζ has simple poles in the upper half plane corresponding to the bound states,
with residues that lie in known subspaces. Then for $\alpha \in \mathbf{R}_+$, one obtains the two
equations

$$\eta = g + [(Q + \mathcal{H})\tilde{y}]p + \mathcal{G}\eta \tag{2}$$

$$Q\eta = -g^\# + [(\mathbf{1} - \mathcal{G}^\#)\tilde{y}]p - \mathcal{H}^\#\eta, \tag{3}$$

where

$$y_m^b(\alpha, \theta, x) := Y_{\kappa_m}^b(-\theta)e^{\kappa_m(\theta \cdot x - \alpha)},$$

$$p^{mb}(x) := \frac{1}{2\kappa} \sum_a d_{ba}^{\kappa_m} u_{\kappa_m}^a(x).$$

Here the functions $u_{\kappa_m}^a(x)$ are the eigenfunctions of the Schrödinger equation,
and the $d_{ba}^{\kappa_m}$ are constants. For $\alpha < 0$ we have

$$\eta(\alpha) = \sum_{m,b} p^{mb} y_m^b(-\alpha).$$

Taking the inner product of equation (3) with a set of linearly independent
eigenfunctions $z^{[n]}$ of $\mathcal{G}^\#$ spanning the eigenspace with the eigenvalue -1 we
obtain (in matrix notation)

$$p = s^{-1}c,$$

in which

$$s_{n,mb} := \langle z^{[n]}, y_m^b \rangle_+, \qquad c^{[n]} := \langle z^{[n]}, g^\# \rangle_+$$

and $\langle \cdot, \cdot \rangle_+$ is the inner product on $L^2(\mathbf{R}_+ \times \mathbf{S}^2)$. Insertion in (2) then leads to the
generalized Marchenko equation: (Here $\tilde{y}$ denotes the transpose of the column
matrix y.)

$$\eta = g + (Q + \mathcal{H})\tilde{y}s^{-1}c + \mathcal{G}\eta. \tag{4}$$

The idea now is to use this equation alone, instead of both (2) and (3), to solve
the Riemann-Hilbert problem. Theorem 2.3.11 of [1] gives the following result.

Lemma 2   *The Riemann-Hilbert problem $H_\sigma^1(S_x)$ has a unique solution if
and only if $\mathcal{G}$ does not have the eigenvalue 1, the operator $\mathcal{G}^\#$ has the eigenvalue -1 with an n_σ-dimensional eigenspace, and the matrix s is invertible. This
solution is obtained from the solution of the generalized Marchenko equation (4)
by Fourier transformation.*

The following results, however, were not known at the time I wrote [1]; they are contained in Lemmas 4.3, 4.4, and 4.6 of [2]. (The proof of Lemma 4.3 in [2] was corrected in [3].)

Lemma 3 *If S is admissible with a potential $V \in W$ and $k = 0$ is not an exceptional point, then the matrix s is invertible and $\|\mathcal{G}\| < 1$. If, furthermore, V causes N bound states of negative energy (counting their multiplicities) then $\dim \operatorname{nul}(1 + \mathcal{G}^\#) = \dim \operatorname{nul}(1 - \mathcal{G}^\#) = N$.*

Note that the last part of this lemma constitutes a new connection between the number of bound states and the scattering data, in addition to the generalized Levinson theorem. It implies, among other things, that a necessary and sufficient condition for a potential to have no bound states is that $\|\mathcal{G}^\#\| < 1$.

The final result, giving a necessary and sufficient condition for a function $\mathbf{R} \times \mathbf{S}^2 \times \mathbf{S}^2$ to be admissible as a scattering amplitude, is the following theorem, which restates Theorem 5.1 of [2] in more complete form. (The "miracle" condition was inadvertently omitted in the statement of Theorem 5.1.)

Theorem B *Let $V \in W$ be a given potential in the Schrödinger equation with the following bound-state properties:*
(*) it causes N bound states (counting their multiplicity) of negative energies $-\kappa_m^2$ with eigenfunctions $p^{mb}(x)$ and characters $Y_{\kappa_m}^b(\theta)$, and $k = 0$ is not an exceptional point.
Let S be the corresponding S matrix; define S_x as before and the operators $\mathcal{G}$ and $\mathcal{G}^\#$ in terms of $A_x = \frac{2\pi i}{k}(1 - S_x)$ (so that they depend parametrically on $x \in \mathbf{R}^3$). Then the following six conditions hold:
(i) $S \in \mathcal{S}$;
(ii) $\operatorname{ind}_L S = N$ (generalized Levinson theorem);
(iii) S satisfies item (viii) of our previous definition (forward analyticity);
(iv) if $N = 0$ then $\|\mathcal{G}^\#\| < 1$; if $N > 0$ then $\mathcal{G}^\#$ has the eigenvalues ± 1 and each of the two corresponding eigenspaces is N-dimensional;
(v) the generalized Marchenko equation:

$$\eta = \mathbf{g} + (Q + \mathcal{H})\mathfrak{P} + \mathcal{G}\eta, \tag{5}$$

where $\mathfrak{P}$ is given by

$$\mathfrak{P}(\alpha, \theta) := \sum_{m,b} y_m^b(\alpha, \theta) p^{mb},$$

has a unique solution and the p^{mb} are the unique solution of the set of linear algebraic equations

$$c^{[n]} = \sum_{m,b} s_{n,mb} p^{mb}; \tag{6}$$

here $s_{n,mb}$, y_m^b, and $c^{[n]}$ are as defined earlier, and the $z^{[n]}$ form a basis in the eigenspace of $\mathcal{G}^\#$ at the eigenvalue -1; in other words, the operator $\mathcal{G}$ does not *have the eigenvalue 1 (so that $\|\mathcal{G}\| < 1$) and the matrix s is invertible; moreover, the potential V has the representation*

$$V(x) = -2\theta \cdot \nabla \left[\eta(\alpha = 0+, \theta, x) - \sum_{m,b} y_m^b(-\theta \cdot x, \theta, x)p^{mb}(x) \right] \qquad (7)$$

for all $\theta \in \mathbf{S}^2$;

(vi) the Jost function with all the required properties exists (see[1]).

Conversely, let S be given and let S_x be defined as before in terms of this function S. If S_x satisfies conditions (i) to (v) for almost all $x \in \mathbf{R}^3$ and the right-hand side of (7) is idependent of θ (this is the "miracle"), then the function ψ defined by

$$\psi(k, \theta, x) = e^{ik\theta \cdot x} + \int_{\theta \cdot x}^{\infty} e^{ik\alpha}\eta(\alpha - \theta \cdot x, \theta, x) + \sum_{m,b} \frac{p^{mb}(x)Y_{\kappa_m}^b(-\theta)}{i(k - i\kappa_m)} e^{i(k - i\kappa_m)\theta \cdot x}$$

in terms of the unique solution η of (5) and p^{mb} of (6) satisfies the Schrödinger equation with the potential given by (7) and the bound-state properties (*). Moreover, ψ satisfies the scattering boundary condition and the function $A = \frac{2\pi i}{k}(\mathbf{1} - S)$ is the corresponding scattering amplitude.

References

1. Newton, R. G.: *Inverse Schrödinger Scattering in Three Dimensions*, Springer Verlag, New York 1989.
2. Newton, R.G.:, Factorizations of the S matrix, J. Math. Phys. **31** (1990) 2414–2424.
3. Newton, R.G.: Uniqueness in some quasi-Goursat problems in 3+1 dimensions and the inverse scattering problem, J. Math. Phys. **32** (1991) 3130–3134.

Inverse Spectral Problems in Riemannian Geometry

Peter A. Perry

Department of Mathematics, University of Kentucky, Lexington, Kentucky 40506-0027, U.S.A.

1 Introduction

Over twenty years ago, Marc Kac posed what is arguably one of the simplest inverse problems in pure mathematics: "Can one hear the shape of a drum?" [19]. Mathematically, the question is formulated as follows. Let Ω be a simply connected, plane domain (the drumhead) bounded by a smooth curve γ, and consider the wave equation on Ω with Dirichlet boundary condition on γ (the drumhead is clamped at the boundary):

$$\Delta u(x,t) = \frac{1}{c^2} u_{tt}(x,t) \text{ in } \Omega,$$

$$u(x,t) = 0 \qquad \text{on } \gamma.$$

The function $u(x,t)$ is the displacement of the drumhead, as it vibrates, at position x and time t. Looking for solutions of the form $u(x,t) = \operatorname{Re} e^{i\omega t} v(x)$ (normal modes) leads to an eigenvalue problem for the Dirichlet Laplacian on Ω:

$$\Delta v(x) + \lambda v(x) = 0 \text{ in } \Omega$$
$$v(x) = 0 \text{ on } \gamma \tag{1}$$

where $\lambda = \omega^2/c^2$. We write the infinite sequence of Dirichlet eigenvalues for this problem as $\{\lambda_n(\Omega)\}_{n=1}^{\infty}$, or simply $\{\lambda_n\}_{n=1}^{\infty}$ if the choice of domain Ω is clear in context. Kac's question means the following: is it possible to distinguish "drums" Ω_1 and Ω_2 with distinct (modulo isometries) bounding curves γ_1 and γ_2, simply by "hearing" all of the eigenvalues of the Dirichlet Laplacian?

Another way of asking the question is this. What is the geometric content of the eigenvalues of the Laplacian? Is there sufficient geometric information to determine the bounding curve γ uniquely? In what follows we will call two domains *isospectral* if all of their Dirichlet eigenvalues are the same. We refer to the set of all domains (modulo rigid motions in the plane) with the same Dirichlet eigenvalues as a given domain Ω as the *isospectral set of* Ω. We would like to characterize the isospectral set of a given domain.

Some surprising and interesting results are obtained by considering the heat equation on Ω with Dirichlet boundary conditions, which gives rise to the same boundary value problem as before. The heat equation is

$$\Delta u(x,t) = u_t(x,t) \quad \text{in } \Omega$$
$$u(x,t) = 0 \qquad \text{on } \gamma$$
$$u(x,0) = f(x)$$

where $u(x,t)$ is the temperature at point x and time t, and $f(x)$ is the initial temperature distribution. This evolution equation has the formal solution

$$u(x,t) = (e^{t\Delta} f)(x)$$

where the operator $e^{t\Delta}$ can be calculated using the spectral resolution of Δ. Indeed, if $\phi_j(x)$ is the normalized eigenfunction of the boundary value problem (1) with eigenvalue λ_j, the operator $e^{t\Delta}$ has integral kernel $K(t,x,y)$ (the "heat kernel") given by

$$K(t,x,y) = \sum_{j=1}^{\infty} e^{-t\lambda_j} \phi_j(x) \phi_j(y). \tag{2}$$

The trace of $K(t,x,y)$ is actually a spectral invariant: by (2), we can compute

$$\int_{\Omega} K(x,x,t)\, dx = \sum_{j=1}^{\infty} e^{-t\lambda_j}. \tag{3}$$

Note that the function (3) determines the spectrum $\{\lambda_n\}_{n=1}^{\infty}$.

To analyze the geometric content of the spectrum, one calculates the same trace by a completely different method: one constructs the heat kernel by perturbation from the explicit heat kernel for the plane, and then one computes the trace explicitly. It turns out that the trace has a small-t asymptotic expansion

$$\int_{\Omega} K(x,x,t)\, dx \sim \frac{1}{4\pi t} \left(a_0 + a_{1/2} t^{1/2} + a_1 t + \cdots \right),$$

where

$$a_0 = \text{area}(\Omega)$$
$$a_1 = \text{length}(\gamma)$$

Although a strict derivation is a bit involved (see Chavel [12] and references therein), there is a simple heuristic argument, due to Mark Kac [19], which shows why a_0 and a_1 should give the area of Ω and the length of γ. The heat kernel in the plane is

$$K_0(x,y,t) = \frac{1}{4\pi t} \exp\left(|x-y|^2 / 4\pi t \right).$$

We expect that, for small times, $K(x, x, t) \simeq K_0(x, x, t)$ (a Brownian particle starting out in the interior doesn't "see" the boundary for a time of order $\sqrt{t}$) so that, to lowest order,

$$\int_\Omega K(x, x, t) \, dx \simeq \int_\Omega K_0(x, x, t) \, dx = \frac{1}{4\pi t} \text{area}(\Omega).$$

For times of order $\sqrt{t}$, boundary effects become important. We can approximate the heat kernel near the boundary locally by the "method of images." Locally, the boundary looks like the line $x_1 = 0$ in the x_1-x_2 plane; letting $x \mapsto x^*$ be the reflection $(x_1, x_2) \mapsto (-x_1, x_2)$, the kernel

$$K_\Omega(x, y, t) = K_0(x, y, t) - K_0(x, y^*, t)$$

vanishes on $x_1 = 0$. Hence,

$$K_\Omega(x, x, t) \simeq \frac{1 - e^{-2\delta^2/t}}{4\pi t}$$

where δ is the distance from x to the boundary. Writing the volume integral for the additional term as an integral over the boundary curve and the distance from the boundary,

$$\int_\gamma \int_0^\infty \frac{1}{4\pi t} e^{-2\delta^2/t} \, d\delta \, ds,$$

we have

$$\int_\Omega K(x, x, t) \, dx \simeq \frac{\text{area}(\Omega)}{4\pi t} - \frac{\text{length}(\gamma)}{4} \frac{1}{\sqrt{2\pi t}} + o\left(\frac{1}{\sqrt{t}}\right).$$

It follows that the isospectral set of a given "drum" Ω contains only drums with the same area and perimeter.

2 Can one hear the shape of a manifold?

Kac's early observations stimulated a flurry of work in which mathematicians generalized Kac's problem, studied the geometric content of the spectrum, and constructed counterexamples to Kac's conjecture as generalized to Riemannian manifolds. Guides to the extensive literature in spectral geometry include the books of of Bérard [2], Berger, Gauduchon, and Mazet [3], and Chavel [12].

Here we will briefly discuss the generalization of Kac's problem and some of the known results. A *Riemannian manifold* of dimension n is a smooth n-dimensional manifold M equipped with a Riemannian metric g which defines the length of tangent vectors and determines distances and angles on the manifold. The metric also determines the Riemann curvature tensor of M. In two dimensions, the Riemann curvature tensor is in turn determined by the scalar curvature, and in three dimensions it is completely determined by the Ricci curvature tensor. If M is compact, the associated Laplacian has an infinite set of discrete eigenvalues $\{\lambda_n\}_n = 1^\infty$. What is the geometric content of the spectrum for a compact Riemannian manifold?

An early paper of Milnor [20] constructs a pair of 16-dimensional torii with the same spectrum. The torii T_1^n and T_2^n are quotients of $\mathbf{R}^n$ by lattices Γ_1 and Γ_2 of translations of $\mathbf{R}^n$. Since the two torii are isometric if and only if their lattices are congruent, it suffices to construct a pair of non-congruent 16-dimensional lattices whose associated torii have the same spectrum.

To understand the analysis involved in Milnor's construction, consider the following simple 'trace formula' for a torus $T^n = \mathbf{R}^n/\Gamma$ which computes the trace of the heat kernel on a torus in terms of the lengths of the lattice vectors of Γ. Using the "method of images," it is easy to see that the heat kernel on the torus is given by the formula

$$K_\Gamma(x, y, t) = \sum_{\omega \in \Gamma} K_0(x + \omega, y, t),$$

where

$$K_0(x, y, t) = \frac{1}{(4\pi t)^{n/2}} e^{-|x-y|^2/4t}$$

is the heat kernel on $\mathbf{R}^n$. It follows that

$$\int_{T^n} K(x, x, t)\, dx = \frac{\text{vol}(T^n)}{4\pi t} \sum_{\omega \in \Gamma} e^{-|\omega|^2/4t}.$$

Milnor noted that there exist non-congruent lattices in 16 dimensions with the same set of "lengths" $\{|\omega| : \omega \in \Gamma\}$, first discovered by Witt [26]. Since the trace of the heat kernel determines the spectrum, and the heat trace is in turn determined by the lengths, it follows that the corresponding non-isometric torii have the same spectrum.

Later, other mathematicians found lower-dimensional examples of pairs of non-isometric Riemannian manifolds with the same spectrum. The construction of these examples involved Riemann surfaces with constant curvature and genus $g \geq 5$, and Riemann surfaces with variable curvature and genus $g \geq 3$. Among mathematicians contributing to this research were Vigneraas [25], Buser [4], Sunada [24], and Brooks-Tse [9]. Sunada [24] proved a simple trace formula that reduces the construction of such examples to an exercise in group theory and used it to construct isospectral surfaces. A very readable exposition of this work may be found in the paper of Brooks [6].

All of these examples showed that it was possible to construct pairs (or more generally, finite families) of Riemannian manifolds with the same spectrum. Later, Gordon and Wilson [18] and DeTurck and Gordon [14, 15, 16] constructed continuous families of isospectral manifolds in sufficiently high dimension ($n \geq 5$). Two major questions remained:

1) Can one show that the isospectral set of a given manifold is at most finite in "low" dimension?
2) Can one find counterexamples for Kac's original problem, i.e., can one construct isospectral, non-congruent planar domains?

As we will see, some partial progress has been made on the first question and decisive progress has been made on the second.

3 Some Positive Results

In 1986, Osgood, Phillips, and Sarnak proved one of the first major positive results on isospectral sets of surfaces and planar domains. Informally, a sequence of planar domains Ω_j converges in C^∞ sense to a limiting, non-degenerate domain Ω if the bounding curves γ_j converge in C^∞ sense modulo rigid motions of the plane and the limit curve encloses a nondegenerate bounded region. Similarly, a set of compact surfaces S_j converges in C^∞ sense to a limiting, non-degenerate surface S if the surfaces are all diffeomorphic to S and the metrics g_j on S_j, pulled back to S, converge in C^∞ sense to a positive definite metric on S. Osgood, Phillips, and Sarnak showed [22]:

Theorem 1. *(i) Let Ω_j be a sequence of isospectral planar domains. There is a subsequence which converges in C^∞ sense to a nondegenerate limiting surface.*

(ii) Let S_j be a sequence of isospectral compact surfaces. There is a subsequence of the S_j converging in C^∞ sense to a non-degenerate surface S.

Later, Osgood, Phillips, and Sarnak [23] generalized the first part of their result to non-simply connected planar domains.

If one can show that the isospectral set of a given domain or surface is *discrete*, the compactness result implies that the isospectral set of the given domain or surface is *finite*. Although no such discreteness results have yet been proved, there has been some progress studying the isospectral sets of Riemannian manifolds in higher dimensions. Brooks, Perry, and Yang [7] and Chang and Yang [10, 11] studied isospectral sets of *conformally equivalent* metrics on a *fixed* compact manifold in three dimensions: that is, metrics on a fixed underlying smooth manifold which define the same angles, but not necessarily the same distances. Later, Anderson [1] and Brooks, Perry, and Petersen [8] studied sets of manifolds with the same spectrum and were able to prove compactness subject to certain a *priori* assumptions on the geometry. Brooks, Perry, and Petersen proved:

Theorem 2. *Suppose $\{M_j\}$ is a sequence of isospectral manifolds such that either*

(i) All of the M_j have negative sectional curvatures, or
(ii) All of the M_j have Ricci curvature bounded below

Then there are finitely many diffeomorphism types, and there is a subsequence which converges in C^∞ to a nondegenerate limiting manifold.

It is expected that the curvature constraints can be removed eventually.

An important role is played in both of these theorems by two kinds of spectral invariants: *local invariants* expressible as integrals of the curvature tensor and its derivatives over the domain or manifold and *non-local invariants* such as the determinant of the Laplacian and the eigenvalues themselves.

Let us consider the case of a compact manifold M without boundary in 2 or 3 dimensions. Typically, the local invariants are the *heat invariants* arising in the small–t asymptotic expansion of the heat trace. If $K_M(x, y, t)$ is the fundamental solution kernel for the heat equation on a Riemannian manifold M, one has the asymptotic expansion

$$\int_M K(x, x, t)\, d\operatorname{vol}(x) \sim (4\pi t)^{-n/2} \sum_{j=0}^{\infty} a_j t^j$$

with the a_j spectral invariants. The term a_0 gives the volume of the Riemannian manifold, and the higher-order terms are integrals over M of the scalar ($n = 2$) or Ricci ($n = 3$) curvatures and their covariant derivatives up to order $j - 2$.

The heat invariants by themselves do not contain sufficient information to prove compactness. An easy way to see this is to note that there are deformations of constant curvature surfaces of genus 2 or higher in which the length of a closed geodesic tends to zero, thereby dividing the manifold into two parts in the limit as $\ell \to 0$ (see figure 1). Since the heat invariants in this case depend on the scalar curvature and its derivatives, a_0 is constant and the remaining invariants are all zero! Geometrically, the quantity which becomes uncontrolled in the limit $\ell \to 0$ is the *diameter* of the manifold, defined as the maximum of $\operatorname{dist}_M(x, y)$ where $\operatorname{dist}_M(\cdot, \cdot)$ is the distance with respect to the Riemannian metric on M.

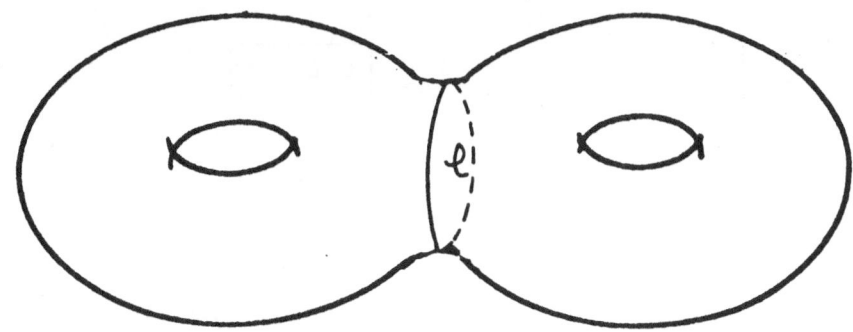

Fig. 1.

What are the nonlocal invariants which prevent degeneration of isospectral manifolds? Osgood, Phillips and Sarnak studied the determinant of the Laplacian, a renormalized product of the nonzero eigenvalues of the Laplacian which is manifestly a spectral invariant. Using results of Wolpert [27], they showed that, for degenerations of the type described above, $\det(-\Delta) \rightarrow 0$ as $\ell \rightarrow 0$, so that fixity of the determinant of the Laplacian rules out such degenerations for isospectral surfaces.

In three dimensions, nonlocal invariants are again used to control the diameter of isospectral manifolds. Cheng's comparison theorem [13] implies that, given a bound on Ricci curvature, one can show that the spectrum bounds the diameter from above (see [8]). Intuitively, if the diameter of the manifold is too large, one can construct test functions in a large number of geodesic balls and obtain a contradiction use the max-min principle. More generally, one expects blow-up of the diameter or other singularity formation to lead to concentration of eigenvalues forbidden by isospectrality.

4 Counterexamples Revisited

Perhaps the most remarkable recent development relating to Kac's problem is the construction, by Gordon, Webb, and Wolpert [17], of isospectral, non-isometric planar domains (figure 2). These domains are obtained from isospectral manifolds-with-boundary constructed using the Sunada technique [24]. One can actually prove isospectrality by the so-called *method of transplantation*: one can take an eigenfunction of the first domain Ω_1, restrict it to each of the triangles A–G, and rearrange the pieces in the second domain as shown. Since the new functions satisfy the Dirichlet boundary conditions and are continuous along each triangle boundary, they are eigenfunctions of the Laplacian on Ω_2. This shows that the Dirichlet spectrum of Ω_1 is contained in the Dirichlet spectrum of Ω_2. A similar transplantation argument shows the reverse inclusion.

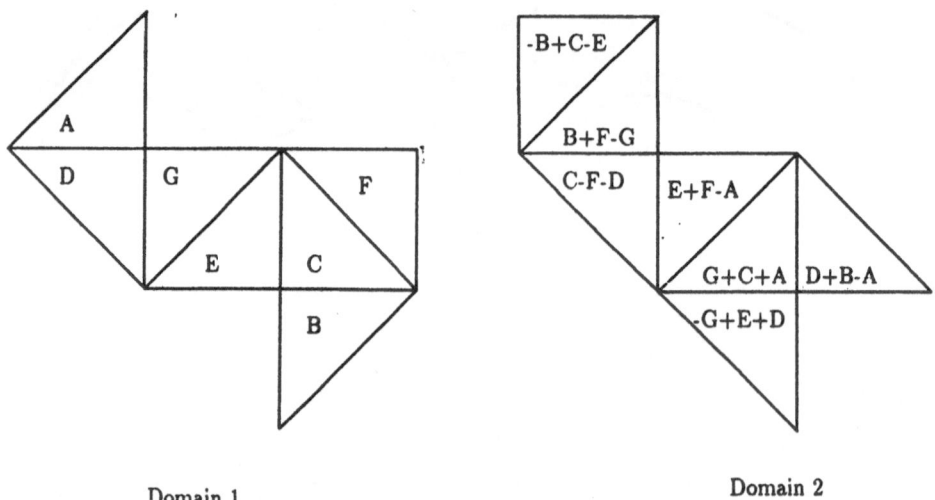

Domain 1   Domain 2

Fig. 2.

More recently, Buser, Conway, Doyle, and Semmler [5] have shown that one can construct pairs of drums which "sound alike" in the stronger sense that there are points $x_1 \in \Omega_1$ and $x_2 \in \Omega_2$ with the property that if Ω_1 is hit at x_1, it makes exactly the same sound as Ω_2 does when hit at x_2. That is, the solutions to the wave equations in Ω_i

$$\Delta u^i - u^i_{tt} = 0$$
$$u^i(x, 0) = 0$$
$$u^i_t(x, 0) = \delta(x - x_i)$$

with Dirichlet boundary conditions satisfy $u^1(x_1, t) = u^2(x_2, t)$ for all $t > 0$.

Thus, one may say, with apologies to Henry David Thoreau, that it is possible to march to the sound of a different drummer, and hear the same beat!

References

1. Anderson, M: Remarks on the compactness of isospectral sets in low dimension. Duke Math. J. **63** (1991) 699–711.

2. Bérard, P.: Spectral geometry : Direct and Inverse Problems. *Lect. Notes in Math.* **1207** (1986).

3. Berger, M., Gauduchon, P., Mazet, E.: Le Spectre d'une Variété Riemannienne. *Lect. Notes in Math.* **194** (1974).

4. Buser, P.: Isospectral Riemann surfaces. Ann. Inst. Fourier Grenoble **36** (1988) 167–192.

5. Buser, P., Conway, J., Doyle, P., Semmler, E.: Preprint, 1992.

6. Brooks, R.: Constructing isospectral manifolds. Amer. Math. Monthly **95** (1988) 823–839.

7. Brooks, R., Perry, P., Yang, P.: Isospectral sets of conformally equivalent metrics. Duke Math. J. **58** (1989) 131–150.

8. Brooks, R., Perry, P., Petersen P. V.: Compactness and finiteness theorems for isospectral manifolds. J. Reine angew. Math **426** (1992) 67–89.

9. Brooks, R., Tse, R.: Isospectral surfaces of small genus. Nagoya J. Math. **107** (1987) 13–24.

10. Chang, S.-Y. A., Yang, P.: Compactness of isospectral conformal metrics on S^3. Comment. Math. Helv. **64** (1989) 363–374.

11. Chang, S.-Y. A., Yang, P.: Isospectral conformal metrics on 3–manifolds. J. American Math. Soc. **3** (1990) 117–145.

12. Chavel, I.: *Eigenvalues in Riemannian Geometry* (Pure and Applied Mathematics, v. 115). New York: Academic Press, 1984.

13. Cheng, S. Y.: Eigenvalue comparison theorems and its geometric applications. Math. Zeits. **143** (1975) 289–297.

14. DeTurck, D., Gordon, C.: Isospectral Riemannian metrics and potentials. Bull. A. M. S. **17** (1987) 137–140.

15. DeTurck, D., Gordon, C.: Isospectral deformations, I. Riemannian structures on two-step nilspaces. Comm. Pure Appl. Math. **40** (1987) 367–387.

16. DeTurck, D., Gordon, C.: Isospectral deformations, II. Trace formulas, metrics, and potentials. Comm. Pure Appl. Math. **42**, (1989) 1067–1095.

17. Gordon, C., Webb, D., Wolpert, S.: You can't hear the shape of a drum. Preprint, 1992.
18. Gordon, C., Wilson, E.: Isospectral deformations of compact solvmanifolds. J. Diff. Geom. **19** (1984) 241–256.
19. Kac. M.: "Can one hear the shape of a drum?" Amer. Math. Monthly **73** (1966) 1–23.
20. Milnor, J.: Eigenvalues of the Laplace operator on certain manifolds. Proc. Nat. Acad. Sci. U. S. A. **51** (1964) 342.
21. Osgood, B., Phillips, R., Sarnak, P.: Extremals of determinants of Laplacians. J. Funct. Anal. **80** (1988) 148–211.
22. Osgood, B., Phillips, R., Sarnak, P.: Compact isospectral sets of surfaces. J. Funct. Anal. **80** (1988) 212–234.
23. Osgood, B., Phillips, R., Sarnak, P.: Moduli space, heights and isospectral sets of plane domains. Ann. of Math. **129** (1989) 293–362.
24. Sunada, T.: Riemannian coverings and isospectral manifolds. Ann. Math. **121** (1984) 169–186.
25. Vignéras, M. F.: Variétés Riemanniennes isospectrales et non isométriques. Ann. Math. **112** (1980) 21–32.
26. Witt, E.: Eine Identität zwischen Modulformen Zweiten Grades. Ab. Math. Sem. Hansischen Univ. **14** (1941) 323–337.
27. Wolpert, S.: Asymptotics of the spectrum and the Selberg zeta function on the space of Riemann surfaces. Commun. Math. Phys. **113** (1987) 283–325.

Local Results for a Two Dimensional Inverse Conductivity Problem

Jeffrey O. Powell

Department of Mathematics, University of Toledo, Toledo, Ohio 43606

1 Introduction

Suppose a region Ω in $\mathbb{R}^n$, $n \geq 2$, contains electrically conductive material, and that its conductivity coefficient is 2 on an unknown subdomain D and 1 on $\Omega \backslash D$. This study is concerned with finding D by injecting a current across the boundary $\partial\Omega$ then measuring the corresponding voltage on $\partial\Omega$. The problem was posed as a mine detection model in [5].

Let D be a bounded domain in the complex plane $\mathbb{C}$, and consider the refraction problem of determining a function u that satisfies

$$\text{div}[(1 + \chi(D))\nabla u] = F \quad \text{in } \mathbb{C}$$
$$u(\infty) = 0. \tag{1.1}$$

$\chi(D) = $ is the indicator function of D; $F(z) = \delta(z - a) - \delta(z - b)$, where δ is Dirac's delta function; and a, b are distinct elements of $\mathbb{C} \backslash \overline{D}$.

Alternatively, let $\Omega \subset \mathbb{R}^n$ be a bounded domain with $C^{2,\alpha}$ boundary and $\overline{D} \subset \Omega$. N will denote an exterior unit normal vector. Consider the problem

$$\text{div}[(1 + \chi(D))\nabla u] = 0 \quad \text{in } \Omega,$$
$$\frac{\partial u}{\partial N} = g \quad \text{on } \partial\Omega \tag{1.2}$$

where $g \not\equiv 0$, $g \in C^{1,\alpha}(\partial\Omega)$, and

$$\int_{\partial\Omega} g = 0.$$

With the notation

$$D^e = \mathbb{C} \backslash \overline{D}, \quad (\text{or } \Omega \backslash D), \quad D^i = D, \quad u^e = u|_{D^e}, \quad u^i = u|_{D^i},$$

(1.1) and (1.2) are equivalent to the following refraction conditions.

$$\Delta u^e = F \quad (\text{or } = 0), \quad \Delta u^i = 0 \quad \text{in } D^e, \ D^i \tag{1.3}$$
$$u^e = u^i, \quad \frac{\partial u^e}{\partial N} = 2\frac{\partial u^i}{\partial N} \quad \text{on } \partial D. \tag{1.4}$$

Existence and uniqueness of the solution u of (1.1) can be shown by using the conditions (1.3), (1.4) and jump relations on ∂D (see [7]) to derive an integral equation which can be solved to determine $v = u - u_1$, where $\Delta u_1 = F$ in $\mathbb{C}$; problem (1.2), with u normalized by

$$\int_{\partial_0 \Omega} u = 0,$$

where $\partial_0 \Omega$ is a compact subset of $\partial \Omega$, has a unique solution $u \in H^1(\Omega) \cap C^\alpha(\overline{\Omega})$ for some $\alpha > 0$. u^e, u^i, and their first derivatives are Hölder continuous in $\overline{D}^e \backslash \{a, b\}$ for (1.1) and $\overline{D}^e$ for (1.2), and $\overline{D}^i$, respectively, if ∂D is $C^{1,1}$; see [9]. The second derivatives of u^e are continuous up to $\partial \Omega$.

2 The inverse problem

Now suppose that D is unknown. Suppose B is a large disk with $a, b \notin \overline{B}$, and that f is a function given on ∂B. Then the inverse problem for (1.1) is to find D, given $u|_{\partial B} = f$. Note that this is equivalent to knowing one Dirichlet–Neumann pair on ∂B, since the problem $\Delta u^e = F$ outside B with $u|_{\partial B} = f$, $u^e(\infty) = 0$ can be solved uniquely.

The inverse problem in Ω is to find D when in addition to g, $u|_{\partial \Omega} = f$ is given.

There are a few global uniqueness results for this problem. Suppose that for some g,

$$\text{div}[(1 + \chi(D_j))\nabla u_j] = 0 \quad \text{in } \Omega, \, j = 1, 2,$$

$$u_1 = u_2, \quad \frac{\partial u_1}{\partial N} = \frac{\partial u_2}{\partial N} \quad \text{on } \partial \Omega.$$

Then $D_1 = D_2$ if the D_j are assumed in advance to be disks in the half–plane Ω; and in the case when the D_j are known to be convex polyhedrons whose distances from $\partial \Omega$ are greater than their diameters; see [6]. In [8], $D_1 = D_2$ was shown when the D_j are finite collections of disks in the half–plane; and for convex cylinders D_j.

3 The local problem

Suppose that D_0 is a known bounded domain and u_0 is the known solution to (1.1) or (1.2) with $D = D_0$. For (1.1), assume $a, b \notin \overline{B}$ for some disk B with $\overline{D}_0 \subset B$; and suppose that a function u^e is "close" on ∂B in some sense to $u_0^e|_{\partial B}$, and satisfies $\Delta u^e = F$ in $\mathbb{C} \backslash \overline{B}$ and $u^e(\infty) = 0$. For (1.2), suppose that a function u^e is close on $\partial \Omega$ to u_0^e, and $\partial u^e / \partial N = g = \partial u_0^e / \partial N$ on $\partial \Omega$. The local inverse problem is to find from this information a domain D "close" to D_0.

Uniqueness of solutions to the local inverse problem has been shown in [1] and [2] for the monotone case, that is, with the hypothesis $D \subset D_0$ or $D_0 \subset D$.

We give an existence–uniqueness result for the non–monotone local inverse problem.

4 The result

We assume that ∂D_0 is analytic and that D_0 is simply connected.

Let $z_0(t)$ be the conformal map of the unit disk to D_0 which satisfies

$$z_0(0) = 0, \quad z_0'(0) > 0,$$

Let u_0 be the solution to (1.1) with $D = D_0$ and a, b outside the closure of some disk B which contains $\overline{D_0}$. Or let u_0 be the solution to (1.2) with $D = D_0$ and g prescribed on $\partial\Omega$ as follows. Let z_1, z_2 be on $\partial\Omega$, with $z_1 \neq z_2$ and Ω strictly convex near the z_j. Parametrized counterclockwise, z on $\partial\Omega$ is $z = z(\tau)$, $\tau \in [0,1]$, with $z_1 = z(0) = z(1)$ and $z_2 = z(\sigma)$ for some σ in $(0,1)$. Choose g so that

$$g > 0 \text{ on } (0,\sigma), \quad g < 0 \text{ on } (\sigma,1)$$
$$g'(0) > 0, \quad g'(\sigma) < 0. \tag{4.1}$$

Suppose data u^e satisfies

$$\Delta u^e = F \text{ on } \mathbb{C}\backslash\overline{B}, \quad u^e(\infty) = 0,$$

or

$$\frac{\partial u^e}{\partial N} = g \text{ on } \partial\Omega$$

and has a harmonic continuation known to lie within some prescribed distance u_0^e in, say, $C^{2+\lambda}$ of a neighborhood of $B\backslash D_0$ or $\Omega\backslash D_0$; it can be shown from (1.4) and the analyticity of ∂D_0 that u_0^e has a harmonic continuation across ∂D_0.

Theorem 1. *There are ϵ_1, $\epsilon_2 > 0$ such that if (a) the measurement u^e satisfies*

$$|u^e - u_0^e|_{2+\lambda}(\partial B) < \epsilon_1$$

or

$$|u^e - u_0^e|_{2+\lambda}(\partial\Omega) < \epsilon_1$$

and (b) ω is from the set $|\omega|_{1+\lambda}(|t| \leq 1) < \epsilon_2$, ω is analytic in $|t| < 1$, and

$$\omega(0) = 0, \quad \text{Im}\,\omega'(0) = 0, \tag{4.2}$$

then for at most one pair u^i, ω, (i) $z = z_0 + \omega$ satisfies $z'(0) > 0$, (ii) u^i is harmonic in $D = \{z \in \mathbb{C} : z(t) = z_0(t) + \omega(t), |t| < 1\}$, and (iii) D in (ii), u^i, and u^e satisfy (1.3), (1.4). In addition, the existence of u^i, ω is assured by fulfillment of the orthogonality condition (5.6), given in section 5.

Uniqueness and solvability of both problems depend on the index of the gradient u_{0z}^e over ∂D_0, the index of a continuous vector field $V = a + ib$ over a closed, piecewise smooth contour Γ being defined as

$$\kappa(V;\Gamma) = \frac{1}{2\pi}[arg(a - ib)]_\Gamma,$$

provided $V \neq 0$ on Γ. Index theory was recently used for an inverse problem in [3].

5 Sketch of proof

If u^e, u^i satisfy (1.3), (1.4) for some D with $\overline{D} \subset B$, B a disk, $a, b \notin \overline{B}$, then (1.4) can be used to show that the conjugate period of u^e over ∂D is zero; so we define

$$U^e = u^e + iv^e \text{ in } D^e \cap B \ (D^e \text{ for } (1.2)), \quad U^i = u^i + iv^i \text{ in } D^i,$$

v^e, v^i being the conjugate harmonic functions, made unique by fixing values at some point z.

Lemma 2. $4U^i(z) = 3U^e(z) + \overline{U^e(z)} - 2i\xi$ for $z \in D$. ξ is a real constant.

Proof. The Cauchy–Riemann conditions and (1.4) imply $\partial v^e / \partial T - 2\partial v^i / \partial T = 0$ on ∂D, where T is the unit tangent $\pi/2$ radians clockwise from N. Integration then gives $v^e = 2v^i + \xi$ on ∂D. Hence, on ∂D, $U^e + \overline{U^e} = U^i + \overline{U^i}$ and $(U^e - \overline{U^e}) - 2(U^i - \overline{U^i}) = 2i\xi$; and substitution for $\overline{U^i}$ proves the lemma; see [4]. $\square$

If $z(t)$ is the conformal map of the unit disk onto D which satisfies

$$z(0) = 0, \quad z'(0) > 0,$$

then this boundary condition becomes

$$4U^i(z(t)) = 3U^e(z(t)) + \overline{U^e(z(t))} - 2i\xi, \quad |t| = 1.$$

Our data has a harmonic conjugate function; the inverse problem is then equivalent to finding a function ϕ^+ analytic in $|t| < 1$ and a conformal map $z(t)$ such that

$$\phi^+(t) = 3U^e(z(t)) + \overline{U^e(z(t))} - 2i\xi, \quad |t| = 1. \tag{5.1}$$

We are searching for $z(t) = z_0(t) + \omega(t)$, so (5.1) becomes the problem of finding ϕ^+, ω analytic in $|t| < 1$ such that

$$\phi^+(t) = 3U^e(z_0(t) + \omega(t)) + \overline{U^e(z_0(t) + \omega(t))} - 2i\xi, |t| = 1. \tag{5.2}$$

From condition (5.2), a linearized problem can be derived. With functions defined in $|t| < 1$ referred to ϕ on the left side, we obtain the following boundary value problem of a type considered by [10]; see also [4]. Find ω_* analytic in $|t| > 1$ and ϕ analytic in $|t| < 1$ such that on $|t| = 1$

$$\phi(t) = \overline{u^e_{0z}(z_0(t))}\omega_*(t) + [u^e_{0z}(z_0(t))z'_0(t)t / \overline{z'_0(t)t}]\overline{\omega_*(t)} + B\omega(t). \tag{5.3}$$

Here, $\omega_*(t) = \overline{\omega(1/\overline{t})}$ and $B\omega = 3B_1\omega + \overline{B_1\omega}$, where $B_1\omega = U^e(z_0 + \omega) - U^e_0(z_0) - u^e_{0z}(z_0)\omega$. For ϵ_1, ϵ_2 sufficiently small, $B\omega$ is a contraction operator.

Denote the conjugate linear part of (5.3) by $A(\phi, \omega)$. Then (5.3) defines an operator equation $A(\phi, \omega) = B\omega$ from the space $Y \times Z$, where Y is the set of functions analytic in $|t| < 1$ which are in $C^{1+\lambda}(|t| \leq 1)$, and Z is the space of ω defined in the hypothesis of Theorem 1, to the space $C^{1+\lambda}(|t| = 1)$. For invertibility, A must have the trivial kernel in $Y \times Z$. We employ the method in [10].

Using (5.3) and its complex conjugate to eliminate ω_* and $\overline{\omega}_*$, we obtain on $|t| = 1$

$$\phi(t) = \overline{[u^e_{0z}(z_0(t))z'_0(t)t/u^e_{0z}(z_0(t))z'_0(t)t]}\overline{\phi(t)} + F(t)$$

(5.4)

$$F(t) = B\omega(t) - \overline{[u^e_{0z}(z_0(t))z'_0(t)t/u^e_{0z}(z_0(t))z'_0(t)t]}\overline{B\omega(t)}$$

Moving $\phi(t)$ to the right side of (5.3) and dividing by the coefficient of $\overline{\omega}_*$, we get

$$\overline{\omega_*(t)} = -[z'_0(t)t/\overline{z'_0(t)t}]\omega_*(t) + G(t) \quad on \ |t| = 1$$

(5.5)

$$G(t) = [\phi(t) - B\omega(t)]z'_0(t)t/\overline{u^e_{0z}(z_0(t))z'_0(t)t}.$$

The coefficients of $\overline{\phi}$, ω_* must be nonzero on $|t| = 1$ and their indices on $|t| = 1$ must be determined in order to solve these Riemann–Hilbert problems; see [11]. If $\kappa(u^e_{0z}; \partial D_0) = 0$, then the indices of (5.4), (5.5) are $\lambda = -2$, $\mu = 2$, respectively. Set $B\omega = 0$. Then according to [11], (5.4) has only the trivial solution $\phi = 0$, which substituted in (5.5) gives $G(t) = 0$. Thus, from [11], homogeneous (5.5) has three linearly independent solutions depending on three real constants. Two of these constants must be zero because $\omega(0) = 0$ and the third may be eliminated by the condition $\text{Im}\,\omega' = 0$, so that (5.5) has only the trivial solution . So the operator A is invertible.

Existence of the solution to the nonhomogeneous problem depends on the solvability of (5.4), which has index $\lambda = -2$. From [11], there is a solution if

$$0 = \int_{|t|=1} \frac{F(t)}{X^+(t)}\,dt,$$

(5.6)

where $X^+(t)$ is the limit as t tends to $|t| = 1$ from inside the unit disk of the fundamental function for the Hilbert problem; see [11].

Details of proofs can be found in [12].

6 Index, nonvanishing of the coefficients

Since ∂D_0 is analytic, u^i_{0z}, u^e_{0z} have analytic continuations to a neighborhood of ∂D_0. Hence, any zeroes on ∂D_0 of u^i_{0z} or u^e_{0z} are isolated. From (1.4),

$$u^i_{0z}(z) = 0 \quad for \ z \in \partial D_0 \ \text{iff} \ u^e_{0z}(z) = 0.$$

(6.1)

Denote by $z_1, \ldots, z_m$ the zeroes of u^i_{0z} on ∂D_0. (6.1) implies $u^e_{0z} \neq 0$ on $\partial D_0 \backslash \{z_1, \ldots, z_m\}$. Let L_ϵ be the Jordan curve formed by arcs of circles $\partial B(z_j; \epsilon)$ centered at z_j which are contained in $\mathbb{C} \backslash D_0$ and

$$\Gamma_\epsilon = \partial D_0 \backslash [\bigcup_{j=1}^{l} B(z_j; \epsilon)].$$

The following lemma and its proof are in [3].

Lemma 3. *For $\epsilon > 0$ sufficiently small, the vector field ∇u_0^e on L_ϵ is homotopic to the vector field ∇u_0^i on L_ϵ.*

This means that $\kappa(u_{0z}^i; L_\epsilon) = \kappa(u_{0z}^e; L_\epsilon)$. Enclose a, b, and L_ϵ and all zeroes and poles of u_{0z}^e except at ∞ in a large disk B_1, and enclose these finitely many isolated poles and zeroes with small disks B_k. If R is the region enclosed by ∂B_1 and L_ϵ and exterior to sufficiently small B_k, $k > 1$, then the argument principle applied to R will show that $\kappa(u_{0z}^e; L_\epsilon) = 0$. Hence, $\kappa(u_{0z}^i; L_\epsilon) = 0$ implies there are no zeroes on ∂D_0. Letting ϵ tend to zero gives the result. For (1.2), determine the index of u_{0z}^e on a slight perturbation of $\partial \Omega$ near the two zeroes of g from the special choice of g; then use the argument principle.

References

1. Alessandrini, G.: Remark on a paper by Bellout and Friedman. Problemi non ben posti ed inversi, **40** (1988) 1–7.
2. Bellout, H., Friedman, A.:Identification problems in potential theory. Arch. Rat. Mech. Anal. **101** (1988) 143–160.
3. Bellout, H., Friedman, A., Isakov, V.: Stability for an inverse problem in potential theory. To appear in Trans. Amer. Math. Soc.
4. Cherednichenko, V.: A problem in the conjugation of harmonic functions and its inverse. Differentsial'nya Uravneniya **18** (1982) 682–699.
5. Friedman, A.:Detection of mines by electronic measurement. SIAM J. Appl. Math. **47** (1987) 201–212.
6. Friedman, A., Isakov, V.: On the uniqueness in the inverse conductivity problem with one measurement. Indiana Univ. Math. J. **38** (1989) 563–579.
7. Isakov, V.:*Inverse Source Problems*, A.M.S. Mathematical Surveys and Monographs 34, 1990.
8. Isakov, V., Powell, J.: On the inverse conductivity problem with one measurement. Inverse Problems **6** (1990) 311–318.
9. Ladyzhenskaya,O., Ural'tseva, N.: *Linear and Quasilinear Elliptic Equations.* Academic Press, New York and London, 1969.
10. Mikhailov, L.: *A New Class of Singular Integral Equations and its Application to Differential Equations with Singular Coefficients.* Wolters–Noordhoff, Groningen, 1970.
11. Muskhelishvili, N.: *Singular Integral Equations.* Noordhoff, Groningen, 1953.
12. Powell, J.: On a small perturbation in the two dimensional inverse conductivity problem. To appear in J. Math. Anal. Appl.

Inverse Scattering at Fixed Energy for Exponentially Decreasing Potentials

A. G. Ramm[1] and P. Stefanov[2]

[1] Mathematics Department, Kansas State University, Manhattan, Kansas 66506, USA
[2] Institute of Mathematics, Bulgarian Academy of Sciences, 1090 Sofia, Bulgaria

1 Introduction

Consider the stationary Schrödinger equation

$$(-\Delta + q(x) - k^2)u = 0 \qquad \text{in } \mathbb{R}^3 \tag{1}$$

with a short-range potential $q(x)$ and let $A(\omega, \theta, k)$ be the related quantum-mechanical scattering amplitude. We are interested in the following inverse problem:

(IP) *Given $A(\omega, \theta, k)$ for a fixed $k > 0$, find $q(x)$.*

It was proven by A.G.Ramm [R1] (see also R.G.Novikov [N]) that IP has a unique solution for the class of compactly supported potentials. The same is true for sufficiently small exponentially decreasing potentials [H-N]. On the other hand, it is known (see [C-S]) that if $q(x)$ decays slowly, then IP may have many solutions. It is interesting to find out the exact decay rate which implies the uniqueness of the solution to IP. While this problem is still open, we prove the following result.

Theorem 1. *Let $q_j = \bar{q}_j \in L^\infty(\mathbb{R}^3)$, $|q_j(x)| \le C_\alpha e^{-\alpha|x|}$ for any $\alpha > 0$, $j = 1, 2$ and denote by $A_j(\omega, \theta, k)$ the corresponding scattering amplitude. Then if $A_1(\omega, \theta, k) = A_2(\omega, \theta, k)$ for some $k > 0$ and all $\omega \in S^2$, $\theta \in S^2$, we have $q_1 = q_2$.*

2 Sketch of the proof

The proof consists of several steps. Assume in what follows k fixed. Without loss of generality we may assume $k = 1$. We will denote below the scattering amplitude for $k = 1$ by $A(\omega, \theta)$.

Step 1 Assume $A_1(\omega, \theta) = A_2(\omega, \theta)$. Then we show that the following orthogonality relation holds:

$$\int ((q_1(x) - q_2(x))u_1(x, \omega)u_2(x, \theta)\, dx = 0 \qquad \text{for all } \omega,\, \theta, \tag{2}$$

where u_j is the scattering solution to (1) which can be defined as the solution of the Lippmann-Schwinger equation

$$u = u_0 - G_0 q u,$$

where $u_0(x, \theta) = e^{ix \cdot \theta}$,

$$[G_0 f](x) = \frac{1}{4\pi} \int \frac{e^{i|x-y|}}{|x-y|} f(y)\, dy.$$

The kernel of G_0 is the outgoing Green's function for the Helmholtz operator $\Delta + 1$. The equality (2) follows directly from the following identity [R2], [R3], [S].

$$A_1(\omega, \theta, k) - A_2(\omega, \theta, k) = -\frac{1}{4\pi} \int ((q_1(x) - q_2(x))u_1(x, \theta, k)u_2(x, -\omega, k)\, dx.$$

Step 2 Denote by X_a the Hilbert space $L^2(\mathbb{R}^3; e^{-2a|x|}dx)$. Next lemma is the key point of the proof of Theorem 1.

Lemma 2. *For any solution ψ of the equation $(-\Delta + q - 1)\psi = 0$ belonging to X_β, there is a sequence ν_j, $j = 1, 2, \ldots$ in $L^2(S^2)$, such that for $\alpha > \beta$*

$$\|\psi - \int_{S^2} u(\cdot, \omega)\nu_j(\omega)\, d\omega\|_{X_\alpha} \to 0, \qquad \text{as } j \to \infty. \tag{3}$$

In other words, we can approximate any exponentially increasing solution to (1) by a linear combination of scattering solutions $u(x, \omega)$, $\omega \in S^2$ in a suitably chosen space X_α.

Proof of Lemma 2 Denote $\psi_0 = (I + G_0 q)\psi$. Using the well-known fact that G_0 is a bounded operator from $L^2_{-\delta}$ to L^2_δ, $\delta > 1/2$, [A] we see that $\psi_0 \in X_\beta$. First we find a sequence ν_j in $L^2(S^2)$, such that

$$\|\psi_0 - \int_{S^2} u_0(\cdot, \omega)\nu_j(\omega)\, d\omega\|_{X_\beta} \to 0, \qquad \text{as } j \to \infty. \tag{4}$$

Then we will show that (3) holds with the same sequence ν_j. Obviously, $(\Delta + 1)\psi_0 = 0$ in $\mathbb{R}^3$. Therefore, we have

$$\psi_0(r\theta) = \sum_{n,m} a_{nm} j_n(r) Y_n^m(\theta), \qquad \sum_{n,m} := \sum_{n=0}^{\infty} \sum_{m=-n}^{n},$$

with some a_{nm}, where $x = r\theta$, $r > 0$, $\theta \in S^2$. Here $Y_n^m(\theta)$ are the spherical harmonics on S^2, while $j_n(r) = \sqrt{\pi/(2r)} J_{n+\frac{1}{2}}(r)$ are the spherical Bessel functions. Since $\psi_0 \in X_\beta$, we obtain

$$\sum_{n,m} |a_{nm}|^2 b_n^2 = \|\psi_0\|_{X_\beta}^2 < \infty,$$

where $b_n^2 = \int_0^\infty |j_n(r)|^2 r^2 e^{-2\beta r} dr < \infty$. Using the expansion

$$e^{ix \cdot \omega} = \sum_{n,m} 4\pi i^n j_n(r) Y_n^m(\theta) \overline{Y_n^m(\omega)}, \qquad x = r\theta,$$

we can write the norm in (4) in the following form

$$\left\| \psi_0 - \int_{S^2} u_0(\,\cdot\,, \omega) \nu_j(\omega)\, d\omega \right\|_{X_\beta}^2 = \sum_{n,m} |a_{nm} - 4\pi i^n \nu_{jnm}|^2 b_n^2,$$

where $\nu_{jnm} = \int_{S^2} \nu_j(\omega) \overline{Y_n^m(\omega)}\, d\omega$. Therefore, it is convenient to set

$$\nu_{jnm} = (4\pi i^n)^{-1} a_{nm}, \quad \text{for } -n \leq m \leq n,\ n \leq j,$$
$$\nu_{jnm} = 0, \qquad\qquad\qquad \text{otherwise},$$

which implies $\nu_j(\omega) = \sum_{n=0}^{j} \sum_{m=-n}^{n} (4\pi i^n)^{-1} a_{nm} Y_n^m(\omega)$. Then (4) holds because

$$\left\| \psi_0 - \int_{S^2} u_0(\,\cdot\,, \omega) \nu_j(\omega)\, d\omega \right\|_{X_\beta}^2 = \sum_{n=j+1}^{\infty} \sum_{m=-n}^{n} |a_{nm}|^2 b_n^2 \to 0,$$

as $j \to \infty$. Now, let us prove that (4) implies (3). The assumptions imposed on q guarantee that $(I + G_0 q)^{-1}$ exists as a bounded operator in $L_{-\delta}^2$, $\delta > 1/2$. Then we can show that

$$\psi = \psi_0 - (I + G_0 q)^{-1} G_0 q \psi_0. \tag{5}$$

Similarly for the solution to the Lippmann-Schwinger equation we have

$$u = u_0 - (I + G_0 q)^{-1} G_0 q u_0. \tag{6}$$

Using (5), (6) and (4) we get (3).

Step 3 Lemma 2 enables us to replace the solutions u_1, u_2 in (2) with any pair of exponentially increasing solutions to (1), so we have

$$\int \big((q_1(x) - q_2(x)) \big) \psi_1(x) \psi_2(x)\, dx = 0 \tag{7}$$

for any $\psi_j \in X_\beta$ with some $\beta > 0$ solutions to (1) with $q = q_j$ respectively. Now we will make use of the special solutions to (1) constructed in [S-U] and [R4]. Namely, there exists $\mu > 0$ such that for any $\zeta \in \mathbb{C}^3$ with the properties $\zeta^2 = 1$, $|\zeta| > \mu > 0$ there exists a solution $\psi(x, \zeta)$ to the equation $(-\Delta_x + q - 1)\psi = 0$ of the form $\psi = e^{ix \cdot \zeta}(1 + \rho(x, \zeta))$ such that $\rho(\,\cdot\,, \zeta) \in L_{-\delta}^2$ with some $\delta > \frac{1}{2}$ and

$\|\rho(\,\cdot\,,\zeta)\|_{L^2_{-\delta}} < C/|\zeta|$. Therefore we see that (7) holds with ψ_1, ψ_2 the special solutions, i.e.

$$\int \big((q_1(x) - q_2(x))\big)\psi_1(x,\zeta_1)\psi_2(x,\zeta_2)\,dx = 0. \tag{8}$$

for $\zeta_1^2 = \zeta_2^2 = 1$, $|\mathrm{Im}\zeta_j| > \mu$. Now we can complete the proof by following a standard argument (see [R5], [S-U]). Let us choose two sequences $\zeta_1(n)$, $\zeta_2(n)$, such that $\zeta_1^2(n) = \zeta_2^2(n) = 1$, $\zeta_1(n) + \zeta_2(n) = -p$ and $|\zeta_1(n)| = |\zeta_2(n)| \to \infty$, as $n \to \infty$, $p \in \mathrm{I\!R}^3$ being fixed. Then (8) with this choice of ζ_1, ζ_2 gives as $n \to \infty$

$$\int e^{-ix\cdot p}\big((q_1(x) - q_2(x))\big)\,dx = 0,$$

which implies $\hat{q}_1(p) = \hat{q}_2(p)$ for all $p \in \mathrm{I\!R}^3$, therefore $q_1 = q_2$.

References

[A]   Agmon, S.: Spectral properties of Schrödinger operators and scattering theory. Ann. Scuola Norm. Sup. Pisa **2** (1975) 151–218.

[C-S]   Chadan, K.,Sabatier, P.: *Inverse problems in Quantum Scattering Theory.* Springer, New York, 1989.

[H-N]   Henkin, G. M., Novikov, R. G.: $\bar{\partial}$-equation in the multi-dimensional inverse scattering problem. Usp. Mat. Nauk **42** (1987) 93–152.

[N]   Novikov, R. G.: Multidimensional inverse spectral problems for the equation $-\Delta\psi + (v(x) - Eu(x))\psi = 0$. Funkt. Analiz i Ego Prilozheniya, **22** (4) (1988), 11–22; Translation in Funct. Anal. and its Appl. **22** (4) (1988) 263–272.

[R1]   Ramm, A. G.: Recovery of the potential from fixed-energy scattering data, Inverse Probllems **4** (1988) 877–886.

[R2]   Ramm, A. G.: Completeness of the products of solutions to PDE and inverse problems. Inverse Problems **5** (1990) 641–664.

[R3]   Ramm, A. G.: Stability estimates in inverse scattering. Acta Appl. Math. **28** (1) (1992) 1–42.

[R4]   Ramm, A. G.: Multidimensional inverse problems and completeness of the products of solutions to PDE. J. Math. Anal. Appl. **136** (1) (1988) 211–253; **136** (1988) 568–574.

[R5]   Ramm, A. G.: *Multidimensional Inverse Scattering Problems*, Longman, New York, 1992 (Expanded Russian edition to appear in Mir, Moscow, 1993)

[S]   Stefanov, P.: A uniqueness result for the inverse back-scattering problem. Inverse Problems **6** (1990) 1055–1064.

[S-U]   Sylvester, J., Uhlmann, G.: Global uniqueness theorem for an inverse boundary value problem. Ann. Math. **125** (1987) 153–169.

Inverse Problems Related to Integrable Nonlinear Partial Differential Equations

Pierre C. Sabatier

Laboratoire de Physique Mathématique, Université MONTPELLIER II, Sciences et Techniques du Languedoc, 34095 - MONTPELLIER Cedex 05, France

Abstract *The so-called inverse scattering method manages solutions of integrable nonlinear partial differential equation as the Fourier method does for linear p.d.e. In its most primitive form, several different inverse problems were involved. It has been shown recently that all the inverse problems of IST can be put in a common structure, that of a unique integral equation for a matrix valued function Ψ which yields the solution $\Psi^{(1)}$ of the n.l.p.d.e and the constraints, if any, which restrict the class of solutions one can obtain. The various integrable n.l.p.d.e are related to the dimension of space and that of matrices. This result enables a classification of integrable n.l.p.d.e. It also clearly shows that the only remaining difficult point is to associate classes of solutions of n.l.p.d.e to classes of kernels of the integral equation. Although this problem here is almost virgin, it is certainly not new in the study of Inverse Problems, where overlooking it is the classical source of illposedness.*

1 Introduction

The so-called "Inverse Method" is a way of solving special nonlinear partial differential equations as linear ones are solved by means of the Fourier transform (FT). As a classical example, the Korteveg-de Vries equation

$$\eta_t + c_0\eta_x + \alpha\eta\eta_x + \beta\eta_{xxx} = 0, \tag{1.1}$$

where α, β, c_0 are numbers, is solved by means of the Direct Scattering Problem (DSP) and the Inverse Scattering Problem (ISP) of the Schrödinger operator on $\mathbb{R}$:

$$L\psi = \left(\frac{\partial^2}{\partial x^2} + \eta \right) \psi(k, x) = -k^2\psi(k, x). \tag{1.2}$$

The spectrum σ of L, which is the set of real values of k plus the N eigenvalues $k = i\kappa_n$, is conserved as η depends on x ,t according to (1.1). If $f^{\pm}(k, x)$ are the solutions of (1.2) that go respectively to $\exp(\pm ikx)$ as $x \rightarrow \pm\infty$, a scattering problem is described by the equality

$$T(k)f_-(k, x) = f_+(-k, x) + R^+(k)f_+(k, x), \tag{1.3}$$

where $R^+(k)$ is the reflection coefficient, $T(k)$ the transmission coefficient. The Cauchy Problem for (1.1) can be solved by the following schematic ansatz :

$$\eta(x,0) \xrightarrow{DSP} \left\{ R^+(k,0), k \in \mathbb{R}; \rho_n(0), 1, 2, ...N \right\}, \tag{1.4}$$

$$\left[\frac{\partial}{\partial t} + i\omega(k) \right] R^+(k,t) = 0 = \left[\frac{\partial}{\partial t} + i\omega(k) \right] \rho_n(t), \tag{1.5}$$

$$[R_+(k,t) \ ; \ \rho_n(t)] \xrightarrow{ISP} \eta(x,t), \tag{1.6}$$

where $\omega(k) = k\left(c_0 - \beta k^2\right)$ and the ρ_n's are proportional to the residues of $T(k)$ at $k = i\kappa_n$. If $\alpha = 0$ (linearized KdV equation) one must only replace DSP by FT and the right hand side of (1.4) by $\tilde{\eta}(k, O)$. The function $\tilde{\eta}(k, t)$ evolves according to (1.5) and its inverse Fourier transform yields $\eta(x, t)$. A similar scheme was obtained for other nonlinear partial differential equations (NLS, etc) in the seventies, starting from the Zacharov Shabat scattering problem and its generalizations. Algebraic methods gave ways to find the "integrable" equations, and one emphasized the solutions obtained from the spectral operator discrete spectrum, i.e. multisolitons. In the eighties, it was rather emphasized that the inverse problems associated to integrable equations rely on $\overline{\partial}$–formulations – for instance, in the KdV problem

$$\frac{\partial}{\partial \overline{k}} \mathcal{F}(k, x) = \mathcal{R}(k)e^{2ikx} \mathcal{F}(-k, x), \tag{1.7}$$

where $\mathcal{F}(k, x)$ is a function readily related to $f_+(k, x)$ and $f_-(k, x)$, and $\mathcal{R}(k)$ is its "defect of holomorphy" on the spectrum support :

$$\mathcal{R}(k) = \frac{1}{2}iR^+(k)\delta(Im \ k) + \sum_{n=1}^{N} \pi\rho_n\delta\left(k - i\kappa_n\right) \quad . \tag{1.8}$$

The $\overline{\partial}$–formulations enabled the study of new integrable equations, including ones with 2-space dimensions (e.g. Davey-Stewartson equations). However, after years two facts unfortunately remain :

(a) No integrable equation with more than two space dimensions is known as yet.

(b) Only one boundary value problem ("Cauchy problem" can be solved for each equation.

They suggest questions : Is the integrability of nonlinear p.d.e. an analytic property (as suggested by (b)) or an algebraic property (as suggested by (a)) ? Can we formulate the inverse method in a way able to go beyond the remarks (a) and (b) ? The formulation I gave recently ([1]) enabled me to demonstrate the algebraic character of remark (a) and to prove that the $\overline{\partial}$–method with one spectral variable cannot give integrable equations with more than two space variables. A slight generalization of it ([2]) enabled me to get integrable equations with N space variables but also with constraints that enforce the evolution in a two-dimensional manifold. I very recently discussed with M. Boiti and F.

Pempinelli ([3]) the relevance of the general scheme to recovering and classi-
fying all known integrable equations, and we were able to show a new one, of
the Davey-Stewartson form. I was also able to extend the method to discrete
variables ([4]).

   In the present lecture, I give the method in a form which is still more general
than what is published as yet ([1], [2], [3], [4]), I sketch how it works, I give the
elements of a classification of integrable nonlinear partial differential equations
and I conclude on some possible extensions. In this approach, what is usually
called "The Inverse Scattering Problem", is in fact the forward problem Equation
(2.1) below.

2 The method

We consider the equation

$$\Psi(k, x, t) = 1 + \int \frac{d\sigma(\lambda, \Lambda)}{-k + \Lambda} T(\lambda, \Lambda, x, t)\Psi(\lambda, x, t), \tag{2.1}$$

or, by short

$$\Psi = 1 + \oint T\Psi, \tag{2.2}$$

where k, λ, $\Lambda \in \mathbf{C}$, $x \in \mathbf{C}^M$, $t \in \mathbf{C}$, Ψ, T have their values in the set $\mathcal{M}_N$ of
$N \times N$ complex-valued matrices, $M \geq 1$, $N > 1$, $d\sigma$ (if T is regular) or at least
the product $T d\sigma$, is a measure in $\mathbf{C}^2$. We call T the spectral data, k the spectral
parameter, x the spatial variable, t the time. Any special dependence of T on x
and t is called a dispersion law. Any function Φ which solves the equation

$$\Phi = \Phi_0 + \oint T\Phi, \tag{2.3}$$

where Φ_0 is a "free term", i.e. a closed function of k, x, t, is called "compatible
with the dispersion relation.

2.1 Assumptions

We assume in the following that for the T's and measures $d\sigma$ we consider,

(a) the equation (2.3) has no more than one solution (uniqueness assumption)
(b) the solution Ψ of (2.1) has the asymptotic expansion as $\mid k \mid \to \infty$:

$$\Psi = 1 + k^{-1}\Psi^{(1)} + k^{-2}\Psi^{(2)} + ...0\left(k^{-n}\right) \tag{2.4}$$

up to the value of n we shall need. These assumptions of course must be checked
on each special choice of T and $d\sigma$.

   Our game is now to find linear operators L acting on Ψ, with coefficients that
may depend on $\Psi^{(1)}$, $\Psi^{(2)}$, etc, if necessary, such that $L\Psi$ is a solution of (2.3)
with a zero free term, and therefore vanishes. Going to the limit $\mid k \mid \to \infty$ yields

nonlinear relations for $\Psi^{(1)}$,... and writing down their consistency conditions gives evolution equations and constraints.

In the simplest case, we introduce a linear differential operator on x and t, called "the evolution operator", $\mathcal{F}$, and a linear differential operator on x, called "the spectral operator", $\mathcal{G}$, both with constant coefficients that are polynomials of degree at most 1 in k. We prove that there exist dispersion relations for T such that $\mathcal{F}\Psi$ and $\mathcal{G}\Psi$ are both compatible with T. More precisely, in this case

$$\mathcal{F}\Psi \equiv (F + kf)\Psi = f\Psi^{(1)} + \oint T\mathcal{F}\Psi, \tag{2.5}$$

$$\mathcal{G}\Psi \equiv (G + kg)\Psi = g\Psi^{(1)} + \oint T\mathcal{G}\Psi, \tag{2.6}$$

where the operators f, g, F, G, do not depend on k. Comparing these equations to (2.1), which can be multiplied on the right by a matrix independent of k, the uniqueness assumption readily yields the nonlinear relations :

$$\mathcal{F}\Psi - \Psi f\Psi^{(1)} = 0, \tag{2.7}$$

$$\mathcal{G}\Psi - \Psi g\Psi^{(1)} = 0. \tag{2.8}$$

The compatibility condition for these relations is

$$[\mathcal{F}, \mathcal{G}]\Psi = \mathcal{F}\left[\Psi g\Psi^{(1)}\right] - \mathcal{G}\left[\Psi f\Psi^{(1)}\right]. \tag{2.9}$$

Expanding Ψ inside (2.9) according to (2.4) first yields the nonlinear evolution equation of $\Psi^{(1)}$ and next either open algorithms to derive other terms or nonlinear constraints on $\Psi^{(1)}$, or both. After dispersion relations are chosen, linear constraints can be sought. Because of the uniqueness assumption, they exist iff there exists a linear differential operator A which is "compatible" with the dispersion relation and a zero free term, so that

$$A\Psi = \oint TA\Psi = 0 = A\Psi^{(1)}. \tag{2.10}$$

We call A a "ghost operator", (2.10) a "ghost equation", just to remind that in a study of integrable nonlinear equations, we have better to discard the dispersion relations that allow them.

In our first paper [1], we studied the case $N = 2$, $M > N$, with $\mathcal{F}$ being of second order in space variables. We constructed the dispersion relations compatible with $\mathcal{F}$, the first order spectral operator compatible with these dispersion relations, and we showed that unless ghost equations are accepted, it is possible to reduce the whole problem to two independent space variables, the $M - 2$ extra components of x being then no more significant than $M - 2$ parameters that label Ψ and are independent of the equation (2.1). In our next paper [2], we studied the case $M = N$ and showed that for $N > 2$, unless there are ghosts or spurious variables as above, the solutions given to a N-dimensional "integrable" nonlinear partial differential equation are restricted by nonlinear constraints to a two-dimensional manifold. In the next section I will show how the method works, essentially on the case $M \leq N$, systematically studied with M. Boiti and F. Pempinelli ([3]), the only novelty of the present paper being the possibility of introducing operators that are polynomials of degree 2 in k.

3 How the method works

A linear differential operator $\mathcal{H}$ acting on Ψ yields

$$\mathcal{H}\Psi = h + \oint T\mathcal{H}\Psi + \oint \text{ additional terms.} \qquad (3.1)$$

The additional terms contain space and time derivatives of T and Ψ. They should vanish for a compatible $\mathcal{H}\Psi$. We call them the parasite terms. Since the elements of Ψ can have independent values, it is always sufficient, and in general it can be proved to be necessary ([1], [2], [3]), that the coefficients of Ψ and its derivatives in the parasite terms should vanish. Hence we obtain the dispersion relations for T that are compatible with $\mathcal{H}$. Now, a short study ([2]) shows that the most general linear differential operator acting on Φ and which is say, degree 2 in k, can be written : as a linear combination of elementary differential operators $\partial^{p_1+p_2+\cdots p_q}/\partial x_1^{p_1}\partial x_2^{p_2}...\partial x_q^{p_q}$ (when it is convenient, we shall write in short ∂^n, with $n = p_1+p_2+...p_q$), with coefficients in the class of endomorphisms of $\mathbf{C}_{N^2}$. The general linear endomorphism E, acting on the matrix Φ, considered as an element of $\mathbf{C}_{N^2}$, can be written ([2]) as

$$E\Phi = \sum_{i,j} E_{ij}(k)\Phi\tau_{ij} = \sum_{i,j} \left(\bar{E}_{ij} + k\dot{E}_{ij} + k^2 \ddot{E}_{ij} \right) \Phi\tau_{ij}, \qquad (3.2)$$

where τ_{ij} is the $N \times N$ matrix with elements

$$[\tau_{ij}]_{\ell m} = \delta_{i\ell}\delta_{jm}. \qquad (3.3)$$

Hence, we can derive $\mathcal{H}\Psi$ in all cases by means of the formula :

$$\sum_{i,j} E_{ij}(k)\partial_n\Psi\tau_{ij} = \oint T \sum_{i,j} E_{ij}(\lambda)\partial^n\Psi\tau_{ij} + \sum_{i,j} E_{ij}(k)\partial^n I\tau_{ij}$$

$$+ \sum_{i,j}\left\{ \left(\dot{E}_{ij} + k\ddot{E}_{ij} \right)\partial^n\Psi^{(1)} + \ddot{E}_{ij}\partial^n\Psi^{(2)} \right\}\tau_{ij} \qquad (3.4)$$

$$+ \oint \sum_{i,j} \left\{ [E_{ij}(\Lambda)T - TE_{ij}(\lambda)]\partial^n\Psi + E_{ij}(\Lambda)[\partial^n(T\Psi) - T\partial^n\Psi] \right\}\tau_{ij}.$$

The second term in the right-hand side of (3.4) vanishes unless $n = 0$, (it is understood that $n = O$ means $\partial^0\Psi = \Psi$).

Now the method given in [3], and which applies to the cases linear in k, readily follows the scheme of Sect. 2. Let us demonstrate it on one example. The most general compatible 2nd order evolution operator, first order in t, and which does not reproduce in addition general cases studied with first order problems, is a priori $\mathcal{F}$:

$$\mathcal{F}\Psi = \sum_{i,j}^{N} \left\{ -\Theta_{ij}\frac{\partial\Psi}{\partial t} - \frac{1}{2}\sum_{\ell,m}^{M} C_{ij}^{\ell m}\frac{\partial^2\Psi}{\partial x_\ell\partial x_m} + \sum_{\ell=1}^{M} B_{ij}^\ell(k)\frac{\partial\Psi}{\partial x_\ell} \right\}\tau_{ij}, \qquad (3.5)$$

where we set $C_{ij}^{\ell m} = C_{ij}^{m\ell}$ (in order to avoid redundancies).

We call "flat matrix" a matrix which is proportional to the identity. Killing the parasite terms containing $\partial\Psi/\partial t$ and $\partial^2\Psi/\partial x_\ell \partial x_m$ enforces that the matrices Θ_{ij} and $C_{ij}^{\ell m}$ be flat — say $\Theta_{ij} = \theta_{ij}1, C_{ij}^{\ell m} = c_{ij}^{\ell m}1$, where θ_{ij} and $c_{ij}^{\ell m}$ (and other lower cases in the following) denote numbers. Killing the parasite terms containing $\partial\Psi/\partial x_\ell$ enforces the N^2 spatial dispersion relations

$$\sum_{m=1}^{M} c_{ij}^{\ell m} \frac{\partial T}{\partial x_m} = B_{ij}^{\ell}(\Lambda)T - TB_{ij}^{\ell}(\lambda). \tag{3.6}$$

They should be consistent with each other. If we want to avoid spurious variables, we are led ([3]) to a generic assumption : There exists at least one couple i, j such that the symmetric $M \times M$ matrix c_{ij} (elements $c_{ij}^{\ell m}$) is inversible. Then it is easily shown that $\mathcal{F}$ must reduce to the form :

$$\mathcal{F}\Psi = \sum_{i,j}^{N} \left\{ -\theta_{ij}\frac{\partial\Psi}{\partial t} + \sum_{\ell,m} c_{ij}^{\ell m} \left(-\frac{1}{2}\frac{\partial^2\Psi}{\partial x_\ell \partial x_m} + \Gamma^m(k)\frac{\partial\Psi}{\partial x_\ell} \right) + \sum_{\ell} \kappa_{ij}^{\ell}\frac{\partial\Psi}{\partial x_\ell} \right\} \tau_{ij}, \tag{3.7}$$

whose spatial dispersion relations are

$$\frac{\partial T}{\partial x_m} = \Gamma^m(\Lambda)T - T\Gamma^m(\lambda). \tag{3.8}$$

Assuming now that T is twice differentiable requires

$$[\Gamma^m(\Lambda), \Gamma^\ell(\Lambda)] = 0. \tag{3.9}$$

With these assumptions, we achieve the vanishing of $\mathcal{F}$ parasite terms by means of the time dispersion relations (which cancel the coefficient of Ψ) :

$$\theta_{ij}\frac{\partial T}{\partial t} = \frac{1}{2}\sum_{\ell,m} c_{ij}^{\ell m} \left[\Gamma^\ell(\Lambda)\Gamma^m(\Lambda)T - T\Gamma^\ell(\lambda)\Gamma^m(\lambda) \right]$$

$$+ \sum_{\ell} \kappa_{ij}^{\ell} \left[\Gamma^\ell(\Lambda)T - T\Gamma^\ell(\lambda) \right]. \tag{3.10}$$

Again, the consistency of these relations requires that $\mathcal{F}$ must be more special :

$$\mathcal{F}\Psi = -\theta\frac{\partial\Psi}{\partial t} + \sum_{\ell,m}^{M} c^{\ell m} \left(-\frac{1}{2}\frac{\partial^2\Psi}{\partial x_\ell \partial x_m} + \Gamma^m(k)\frac{\partial\Psi}{\partial x_\ell} \right) + \sum_{\ell=1}^{M} \kappa^\ell\frac{\partial\Psi}{\partial x_\ell}, \tag{3.11}$$

which is compatible with the dispersion relations (3.9), and those obtained by suppressing the indices i, j in (3.10). If we wish also avoiding redundant results, we must get a standard form by using for instance new coordinates to diagonalize the symmetric matrix $c^{\ell m}$, then reducing the second order term to Δ by rescaling the variables — using if necessary imaginary scales.

The result is the standard form of the evolution operator

$$\mathcal{F}\Psi = -\theta \frac{\partial \Psi}{\partial t} - \frac{1}{2}\Delta \Psi + (\Gamma(k) + \kappa).\mathrm{grad}\ \Psi \qquad (3.12)$$

and its dispersion relations :

$$\mathrm{grad}\ T = \Gamma(\Lambda)T - T\Gamma(\lambda), \qquad (3.13)$$

$$\theta \frac{\partial T}{\partial t} = \Gamma(\Lambda).\Gamma(\Lambda)T - T\Gamma(\lambda).\Gamma(\lambda) + \kappa.\mathrm{grad}\ T. \qquad (3.14)$$

We finally notice also that the equation (2.1) is invariant under the similarity transformation

$$T \longrightarrow STS^{-1}\ ;\ \Psi \longrightarrow S\Psi S^{-1}, \qquad (3.15)$$

where S is a (x, t dependent) matrix. The evolution equations generated by two dispersion relations related by such a transformation must be related by the same one and can be considered equivalent. Thanks to this remark, we can decide that the $\dot{\Gamma}^\ell$ in the standard form (3.12) are diagonal, and the $\bar{\Gamma}^\ell$ off-diagonal.

We can now seek the spectral operators compatible with (3.13) and (3.14). Again we start from the most general first order differential operator, and the analysis is quite similar to that above, giving

$$\mathcal{G}(\Psi) = \{\mathrm{grad}\ \Psi + [\Psi, \Gamma(k)]\}.\Omega, \qquad (3.16)$$

where Ω must satisfy

$$\Gamma.\Omega = 0. \qquad (3.17)$$

Let us pursue the analysis with the simplifying assumption $\bar{\Gamma} = 0$, and set $\dot{\Gamma} = \Gamma$, with

$$\Gamma^\ell = \sum_{i=1}^{N} g_i^\ell \eta_i \qquad \ell = 1, ..., M, \qquad (3.18)$$

where $\eta_i = \tau_{ii}$. Using an obvious vector notation as g_i in $\mathbf{C}^N$ or g^ℓ in $\mathbf{C}^M$, we also set

$$\Omega^\ell = \sum_{i=1}^{N} \omega_i^\ell \eta_i\ ,\quad \omega_i.g_i = 0 \qquad (3.19)$$

in agreement with (3.17). It is easy to prove ([3]) that the antighost condition is

$$\mathrm{rank}\ (g^1, g^2, ..., g^N) = M. \qquad (3.20)$$

From (2.9), (3.12), and (3.16), setting $\Phi = \Psi^{(1)}$, we get a nonlinear evolution equation for $\Phi_N = \mathrm{off\ diag}(\Phi)$, which is coupled to the diagonal part Φ_D

$$-\theta\Gamma \frac{\partial \Phi_N}{\partial t}.\Omega - \frac{1}{2}\Gamma\Delta\Phi_N.\Omega + \Gamma.\mathrm{grad}\ (\mathrm{grad}\ \Phi_N.\Omega)$$

$$+\Gamma.\mathrm{grad}\ [\Phi\Gamma\Phi_N.\Omega]_N - \Gamma[\Phi\Gamma.\ \mathrm{grad}\ \Phi].\Omega = 0, \qquad (3.21)$$

and a constraint equation for Φ_D :

$$\text{grad } \Phi_D.\Omega + (\Phi\Gamma\Phi.\Omega)_D = 0. \tag{3.22}$$

These equations are generalized Davey Stewartson equations. For $M = N = 2$, they give DSI, DSII, and a new integrable equation([3]), that we called DSIII. For $M = N > 2$, we get generalizations plus constraints ([2]) that confine the solutions to manifolds with two space dimensions.

Let us now give a flavour of the method when the operators are allowed k^2 terms.

The evolution operator $\mathcal{F}$:

$$\mathcal{F}\Psi = -\theta\frac{\partial\Psi}{\partial t} - \frac{1}{3}\frac{\partial^3\Psi}{\partial x^3} + k\Gamma\frac{\partial^2\Psi}{\partial x^2} - k^2\Gamma^2\frac{\partial\Psi}{\partial x} \tag{3.23}$$

is compatible with the spatial relations

$$\frac{\partial T}{\partial x} = \Lambda\Gamma T - \lambda T\Gamma, \tag{3.24}$$

$$\theta\frac{\partial T}{\partial t} = -\frac{1}{3}\left(\Lambda^3\Gamma^3 T - \lambda^3 T\Gamma^3\right), \tag{3.25}$$

but the free term depends on k, just so :

$$\mathcal{F}\Psi = -k\Gamma^2\frac{\partial\Psi^{(1)}}{\partial x} - \Gamma^2\frac{\partial\Psi^{(2)}}{\partial x} + \Gamma\frac{\partial^2\Psi^{(1)}}{\partial x} + \oint T\mathcal{F}\Psi. \tag{3.26}$$

One can nevertheless derive a nonlinear relation for $\mathcal{F}\Psi$, since it follows from (2.1), (3.4) and (3.24) that the function

$$\Psi_0 = \left(-\frac{\partial\Psi}{\partial x} + k\Gamma\Psi\right)\alpha + \Psi\beta \tag{3.27}$$

satisfies

$$\Psi_0 = k\Gamma\alpha + \Gamma\Psi^{(1)}\alpha + \beta + \oint T\Psi_0. \tag{3.28}$$

We can identify (3.28) to (3.26) by setting $\alpha = -\Gamma(\partial\Psi^{(1)}/\partial x)$ and β accordingly. The uniqueness theorem yields then :

$$\mathcal{F}\Psi = \left(\frac{\partial\Psi}{\partial x} - k\Gamma\Psi\right)\Gamma\frac{\partial\Psi^{(1)}}{\partial x} + \Psi\left[\Gamma\Psi^{(1)}\Gamma\frac{\partial\Psi^{(1)}}{\partial x} - \Gamma^2\frac{\partial\Psi^{(2)}}{\partial x} + \Gamma\frac{\partial^2\Psi^{(1)}}{\partial x^2}\right]. \tag{3.29}$$

The consistency of this nonlinear evolution relation with a spectral evolution relation yields nonlinear evolution equation and constraints. A first order spectral operator compatible with (3.24) is also easily constructed :

$$\mathcal{G}\Psi = -\frac{\partial^2\Psi}{\partial x^2} + \frac{\partial\Psi}{\partial x}M - \frac{\partial\Psi}{\partial y} + k\left(2\Gamma\frac{\partial\Psi}{\partial x} - \Gamma\Psi M\right) = 0 \tag{3.30}$$

subject to the constraint

$$\Gamma M = 0. \tag{3.31}$$

$\mathcal{G}$ is compatible with the dispersion relations (3.24) and (3.25) and in addition

$$\frac{\partial T}{\partial y} = \Lambda^2 \Gamma^2 T - \lambda^2 T \Gamma^2, \tag{3.32}$$

and, as in (2.8), after setting $b = 0$ for the sake of simplicity :

$$\mathcal{G}\Psi = 2\Psi \Gamma \frac{\partial \Psi^{(1)}}{\partial x}. \tag{3.33}$$

Introducing $\mathcal{H} = \mathcal{F} + (k/2)\Gamma\mathcal{G}$, we obtain from (3.29) and (3.33), after eliminating $\Psi^{(2)}$ by identifying the two first asymptotic terms :

$$\mathcal{H}\Psi \equiv - \left(\theta \frac{\partial \Psi}{\partial t} + \frac{1}{3}\frac{\partial^3 \Psi}{\partial x^3} \right) + \frac{1}{2}k\Gamma \left(\frac{\partial^2 \Psi}{\partial x^2} - \frac{\partial \Psi}{\partial y} \right)$$

$$= \frac{\partial \Psi}{\partial x}\Gamma\frac{\partial \Psi^{(1)}}{\partial x} + \frac{1}{2}\Psi\Gamma \left(\frac{\partial^2 \Psi^{(1)}}{\partial x^2} - \frac{\partial \Psi^{(1)}}{\partial y} \right). \tag{3.34}$$

This equality is not as (2.7) ; $\mathcal{H}$ is not compatible. The consistency between (3.33) and (3.34) give the nonlinear evolution equation for $\Phi = \Psi^{(1)}$:

$$\frac{\partial}{\partial x}\left[2\theta\frac{\partial \Phi}{\partial t} + \frac{1}{6}\frac{\partial^3 \Phi}{\partial x^3} + \frac{\partial \Phi}{\partial x}\Gamma\frac{\partial \Phi}{\partial x} \right] + \frac{1}{2}\frac{\partial^2 \Phi}{\partial y^2} = \frac{\partial \Phi}{\partial x}\Gamma\frac{\partial \Phi}{\partial y} - \frac{\partial \Phi}{\partial y}\Gamma\frac{\partial \Phi}{\partial x}. \tag{3.35}$$

KP equations can be obtained from (3.35). KdV and m KdV are easier to obtain, without k^2 terms. Generalizations are now studied.

4 Classification of integrable nonlinear equations

The method suggests a classification of integrable nonlinear equations according to the following figures (we stick at the case of a first order time differential operator and second order space differential operator) :

(1) M, N

(2)   Order of the space differential operator in $\mathcal{F}$ and $\mathcal{G}$

(3)   Degree in k of the operators $\mathcal{F}$ and $\mathcal{G}$

(4) λ and Λ independent or related (e.g. $\delta(\lambda\text{-}\Lambda)$ inside T)

As an example, for order 2 in $\mathcal{F}$, 1 in $\mathcal{G}$, we obtain the Zacharov Shabat or generalized Z.S. spectral problem in all cases, and

(a) <u>for λ and Λ independent</u>

$[N = 2, M = 2]$ Davey-Stewartson I, II, III

$[N > 2, 2 \leq M \leq N]$ generalized DS + constraints

$[M > N \geq 2]$ ghosts or M-2 spurious variables

(b) <u>for $\lambda = \Lambda$</u>

$[N = 2, M = 1]$ Non linear Schrödinger

$[N > 2, M \leq N - 1]$ DS-type equations + constraints

5 Beyond the state of the art

We currently study two problems

(a) Disentangle the correspondence between $d\sigma$ and the boundary value problems to be solved for any given nonlinear integrable equation. This is in some way the old question of inverse problems associated to the Cauchy Problem. But there has been very little progress in the last 20 years. The facts that integrable equations are obtained here on purely algebraic grounds and with the same "inverse problem" — which is (2.11) and that $d\sigma$ should therefore describe several cases give some hope.

(b) extend the method to more than one spectral variable, for instance by dealing with the equation

$$\Psi(k_1, k_2, .) = I + \int \frac{d\sigma(\Lambda_1, \Lambda_2; \lambda_1, \lambda_2)}{(-k_1 + \Lambda_1)(-k_2 + \Lambda_2)} S(\Lambda_1, \Lambda_2, \lambda_1, \lambda_2, .) \Psi(\lambda_1, \lambda_2, .)$$

(5.1)

or its $\Lambda_1 = \lambda_1, \Lambda_2 = \lambda_2$ form. Preliminary results are very complicated.

References

1. Sabatier P.C., 1992, Inverse Problems **8** 263.
2. Sabatier P.C., 1992, Phys. Lett. A **161** 345.
3. Boiti M., Pempinelli F., Sabatier P.C., 1993, Inverse Problems **9** 1.
4. Sabatier P.C., 1992, Inverse Problems **8** L29.

Some Estimates of Green Function and Applications in Inverse Scattering Theory for the Schrödinger Operator with a Singular Potential

Valeri S. Serov

Department of Computational Mathematics and Cybernetics, Moscow State University, Moscow, Russia

1 Introduction

We propose to discuss here some inverse problems of the scattering theory with fixed energy concerning the Schrödinger operator $-\Delta + q(x)$ in $\mathbb{R}^N$ for $N = 2$ and 3. Assume that the potential $q(x)$ belongs to $L^p_\delta(\mathbb{R}^N)$ for some p, $N/2 < p \leq 2$, and for $\delta > N/p'$, where $1/p + 1/p' = 1$ and $L^p_\delta(\mathbb{R}^N)$, $\delta \in \mathbb{R}$, $1 \leq p \leq \infty$, denotes the weighted L^p spaces defined as

$$L^p_\delta = \left\{ f \in L^p_{\text{loc}} : \|f\|_{p,\delta} = \left(\int_{\mathbb{R}^N} (1 + |x|)^{\delta p} |f(x)|^p dx \right)^{1/p} < \infty \right\} \tag{1.1}$$

2 Green-Faddeev's function

Let $x \in \mathbb{R}^N$, $N \geq 2$, and $z \in \mathbb{C}^N$ be a complex vector with $z \cdot z = k^2$, where k is fixed and $k \geq 0$. Define in the sense of distribution theory

$$g_z(x) = \frac{1}{(2\pi)^N} \int_{\mathbb{R}^N} \frac{e^{ix \cdot \xi} d\xi}{\xi^2 + 2z \cdot \xi} \tag{2.1}$$

It is easy to check that $g_z(x)$ is a tempered distribution satisfying

$$(-\Delta_x - 2iz \cdot \nabla_x) g_z(x) = \delta(x), \tag{2.2}$$

where Δ_x is the Laplacian in $\mathbb{R}^N$, $\nabla_x = (\partial/\partial x_1, \ldots, \partial/\partial x_N)$ and $\delta(x)$ is the δ-function. The distribution $G_z(x) = e^{ix \cdot z} g_z(x)$ is then a fundamental solution for the operator $-\Delta - k^2$ in $\mathbb{R}^N$, i.e.,

$$-(\Delta + k^2) G_z(x) = \delta(x) \tag{2.3}$$

In his work on quantum scattering, Faddeev introduced this function $G_z(x)$ in 1965 (see [1]). It follows that $G_z(x)$ differs from Green's function of classical scattering theory by a global metaharmonic function on $\mathbb{R}^N$:

$$G_z(x) = G_k^+(x) + H_z(x), \quad z \cdot z = k^2, \quad (\Delta + k^2) H_z(x) = 0 \tag{2.4}$$

where

$$G_k^+(x) = \frac{1}{(2\pi)^N} \lim_{\varepsilon \to +0} \int_{\mathbb{R}^N} \frac{e^{ix\cdot\xi}d\xi}{\xi^2 - k^2 - i\varepsilon} = \frac{i}{4}\left(\frac{k}{2\pi|x|}\right)^{\frac{N-2}{2}} H_{\frac{N-2}{2}}^{(1)}(k|x|) \quad (2.5)$$

is the outgoing Green's function.

Theorem 1. *Let $N = 2$ or 3, $f \in L_\delta^p(\mathbb{R}^N)$ for some p, $N/2 < p \le 2$, and $\delta > N/p'$, where $1/p + 1/p' = 1$. Let also $k \ge 0$ be fixed, $z \in \mathbb{C}^N$ – complex vector, $z \cdot z = k^2$ and $\mathrm{Im}\, z \ne 0$. Then for $|z|$ sufficiently large, the convolution map $f \to g_z * f$ is bounded from $L_\delta^p(\mathbb{R}^N)$ to $L^\infty(\mathbb{R}^N)$ and satisfies the estimate*

$$\|g_z * f\|_{\infty,0} \le \frac{c}{|z|^\beta}\|f\|_{p,\delta} \qquad\qquad 2.6$$

where $\beta = \alpha/(\alpha + 1)$ for $N = 3$, $\beta = 2\alpha/(\alpha + 1)$ for $N = 2$ and $0 < \alpha < 1 - N/(2p)$.

Corollary 2. *Let potential $q(x) \in L_\delta^p(\mathbb{R}^N)$, $N = 2$ or 3, for some p, $N/2 < p \le 2$, and $\delta > N/p'$, where $1/p + 1/p' = 1$. Then there exists a constant $c_0 > 0$ such that for $|z| > c_0$, there exists a unique solution of $(-\Delta + q - k^2)u = 0$ in $\mathbb{R}^N$ of the form*

$$u(x, z) = e^{ix\cdot z}(1 + R(x, z)), \quad z \cdot z = k^2 \qquad\qquad (2.7)$$

with $R \in L^\infty(\mathbb{R}^N)$. Further more the following estimate holds

$$\|R\|_{\infty,0} \le \frac{c}{|z|^\beta}, \qquad\qquad (2.8)$$

where β is as in Th. 1.

Sylvester and Uhlmann were the first to prove a kind of weighted L^2 estimate, i.e. the convolution map $f \to g_z * f$ is bounded from $L_\delta^2(\mathbb{R}^N)$ to $L_{\delta+1}^2(\mathbb{R}^N)$ for δ, $-1 < \delta < 0$ (see [2], [3]). In this papers they proved that there exists a unique solution of $(-\Delta + q)u = 0$ in $\mathbb{R}^N$ of the form (2.7) for potential $q \in L^\infty(\mathbb{R}^N)$ and with compact support. Besides, they proved that function $R(x, z)$ from (2.7) belongs to $L_\delta^2(\mathbb{R}^N)$ and the following estimate holds

$$\|R\|_{2,\delta} \le \frac{c}{|z|}$$

After that Sylvester and Uhlmann applied this result for proving a uniqueness theorem for an inverse boundary value problem in electrical prospection and a global uniqueness theorem for an inverse boundary value problem.

3 Applications in scattering theory

Let us consider the bounded solutions $u(x, k)$ of the equation

$$(-\Delta + q - k^2)u(x, k) = 0 \qquad (3.1)$$

It is well-known that this solutions satisfy Sommerfeld's radiation condition at the infinity and are the unique solution of the Lippmann-Schwinger equations

$$u(x, k, \theta) = e^{ikx\cdot\theta} - \int_{\mathbb{R}^N} G_k^+(x - y)q(y)u(y, k, \theta)dy, \qquad (3.2)$$

where $x \in \mathbb{R}^N$, $k \in \mathbb{R}$ and $\theta \in S^{N-1}$ (the unit sphere in $\mathbb{R}^N$). Asymptotically, $u(x, k, \theta)$ admits the expansion

$$u(x, k, \theta) = e^{ikx\cdot\theta} + c_N \frac{e^{ik|x|}k^{(N-3)/2}}{|x|^{(N-1)/2}} A(k, \theta', \theta) + o\left(\frac{1}{|x|^{(N-1)/2}}\right) \qquad (3.3)$$

where $|x|$ tends to infinity, $\theta' = \frac{x}{|x|}$ and the scattering amplitude $A(k, \theta', \theta)$ is defined as

$$A(k, \theta', \theta) = -\int_{\mathbb{R}^N} e^{-ik\theta'\cdot y}q(y)u(y, k, \theta)dy \qquad (3.4)$$

We study the inverse scattering problem in potential scattering at fixed energy, i.e. k^2 be fixed and $k \geq 0$. The question is: Does $A(k, \theta', \theta)$ given for a fixed $k \geq 0$ and for all $\theta', \theta \in S^{N-1}$ determine uniquely q? For potentials q of compact support and $q \in L^\infty$ it was proven by Ramm (see [4]) that the answer is affirmative. The same is true for sufficiently small exponentially decreasing potential (see [5]). Weder (see [6]) proved that if a potential decreasing like $c(1 + |x|)^{-N-\varepsilon}$, $\varepsilon > 0$, and from $L_{loc}^\infty(\mathbb{R}^N)$ is already known outside a compact set, then it can be determined completely. The proof of the uniqueness reduces this inverse problem to the problem of finding $q(x)$ from the Dirichlet-to-Neumann map (see [3], [7], [8]). On the other hand in [9] another approach is proposed based on the completeness of the scattering solutions (2.7). In this paper [9] was proved that the uniqueness holds for potentials of non-compact support in $\mathbb{R}^3$ that decrease faster than any exponential $ce^{-a|x|}$, $a > 0$, as $|x| \to +\infty$ and $q \in L_{loc}^\infty(\mathbb{R}^3)$. Sun and Uhlmann (see [10]) considered the problem of determining the strength and location of the singularities of the potential q in $\mathbb{R}^2$. They assumed that $q \in L^p(\Omega)$, $2 < p \leq \infty$, and Ω is a bounded domain with smooth boundary and $q \equiv 0$ in $\mathbb{R}^2 \setminus \Omega$. The following theorem is true.

Theorem 3. *1) Let $q_1(x)$, $q_2(x) \in L^p(\mathbb{R}^3)$, $p > 3/2$, with compact support. If $A_{q_1}(k, \theta', \theta) = A_{q_2}(k, \theta', \theta)$ for all θ', $\theta \in S^2$ and for a fixed $k \geq 0$, then $q_1 = q_2$ in the sense of L^p. 2) Let $q_1(x)$, $q_2(x) \in L_\delta^p(\mathbb{R}^2)$, $1 < p \leq 2$, $\delta > 2/p'$, $1/p + 1/p' = 1$. If $A_{q_1}(k, \theta', \theta) = A_{q_2}(k, \theta', \theta)$ for all θ', $\theta \in S^1$ and for a fixed $k \geq 0$, then*

$$q_1 = q_2 \pmod{H^t(\mathbb{R}^2)}$$

where H^t is the Sobolev space and $t < 3 - 2/p - 2p/(2p - 1)$.

Remark. In the the case $N = 2$ it is easy to check that for any p, $(7 + \sqrt{17})/8 < p \leq 2$, $H^t(\mathbb{R}^2) \subset L^r(\mathbb{R}^2)$, where $r > p$. It means that the potentials $q_1(x)$ and $q_2(x)$ have the same singularities.

Corollary 4. *Let $\Omega \subset \mathbb{R}^2$ be a bounded domain with smooth boundary. Let $q_1(x)$, $q_2(x) \in L^p(\Omega)$ for some p, $(7 + \sqrt{17})/8 < p \leq 2$, and $q_1(x) \equiv 0$, $q_2(x) \equiv 0$ in $\mathbb{R}^2 \setminus \Omega$. Assume also that $\{0\}$ is not a Dirichlet eigenvalue of $-\Delta + q_i(x)$, $i = 1, 2$. If $\Lambda_{q_1} = \Lambda_{q_2}$, where Λ_{q_i} is the Dirichlet–to–Neumann map (see [3]), then*

$$q_1(x) - q_2(x) \in L^r(\mathbb{R}^2)$$

for some $r > p$.

References

1. Faddeev, L. D.: Growing solutions of the Schrödinger equation. Dokl. Akad. Nauk SSSR. **165** (1965) 514–517 (Transl.: Sov. Phys. Dokl. **10** 1033).
2. Sylvester, J., Uhlmann, G.: A uniqueness theorem for an inverse boundary value problem in electrical prospection. Comm. Pure Appl. Math. **39** (1986) 91–112.
3. Sylvester, J., Uhlmann, G.: A global uniqueness theorem for an inverse boundary value problem. Annals of Math. **125** (1987) 153–169.
4. Ramm, A. G.: Recovery of the potential from fixed-energy scattering data. Inverse Probl. **4** (1988) 877-886.
5. Henkin, G .M., Novikov, R. G.: $\bar{\partial}$-equation in the multi-dimensional inverse scattering problem. Usp. Mat. Nauk. **42** (187) 93–152.
6. Weder, R.: Global uniqueness at fixed-energy in multidimensional inverse scattering theory. Inverse Probl. **7** (1991) 927–938.
7. Sylvester, J., Uhlmann, G.: The Dirichlet-to-Neumann map and applications in inverse problems for partial differential equations. SIAM, Philadelphia, 99–139 (1990).
8. Nachman, A.: Reconstruction from boundary measurements. Annals of Math. **128** (1988) 531–587.
9. Ramm, A. G., Stefanov, P. D.: Fixed energy inverse scattering for non-compactly supported potentials (Preprint).
10. Sun, Z., Uhlmann, G.: Recovery of singularities for formally determined inverse problems (Preprint).

Reconstruction of Electromagnetic Parameters from Boundary Measurements

E. Somersalo[1], P. Ola[2] and L. Päivärinta[2]

[1] Rolf Nevanlinna Institute, P.O. Box 26 (Teollisuuskatu 23), 00014 UNIVERSITY OF HELSINKI, Finland
[2] Department of Mathematics, University of Oulu, Linnanmaa, 90570 Oulu, Finland

The inverse boundary value problem discussed in this article is to determine the electric and magnetic material parameters in a bounded domain from field measurements at the boundary of the body.

Let $\Omega \subset \mathbb{R}^3$ be a bounded $C^{1,1}$ domain with a connected complement $\mathbb{R}^3 \setminus \overline{\Omega}$. The electric permittivity ε, conductivity σ and magnetic permeability μ are assumed to be C^3 functions in $\mathbb{R}^3$, and for some constants ε_m, ε_M, σ_M, μ_m and μ_M,

$$0 < \varepsilon_m \le \varepsilon(x) \le \varepsilon_M, \quad 0 \le \sigma(x) \le \sigma_M, \quad 0 < \mu_m \le \mu(x) \le \mu_M.$$

Furthermore, it is assumed that in $\mathbb{R}^3 \setminus \Omega$,

$$\varepsilon(x) = \varepsilon_0, \quad \sigma(x) = \sigma_0, \quad \mu(x) = \mu_0,$$

for some constants $\varepsilon_0 > 0$, $\sigma_0 \ge 0$ and $\mu_0 > 0$. The time harmonic electric and magnetic fields E and H satisfy Maxwell's equations

$$\nabla \wedge E = i\omega\mu H, \quad \nabla \wedge H = -i\omega\gamma E, \tag{1}$$

where $\omega > 0$ is a fixed frequency and $\gamma = \varepsilon + i\sigma/\omega$. We assume that the fields satisfy the electric boundary condition

$$n \wedge E\big|_{\partial\Omega} = F, \tag{2}$$

where $n = n(x)$ is the outward unit normal vector to the boundary $\partial\Omega$. Assuming that ω is not a resonance frequency (i.e. the boundary value problem has a unique solution), we may define the boundary map

$$\Lambda : F \mapsto n \wedge H\big|_{\partial\Omega}.$$

The inverse boundary value problem is to retrieve the functions ε, σ and μ from the knowledge of the boundary map Λ.

The problem in its present form was formulated in the article [9]. The starting point was an equation that relates the boundary values of the fields to the material parameters in the interior of Ω. Let E_0 and H_0 satisfy the homogenous space equations

$$\nabla \wedge E_0 = i\omega\mu_0 H_0, \quad \nabla \wedge H_0 = -i\omega\gamma_0 E_0 \tag{3}$$

in Ω, where $\gamma_0 = \varepsilon_0 + i\sigma_0/\omega$. Then equations (1) and (3) together with Gauss' formula imply that

$$\int_{\partial\Omega} (F \cdot H_0 + \Lambda(F) \cdot E_0)dS = \int_{\partial\Omega} n \cdot (E \wedge H_0 - E_0 \wedge H)dS$$

$$= i\omega \int_\Omega \big((\mu - \mu_0)H \cdot H_0 - (\gamma - \gamma_0)E \cdot E_0 \big)dx. \tag{4}$$

Following the article [9], consider first the linearized problem. Denote

$$\delta\gamma = \gamma - \gamma_0, \quad \delta\mu = \mu - \mu_0,$$

and assume that these perturbations off the background constant values are small. The exact conditions for the smallness are too tedious to state here. We are looking for the solutions E and H to the problem (1)–(2) in a special form,

$$E = e^{i\zeta_1 \cdot x}(\eta_1 + R), \quad H = e^{i\zeta_1 \cdot x}(\nu_1 + Q),$$

where ζ_1, η_1 and ν_1 are complex vectors in $\mathbb{C}^3$. The boundary condition for E is fixed as

$$n \wedge R\big|_{\partial\Omega} = 0.$$

Furthermore, a special choice for the fields E_0 and H_0 is made as

$$E_0 = e^{i\zeta_2 \cdot x}\eta_2, \quad H_0 = e^{i\zeta_2 \cdot x}\nu_2.$$

The complex wave vectors ζ_j and polarizations η_j and ν_j are related to each others through the algebraic conditions

$$\zeta_j \wedge \eta_j = \omega\mu_0\nu_j, \quad \zeta_j \wedge \nu_j = -\omega\gamma_0\eta_j, \quad j = 1, 2,$$

and

$$\zeta_j \cdot \zeta_j = k^2 = \omega^2\mu_0\gamma_0.$$

For a fixed $\xi \in \mathbb{R}^3$, one may choose the wave vectors in such a way that an extra condition

$$\zeta_1 + \zeta_2 = \xi$$

holds. It is perhaps not surprising that with these choices, the fields R and Q are small to first order in $\delta\gamma$ and $\delta\mu$, and thus

$$\int_{\partial\Omega} (F \cdot H_0 + \Lambda(F) \cdot E_0)dS = i\omega \int_\Omega (\delta\mu H \cdot H_0 - \delta\gamma E \cdot E_0)dx$$

$$= i\omega \int_\Omega (\delta\mu(\nu_1 \cdot \nu_2) - \delta\gamma(\eta_1 \cdot \eta_2))e^{i\xi \cdot x}dx + 0(\delta\mu^2, \delta\gamma^2). \tag{5}$$

In the above formula, the left hand side is completely determined by the knowledge of the mapping Λ, while the right hand side is the Fourier transform of a linear combination of the unknown material parameters plus a higher order perturbation. It was shown in [9], that under certain smallness conditions on $\delta\gamma$

and $\delta\mu$, these perturbations can be obtained up to an error that is small to order more than one in $\delta\gamma$ and $\delta\mu$.

The linearized theory is in a way a natural extension of Calderón's work on the impedance imaging problem ([1]) to electrodynamics. In view of the recent developement on the impedance imaging problem (see [3], [6], [8], [10]), it is natural to ask whether one can choose the fields in such a manner that the residual term in (5) tends to zero as $|\zeta_j| \to \infty$, $j = 1, 2$. The answer is, unfortunately, negative. (This can be understood in the light of the article [4] to be a consequence of the fact that Maxwell's equations form a first order system and the regular fundamental solutions behave like $O(|\xi|^{-1})$ at infinity.) however, the idea of using the exponentially growing solutions to analyze the fully non–linear problem turns out to be very useful after some modifications. In [2], these ideas were used succesfully in the inverse scattering problem of electrodynamics. It was shown in the cited article that when the magnetic permeability of a scatterer is constant, the inverse scattering problem of solving ε and σ from the far field measurements has a unique solution. Some of the techniques developed in that article is seen in the discussion below.

In the following, the ideas of the article [7] are briefly sketched. The discussion consists of three parts: (A) The construction of certain exponentially growing solutions to Maxwell's equations, (B) recovery of the boundary values of these solutions starting from the boundary map Λ, and (C) asymptotics of the integral identity (4) when ζ grows.

We start by explaining the ideas needed in (A). First, we recall the exponentially growing fundamental solution for the Helmholtz equation. Let $\zeta \in \mathbb{C}^3$, $\zeta \cdot \zeta = k^2$. Define

$$G_\zeta(x) = e^{i\zeta \cdot x} g_\zeta(x),$$

where

$$g_\zeta(x) = \left(\frac{1}{2\pi}\right)^3 \int \frac{e^{i\xi \cdot x}}{|\xi|^2 + 2\xi \cdot \zeta} d\xi.$$

Then G_ζ satisfies the equation

$$(-\Delta - k^2)G_\zeta = \delta.$$

The remarkable property of the function g_ζ, proved in [10], is that it defines a convolution operator with a decreasing norm with respect to $|\zeta|$ in appropriate weighted L^p–spaces. More precisely, let

$$L_\delta^2 = \{f \in L_{\text{loc}}^2 : \|f\|_\delta = \left(\int (1 + |x|^2)^\delta |f(x)|^2 dx\right)^{1/2} < \infty\}.$$

We have

Proposition 1 (Sylvester–Uhlmann). *The convolution with the kernel g_ζ defines a continuous operator*

$$g_\zeta * : L_\delta^2 \to L_{-\delta}^2, \quad 1/2 < \delta < 1,$$

and

$$\|g_\zeta * f\|_{-\delta} \le \frac{C}{|\zeta|}\|f\|_\delta.$$

To find an appropriate counterpart to this operator for Maxwell's equations, define first the operators

$$L = \begin{pmatrix} 0 & \dfrac{i}{\gamma_0}\nabla\wedge \\ -\dfrac{i}{\mu_0}\nabla\wedge & 0 \end{pmatrix}, \quad W = \begin{pmatrix} \tilde{\gamma}1_{3\times3} & 0 \\ 0 & \tilde{\mu}1_{3\times3} \end{pmatrix},$$

where

$$\tilde{\gamma} = \frac{\gamma}{\gamma_0} - 1, \quad \tilde{\mu} = \frac{\mu}{\mu_0} - 1,$$

and $1_{3\times3}$ is a 3×3 unit matrix. With these notations, Maxwell's equations (1) read as

$$(L - \omega)\begin{pmatrix} E \\ H \end{pmatrix} = W\begin{pmatrix} E \\ H \end{pmatrix}.$$

Furthermore, if we define the differential operator B as

$$B = \frac{k^2}{\omega}\begin{pmatrix} 1_{3\times3} + \dfrac{\nabla\nabla\cdot}{k^2} & \dfrac{i}{\omega\gamma_0}\nabla\wedge \\ \dfrac{-i}{\omega\mu_0}\nabla\wedge & 1_{3\times3} + \dfrac{\nabla\nabla\cdot}{k^2} \end{pmatrix},$$

a straightforward calculation shows that,

$$B(L - \omega) = (L - \omega)B = (-\Delta - k^2)1_{6\times6}.$$

In the light of this identity, it is natural to look for a solution to Maxwell's equations in the form

$$\begin{pmatrix} E \\ H \end{pmatrix} = e^{i\zeta\cdot x}\begin{pmatrix} \eta \\ \nu \end{pmatrix} + BG_\zeta * \left(W\begin{pmatrix} E \\ H \end{pmatrix}\right), \tag{6}$$

where the wave vector and the polarizations satisfy the same algebraic conditions as before.

The striking feature of the above equation is that the integral operator is of zeroth order and, consequently, there is no hope of getting nice asymptotic behaviour with large ζ. To reduce the order of the operator, we use first a rather standard trick: One can commute the operators B and $G_\zeta *$ and use Maxwell's equations. (This idea, applied to the classical scattering problem, can be found e.g. in the monograph of Müller [5].) We find that the integral equation is equivalent to

$$\begin{pmatrix} E \\ H \end{pmatrix} = e^{i\zeta\cdot x}\begin{pmatrix} \eta \\ \nu \end{pmatrix} + G_\zeta * \left((V + S)\begin{pmatrix} E \\ H \end{pmatrix}\right), \tag{7}$$

where

$$V = \begin{pmatrix} \omega^2(\mu\gamma - \mu_0\gamma_0) + (\nabla\alpha)^t\cdot & \dfrac{i\omega}{\gamma}\nabla(\mu\gamma)\wedge \\ -\dfrac{i\omega}{\mu}\nabla(\mu\gamma)\wedge & \omega^2(\mu\gamma - \mu_0\gamma_0) + (\nabla\beta)^t\cdot \end{pmatrix},$$

with

$$\alpha = \frac{\nabla\gamma}{\gamma}, \quad \beta = \frac{\nabla\mu}{\mu},$$

superindex "t" denoting transpose, and

$$S = \begin{pmatrix} \alpha \cdot \nabla & 0 \\ 0 & \beta \cdot \nabla \end{pmatrix}.$$

The next step is to remove the first order term. To this end, denote

$$M = \begin{pmatrix} \gamma^{1/2}\mathbf{1}_{\times 3} & 0 \\ 0 & \mu^{1/2}\mathbf{1}_{3\times 3} \end{pmatrix}.$$

By direct calculation, one can verify the commutator relation

$$[\Delta, M] = M(S + q),$$

where

$$q = \begin{pmatrix} \dfrac{\Delta\gamma^{1/2}}{\gamma^{1/2}}\mathbf{1}_{3\times 3} & 0 \\ 0 & \dfrac{\Delta\mu^{1/2}}{\mu^{1/2}}\mathbf{1}_{3\times 3} \end{pmatrix}.$$

Assume now that E and H are solutions to the integral equation (7). If we apply the Helmholtz operator to the equation, we find that

$$(-\Delta - k^2)\begin{pmatrix} E \\ H \end{pmatrix} = (V + S)\begin{pmatrix} E \\ H \end{pmatrix}.$$

Therefore,

$$(-\Delta - k^2)M\begin{pmatrix} E \\ H \end{pmatrix} = M(V + S)\begin{pmatrix} E \\ H \end{pmatrix} - [\Delta, M]\begin{pmatrix} E \\ H \end{pmatrix}$$

$$= M(V - q)\begin{pmatrix} E \\ H \end{pmatrix}.$$

A subsequent application of the operator $G_\zeta *$ yields, after rearrenging the terms, the integral equation

$$\begin{pmatrix} R \\ Q \end{pmatrix} = \begin{pmatrix} J \\ K \end{pmatrix} + M^{-1}g_\zeta * \left(M(V - q)\begin{pmatrix} R \\ Q \end{pmatrix} \right), \tag{8}$$

where

$$\begin{pmatrix} J \\ K \end{pmatrix} = (M^{-1}M_0 - 1)\begin{pmatrix} \eta \\ \nu \end{pmatrix} + M^{-1}g_\zeta * \left(M(V - q)\begin{pmatrix} \eta \\ \nu \end{pmatrix} \right).$$

and

$$M_0 = \begin{pmatrix} \gamma_0^{1/2}\mathbf{1}_{3\times 3} & 0 \\ 0 & \mu_0^{1/2}\mathbf{1}_{3\times 3} \end{pmatrix}.$$

Note that equation (8) contains the integral operator $g_\zeta *$, which has a small norm for large $|\zeta|$ by Proposition 1. Consequently, equation (8) is solvable by Neumann series in the weighted L^2–spaces. we have the following result:

Theorem A *For sufficiently large* $|\zeta|$, *the integral equation (6) has a unique solution* (E, H) *with*

$$e^{i\zeta \cdot x} E - \eta, \quad e^{i\zeta \cdot x} H - \nu \in L^2_{-\delta}.$$

The solutions can be found by solving the equivalent integral equation (8).

What is more, we get the asymptotics of the special solutions by iterating equation (8). This important feature is used in step (C), when we calculate the asymptotics of the key integral equation (4).

Now, we turn our attention to the problem (B): Given the boundary map Λ, how does one calculate the left hand side of equation (4) when E and H are the special solutions of the integral equation (6)? This part of the work uses similar ideas to those employed in the article [6] of Nachman.

Consider first an exterior problem: Assume that in $R^3 \setminus \overline{\Omega}$, E and H satisfy

$$(L - \omega)\begin{pmatrix} E \\ H \end{pmatrix} = 0.$$

Let B_R be a ball with radius R large enough so that Ω is contained in B_R. Integration by parts implies for $x \in B_R \setminus \overline{\Omega}$, that

$$\begin{pmatrix} E(x) \\ H(x) \end{pmatrix} = \int_{B_R \setminus \overline{\Omega}} (L - \omega) B \left(G_\zeta(x - y) \begin{pmatrix} E(y) \\ H(y) \end{pmatrix} \right) dy$$

$$= \left(\int_{\partial B_R} - \int_{\partial \Omega} \right) \begin{pmatrix} \nabla \wedge & \frac{i}{\omega\gamma_0}(\nabla\wedge)^2 \\ \frac{-i}{\omega\mu_0}(\nabla\wedge)^2 & \nabla\wedge \end{pmatrix} G_\zeta(x - y) \begin{pmatrix} n \wedge E(y) \\ n \wedge H(y) \end{pmatrix} dS(y)$$

$$= \left[\begin{pmatrix} K_{B_R} & \frac{i}{\omega\gamma_0} D_{B_R} \\ \frac{-i}{\omega\mu_0} D_{B_R} & K_{B_R} \end{pmatrix} - \begin{pmatrix} K_{R^3\setminus\overline{\Omega}} & \frac{i}{\omega\gamma_0} D_{R^3\setminus\overline{\Omega}} \\ \frac{-i}{\omega\mu_0} D_{R^3\setminus\overline{\Omega}} & K_{R^3\setminus\overline{\Omega}} \end{pmatrix} \right] \begin{pmatrix} n \wedge E(y) \\ n \wedge H(y) \end{pmatrix},$$

where a shorthand notation for the potential operators was used. As in usual scattering theory, we want to push the radius of the sphere B_R towards infinity to get rid of the integration over the outer surface. In order to do that, one has to have a counterpart of the standard Silver–Müller radiation condition at infinity. Following Nachman's terminology, we say that the fields E and H satisfy the *weak radiation condition*, if

$$\lim_{R \to \infty} \begin{pmatrix} K_{B_R} & \frac{i}{\omega\gamma_0} D_{B_R} \\ \frac{-i}{\omega\mu_0} D_{B_R} & K_{B_R} \end{pmatrix} \begin{pmatrix} n \wedge E(y) \\ n \wedge H(y) \end{pmatrix} = 0$$

uniformly in compact sets.

The following proposition is a counterpart of the classical Stratton–Chu representation formula for exterior domains.

Proposition 2. *Let E and B be given by*

$$\begin{pmatrix} E \\ H \end{pmatrix} = \begin{pmatrix} K_{\mathbb{R}^3 \setminus \overline{\Omega}} & \dfrac{i}{\omega\gamma_0} D_{\mathbb{R}^3 \setminus \overline{\Omega}} \\ \dfrac{-i}{\omega\mu_0} D_{\mathbb{R}^3 \setminus \overline{\Omega}} & K_{\mathbb{R}^3 \setminus \overline{\Omega}} \end{pmatrix} \begin{pmatrix} n \wedge u \\ n \wedge v \end{pmatrix}.$$

Then E and H satisfy Maxwell's equations and the weak radiation condition in the exterior domain. The converse is true with $u = E$, $v = H$.

To proceed, we need the boundary values of the potential operators appearing in the previous proposition. In general, when the point x goes to the boundary $\partial\Omega$ from the exterior domain, we have for a tangential field F the following traces:

$$K_{\mathbb{R}^3 \setminus \overline{\Omega}} F\big|_{\partial\Omega} = KF + \frac{1}{2} n \wedge F,$$

and

$$D_{\mathbb{R}^3 \setminus \overline{\Omega}} F\big|_{\partial\Omega} = DF + \frac{1}{2} n \mathrm{Div} F,$$

where Div is the tangential divergence, and K and D are the vector layer potentials defined as

$$KF(x) = \mathrm{p.v.} \int_{\partial\Omega} \nabla_x G_\zeta(x - y) \wedge F(y) dS(y), \quad x \in \partial\Omega,$$

and

$$DF(x) = \nabla V(\mathrm{Div} F)(x) + k^2 V F(x), \quad x \in \partial\Omega,$$

the operator V being

$$VF(x) = \int_{\partial\Omega} G_\zeta(x - y) F(y) dS(y), \quad x \in \partial\Omega.$$

Consider now the exterior problem of finding fields E and H satisfying Maxwell's equations

$$\nabla \wedge E = i\omega\mu_0 H, \quad \nabla \wedge H = -i\omega\gamma_0 E$$

in $\mathbb{R}^3 \setminus \overline{\Omega}$ with the boundary condition

$$n \wedge H = \Lambda(n \wedge E)$$

at the boundary $\partial\Omega$, and furthermore, it is required that the fields

$$E - e^{i\zeta \cdot x}, \quad H - e^{i\zeta \cdot x}$$

satisfy the weak radiation condition at infinity. If a solution to this exterior problem exists, we must have by Proposition 2,

$$\begin{pmatrix} E \\ H \end{pmatrix} = e^{i\zeta \cdot x} \begin{pmatrix} \eta \\ \nu \end{pmatrix} + \begin{pmatrix} K_{\mathbb{R}^3 \setminus \overline{\Omega}} & \dfrac{i}{\omega\gamma_0} D_{\mathbb{R}^3 \setminus \overline{\Omega}} \\ \dfrac{-i}{\omega\mu_0} D_{\mathbb{R}^3 \setminus \overline{\Omega}} & K_{\mathbb{R}^3 \setminus \overline{\Omega}} \end{pmatrix} \begin{pmatrix} n \wedge E \\ n \wedge \Lambda(E) \end{pmatrix}.$$

Let x go to the boundary $\partial\Omega$ and read off the equation for the field $n \wedge E$ at the boundary. We find that $F = n \wedge E$ must satisfy the integral equation

$$\frac{1}{2}F = n \wedge \eta e^{i\zeta \cdot x} + n \wedge KF + \frac{i}{\omega\gamma_0} n \wedge D\Lambda F \qquad (9)$$

By analyzing the correspondence of this integral equation to boundary value problems for Maxwell's equations, one can prove the following result.

Theorem B *If $\omega > 0$ is not a resonance frequency, the integral equation (9) is a Fredholm equation (in appropriate function spaces). Furthermore, if $|\zeta|$ is large enough, it has a unique solution.*

The crux of the above theorem is that the solution to the boundary integral equation (9) allows us to reconstruct the boundary values of the special solutions of Theorem A. Indeed, define the fields E_e and H_e in the exterior domain $\mathbb{R}^3 \setminus \overline{\Omega}$ as

$$E_e = \eta e^{i\zeta \cdot x} + (K_{\mathbb{R}^3\setminus\overline{\Omega}} + \frac{i}{\omega\gamma_0}D_{\mathbb{R}^3\setminus\overline{\Omega}}\Lambda)F, \qquad H_e = -\frac{i}{\omega\mu_0}\nabla \wedge E_e,$$

where F is the solution of (9). Define E_i and H_i in the interior domain Ω as

$$\nabla \wedge E_i = i\omega\mu H_i, \qquad \nabla \wedge H_i = -i\omega\gamma E_i,$$

with the boundary condition

$$n \wedge E_i|_{\partial\Omega} = n \wedge E_e|_{\partial\Omega}.$$

Then, an integration by parts argument shows that the field E defined as

$$E(x) = \begin{cases} E_i(x), & x \in \Omega \\ E_e(x), & x \notin \Omega \end{cases}$$

satisfies the integral equation (6) and thus coincides with the special solution of Theorem A. This argument shows that not only does Λ determine the tangential boundary values of the special solution E, but there is a *constructive* method of obtaining the left hand side of equation (4) for large $|\zeta|$.

Finally, we turn to the problem (C), the large $|\zeta|$ asymptotics of equation (4). As it was indicated after Theorem A, the asymptotics of the special solutions can be obtained by solving the integral equation (8) by Neumann series. Rather tedious caculations that we omit here give the following quite compact result.

Theorem C *With an appropriate choice of the polarization vectors η_j and ν_j,*

$$\lim_{|\zeta|\to\infty} \int_\Omega ((\mu - \mu_0)H_\zeta \cdot H_0 - (\gamma - \gamma_0)E_\zeta \cdot E_0)dx = \frac{i}{\omega\mu_0}((\Delta u + F(u,v))/u)\hat{}(\xi),$$

where the hat denotes the Fourier transform and

$$u = \left(\frac{\mu}{\mu_0}\right)^{1/2}, \qquad v = \left(\frac{\gamma}{\gamma_0}\right)^{1/2},$$

and

$$F(u,v) = k^2 u(1 - u^2 v^2).$$

By swapping ν and η, similar equation follows with u and v interchanged. Thus, at the large $|\zeta|$ limit, we obtain an elliptic semilinear system

$$\begin{cases} \triangle u + F(u,v) = pu, \\ \triangle v + F(v,u) = qv, \end{cases}$$

where the coefficients p and q depend solely on the data. Furthermore, we know that $u = v = 1$ outside the domain Ω. The Unique Continuation Principle for elliptic equation implies then that the system has a unique solution.

As a conclusion, we have shown that for the system of Maxwell's equations, there is a quite natural counterpart to the exponentially growing special solutions used in the literature for solving the impedance imaging problem in three dimensions. Furthermore, there is a constructive way of obtaining the boundary values of these solutions. These boundary values in turn determine asymptotically the coefficients of a uniquely solvable system of semilinear elliptic equations, whose solutions are the sought electromagnetic parameters within the body.

References

1. Calderón, A. P.: On an inverse boundary value problem. In: Meyer, W. H., Raupp, M. A. (eds.): *Seminar on Numerical Analysis and its Applications to Continuum Physics.* Brazilian Math Society, Rio de Janeiro 1980, 65-73.
2. Colton,D., Päivärinta, L.: The uniqueness of a solution to an inverse scattering problem for electromagnetic waves. Arch. Rat. Mech. Anal. **119** (1992) 59-70.
3. Henkin, G., Novikov, R.: A multidimensional inverse problem in quantum and acoustic scattering. Inverse Problems **4** (1988) 103-121.
4. Isakov, V.: Completeness of products of solutions and inverse problems for some PDO's. J. Diff. Equations **92** (1991) 305-316.
5. Müller, C.: *Foundations of the Mathematical Theory of electromagnetic Waves.* Springer, New York 1969.
6. Nachman, A.: Reconstructions from boundary measurements. Annals of Math. **128** (1988) 531-576.
7. Ola, P., Päivärinta, L., Somersalo, E.: An inverse boundary value problem in electrodynamics. To appear in Duke J. Math.
8. Ramm, A. G.: Recovery of the potential from fixed energy scattering data. Inverse Problems **4** (1988) 877-886.
9. Somersalo, E., Isaacson, D., Cheney, M.: A linearized inverse boundary value problem for Maxwell's equations. J. Comput. Appl. Math. **42** (1992) 123-136.
10. Sylvester, J., Uhlmann, G.: A global uniqueness theorem for an inverse boundary value problems. Annals of Math. **125** (1987) 153-169.

Inverse Boundary Value Problems for Schrödinger Operators

Ziqi Sun[*]

Department of Mathematics and Statistics, Wicita State University, Wichita, KS 67260, USA

In this note we shall consider the problem of determining potential functions of a Schrödinger operator from its Dirichlet to Neumann map on a bounded domain $\Omega \in \mathbb{R}^n$, $n \geq 2$. This problem is closely related to both the problem of determining the spatially dependent conductivity of a body from steady state direct currrent measurements at the boundary [C] and the scattering problems at a fixed energy, in which one attempts to infer the potential from the related quantum-mechanical scattering amplitude. [No-H]

1. Description of results

We consider the Schrödinger operator with a magnetic (vector) potential $\vec{A}(x)$ and a electric (scalar) potential $q(x)$ in $\mathbb{R}^n$, $n \geq 2$,

$$(1.1) \qquad H_{\vec{A},q} = \sum_{j=1}^{n}\left(-i\frac{\partial}{\partial x_j} + A_j(x)\right)^2 + q(x),$$

where $x = (x_1, x_2, ..., x_n) \in \mathbb{R}^n$, $i = \sqrt{-1}$ and $\vec{A} = (A_1, A_2, ..., A_n)$. We assume that both $\vec{A}$ and q are real valued functions and $A_j \in W^{1,\infty}(\mathbb{R}^n)$, $1 \leq j \leq n$ and $q \in L^{\infty}(\mathbb{R}^n)$.

Let Ω be a bounded domain in $\mathbb{R}^n$ with smooth boundary. We assume that zero is not a Dirichlet eigenvalue of (1.1) on Ω (This condition is always satisfied in many applications). Then for any boundary value $f \in H^{1/2}(\partial\Omega)$ there exists a unique solution $u \in H^1(\Omega)$ which solves

$$(1.2) \qquad H_{\vec{A},q}u = 0 \quad in \ \Omega \quad and \quad u\,|\,_{\partial\Omega} = f.$$

---
[*] Partly supported by NSF DMS 9123742

The Dirichlet to Neumann map of (1.1) on Ω is the operator $\Lambda_{\vec{A},q}$, mapping $H^{1/2}(\partial\Omega)$ into $H^{-1/2}(\partial\Omega)$, defined as follows.

(1.3) $$\Lambda_{\vec{A},q}: f \to \frac{\partial u}{\partial N}\big|_{\partial\Omega} + i(\vec{A}\cdot N)f, \quad f \in H^{1/2}(\partial\Omega)$$

with u solution of (1.2) and N the outer normal on $\partial\Omega$. It is a straightforward computation to check that the "energy" associated to the solution u of (1.2),

$$I_{\vec{A},q}(u) = \int_{\Omega}\left(\nabla u \nabla\overline{u} + (\vec{A}^2 + q)u\overline{u} + i\vec{A}\cdot(u\nabla\overline{u} - \overline{u}\nabla u)\right)dx,$$

can be expressed in terms of the boundary value f through $\Lambda_{\vec{A},q}$:

$$I_{\vec{A},q}(u) = \int_{\partial\Omega}\overline{f}\Lambda_{\vec{A},q}(f)ds.$$

In this paper we assume that $supp\ \vec{A}$, $supp\ q \in \overline{\Omega}$. Thus in this case

(1.4) $$\Lambda_{\vec{A},q}(f) = \frac{\partial u}{\partial N}\big|_{\partial\Omega}.$$

When Ω is given, the Dirichlet to Neumann map $\Lambda_{\vec{A},q}$ is uniquely determined by $H_{\vec{A},q}$, i.e. by potentials $\vec{A}$ and q. The problem under discussion in this paper is whether the converse is true. More specifically, we ask whether the potentials $\vec{A}$ and q are uniquely determined by $\Lambda_{\vec{A},q}$.

In recent years significant progress has been made on this problem in the case of $\vec{A} = 0$. In this case global uniqueness holds in dimension $n \geq 3$ [S-U,I][N-S-U][L-N] and local and generic uniqueness hold in dimension $n = 2$ [S-U,II][Su-U].

In the case of $\vec{A} \neq 0$, however, there is an obstruction to uniqueness. In fact a change of the magnetic potential $\vec{A}$ to its gauge equivalence $\vec{A}' = \vec{A} + \nabla g$ for some $g \in W^{1,\infty}$ with $g = \frac{\partial g}{\partial N} = 0$ on $\partial\Omega$ would not change the Dirichlet to Neumann map $\Lambda_{\vec{A},q}$. Indeed, it is a straightforward computation to show that replacing $\vec{A}$ by $\vec{A}'$ in (1.1) is equivalent to replacing the solution u in (1.2) by $u' = ue^{-ig}$. Since u' carries the same boundary value and the normal derivative as u, it follows that $\Lambda_{\vec{A}',q} = \Lambda_{\vec{A},q}$.

It is easy to see that the above gauge transformation $\vec{A} \to \vec{A}'$ preserves the $rot\ (\vec{A})$ in dimension $n \geq 3$ and preserves $curl\ (\vec{A})$ in dimension $n = 2$, where

$$rot\ (\vec{A}) = \sum_{j,l=1}^{n}\left(\frac{\partial A_l}{\partial x_j} - \frac{\partial A_j}{\partial x_l}\right)dx_j \wedge dx_l, \quad curl\ (\vec{A}) = \frac{\partial A_1}{\partial x_2} - \frac{\partial A_2}{\partial x_1}.$$

Physically, $rot\,(\vec{A})$ (or $curl\,(\vec{A})$) is the magnetic field associated to (1.1).

The inverse boundary value problem for (1.1) is now the problem of determining both $rot\,(\vec{A})$ (or $curl\,(\vec{A})$) and q from knowledge of $\Lambda_{\vec{A},q}$. One can show that in general this is the best one can expect in the case of $\vec{A}\neq 0$. We have obtained two uniqueness theorems concerning this problem. (See Theorem 1 and Theorem 2 below.) Theorem 1 deals with this problem in the case of $n\geq 3$. It shows that the Dirichlet to Nuemann map $\Lambda_{\vec{A},q}$ determines uniquely $rot\,(\vec{A})$ and q provided $rot\,(\vec{A})$ is small. Theorem 2 is a generic local uniqueness theorem, which shows that $\Lambda_{\vec{A},q}$ determines uniquely $curl\,(\vec{A})$ and q provided $curl\,(\vec{A})$ is small and q is close to "most" scalar potentials. In what follows we use $W_{\Omega}^{m,\infty}$ to denote the space of functions f in $W^{m,\infty}(\mathbb{R}^n)$ with $supp\,f\subset\bar{\Omega}$.

Theorem 1 $(n\geq 3)$. *Let* $\vec{A}_j\in W_{\Omega}^{2,\infty}$, $q_j\in L^{\infty}(\Omega)$, $j=1,2$. *Assume that zero is not a Dirichlet eigenvalue of* $H_{\vec{A}_j,q_j}$, $j=1,2$. *Then there exists a positive constant* $\varepsilon=\varepsilon(\Omega)$ *such that if*

$$\|\,rot\,(\vec{A}_j)\,\|_{L^{\infty}(\Omega)}<\varepsilon,\ j=1,2\quad and\quad \Lambda_{\vec{A}_1,q_1}=\Lambda_{\vec{A}_2,q_2},$$

then

$$rot\,(\vec{A}_1)=rot\,(\vec{A}_2)\quad and\quad q_1=q_2,$$

Theorem 2 $(n=2)$. *There exists an open and dense set* $\mathcal{O}\subset W_{\Omega}^{1,\infty}$ *such that, for every* $q\in\mathcal{O}$, *there are a* $W_{\Omega}^{1,\infty}$ *neighborhood* $\mathcal{O}_q$ *of* q *and a positive constant* $\varepsilon_q=\varepsilon_q(\Omega)$, *such that if*

$$\|\,curl\,(\vec{A}_j)\,\|_{W_{\Omega}^{2,\infty}}<\varepsilon_q\ and\ q_j\in\mathcal{O}_q,\ j=1,2\quad and\quad \Lambda_{\vec{A}_1,q_1}=\Lambda_{\vec{A}_2,q_2},$$

then

$$curl\,(\vec{A}_1)=curl\,(\vec{A}_2)\quad and\quad q_1=q_2,$$

Theorem 1 and 2 can be viewed as extensions of the uniqueness results proven in [N-S-U] and [Su-U]. The detailed proofs of Theorem 1 and 2 have been presented in [Su,I] and [Su,II]. In the next two sections we shall review the proofs, focusing mainly on the ideas and motivation behind the analysis.

2. Proof of Theorem 1

We begin with an identity which relates potential functions $\vec{A}$ and q to the Dirichlet to Neumann map $\Lambda_{\vec{A},q}$.

Propositon 2.1. *Let* $\vec{A}_j$ *and* q_j, $j=1,2$, *be potential functions. Then*

$$i\int_{\Omega}(\vec{A}_1-\vec{A}_2)\cdot(u_1\nabla\overline{u}_2-\overline{u}_2\nabla u_1)dx + \int_{\Omega}(\vec{A}_1^2-\vec{A}_2^2+q_1-q_2)u_1\overline{u}_2dx$$

$$= -\int_{\partial\Omega}\overline{u}_2(\Lambda_{\vec{A}_1,q_1}-\Lambda_{\vec{A}_2,q_2})u_1 ds.$$

holds for arbitrary u_j solution of $H_{\vec{A}_j,q_j}u_j = 0$, $j = 1,2$.

The proof of this identity follows an integration by parts argument and the definition of $\Lambda_{\vec{A},q}$. Following the idea of the geometric optics construction of solutions to hyperbolic equations, we look for exponentially growing solutions of the form given below in the null space of $H_{\vec{A},q}$.

(2.1) $$u(x,\xi) = e^{\xi x + \phi(x,\xi)}(1 + \omega(x,\xi)),$$

where $\xi \in \mathbf{C}^n$ is a complex vector satisfying $\xi\cdot\xi = 0$ and the function $\omega(x,\xi)$ behaves like $|\xi|^{-1}$ as $|\xi|$ tends to ∞ in an appropriate function space. We shall show that it is always possible to construct such solutions provided $\| rot\,(\vec{A})\|_{L^\infty(\Omega)}$ is sufficiently small and $|\xi|$ is sufficiently large.

Substituting (2.1) into the equation $H_{\vec{A},q}u = 0$, we get two equations

(2.2) $$\xi\cdot\nabla\phi = -i\xi\cdot\vec{A},$$

(2.3) $$\Delta\omega + 2(\xi+\nabla\phi+i\vec{A})\cdot\nabla\omega - G\omega = G,$$

where

(2.4) $$G = \vec{A}^2 - i\nabla\cdot\vec{A} + q - 2i\vec{A}\cdot\nabla\phi - \nabla\phi\cdot\nabla\phi - \Delta\phi.$$

From now on we assume

(2.5) $$|\xi| \geq 1, \quad \vec{A}\in W_{\overline{\Omega}}^{2,\infty}, \quad q \in L^\infty(\mathbf{R}^n), \quad supp\,q \subset \overline{\Omega}.$$

We construct ϕ by Fourier transforming (2.2):

(2.6) $$\phi(x,\xi/|\xi|) = \left(\frac{\xi\cdot\vec{A}^\wedge(\eta)}{\xi\cdot\eta}\right)^\vee = (2\pi)^{-n}\int_{\mathbf{R}^n}e^{-ix\eta}\left(\frac{\xi\cdot\vec{A}^\wedge(\eta)}{\xi\cdot\eta}\right)d\eta,$$

where $\eta = (\eta_1,\eta_2,\,...,\,\eta_n)$ is the dual coordinates and $\wedge$ denotes the Fourier transform with respect to x. We denote by $\vee$ its inverse. Then one can show the following result.

Proposition 2.2. *The solution $\phi(x,\xi)$ defined by (2.6) has the following three properties.*

(2.7)
$$\| \phi(\cdot, \xi / | \xi |) \|_{W^{2,\infty}(\Omega)} \le C \| \vec{A} \|_{W_H^{2,\infty}},$$

(2.8)
$$\| \nabla \phi + i \vec{A} \|_{L^\infty(\Omega)} \le C \| rot\, (\vec{A}) \|_{L^\infty(\Omega)},$$

(2.9)   If $\xi(s): (a,b) \to \mathbf{C}^n$ is a differentiable map with $\xi(s) \cdot \xi(s) = 0$ and $|\xi(s)| \ge 1$ for all s, then $s \to \phi(\cdot, \xi(s)/|\xi(s)|)$ is differentiable as a map from (a,b) to $L^\infty(\Omega)$.

The constant C involved in (2.7)-(2.9) depends only on Ω.

To construct ω we set $\widetilde{G} = G\chi_\Omega$ and $\widetilde{\phi} = \phi\chi_\Omega$, where χ_Ω is the indicator function of Ω. (2.3) can be written as

(2.10)
$$(L_\xi + 2(\nabla\widetilde{\phi} + i\vec{A}) \cdot \nabla - \widetilde{G})\omega = \widetilde{G}.$$

where

$$L_\xi = \Delta + 2\xi \cdot \nabla.$$

For $\xi \neq 0$, the operator L_ξ admits a bounded inverse $L_\xi^{-1}: L^2(\Omega) \to H^1(\Omega)$ [S-U,I]. Applying L_ξ^{-1} to both sides of (2.10) we get the following integral equation for ω.

(2.11)
$$(I + 2F_1 + F_2)\omega = L_\xi^{-1}\widetilde{G},$$

where

$$F_1 = L_\xi^{-1} \circ (\nabla\widetilde{\phi} + i\vec{A}) \circ \nabla, \qquad F_2 = L_\xi^{-1} \circ (\widetilde{G}).$$

Using properties of the inverse operator L_ξ^{-1} as well as properties of ϕ, one can show that the operator $I + 2F_1 + F_2$ is invertible provided that $|\xi|$ is large and $\| rot\, (\vec{A_j}) \|_{L^\infty(\Omega)}$ is small. This gives us the following

Proposition 2.3. *Let $\vec{A}$ and q be potential functions satisfying (2.5). Then there exist positive constants $\delta = \delta(\Omega)$ and K such that if*

$$\| rot\, (\vec{A_j}) \|_{L^\infty(\Omega)} < \epsilon, \quad and \quad |\xi| > K,$$

then the equation (2.3) has a solution $\omega \in H^1(\Omega)$. Moreover,

(2.12)
$$\| \omega \|_{L^2(\Omega)} \le C |\xi|^{-1}$$

and

$$(2.13) \qquad \|\, \nabla \omega \,\|_{L^2(\Omega)} \leq C,$$

where K and C depend only on Ω, $\|\, \vec{A} \,\|_{W^{2,\infty}(\Omega)}$ and $\|\, q \,\|_{L^\infty(\Omega)}$.

Proposition 2.1 shows that if $\Lambda_{\vec{A}_1, q_1} = \Lambda_{\vec{A}_2, q_2}$, then

$$(2.14) \qquad i \int_\Omega (\vec{A}_1 - \vec{A}_2) \cdot (u_1 \nabla \overline{u}_2 - \overline{u}_2 \nabla u_1) dx + \int_\Omega (\vec{A}_1^2 - \vec{A}_2^2 + q_1 - q_2) u_1 \overline{u}_2 dx = 0$$

holds for arbitrary u_j solution of $H_{\vec{A}_j, q_j} u_j = 0$, $j = 1, 2$. We now choose u_1 and u_2 to be exponentially growing solutions.

Let k, γ_1 and γ_2 be three mutually orthogonal vectors in $\mathbb{R}^n$ with $|\gamma_1| = |\gamma_2| = 1$. Let ζ, $\xi \in \mathbb{C}^n$ be given by $\zeta = \gamma_1 + i\gamma_2$ and $\xi = s\zeta + g(s, k)\gamma_1$, where s is a positive real parameter and

$$(2.15) \qquad g(s, k) = 2^{-1} |k|^2 ((|k|^2 + 4s^2)^{1/2} + 4s)^{-1}.$$

Then we set

$$(2.16) \qquad \xi_1 = \frac{ik}{2} + \xi, \qquad \overline{\xi}_2 = \frac{ik}{2} - \xi.$$

One checks that

$$(2.17) \qquad \xi_1 \cdot \xi_1 = \xi_2 \cdot \xi_2 = 0, \quad \xi_1 + \overline{\xi}_2 = ik, \quad \xi_1 - \overline{\xi}_2 = 2\xi,$$

$$(2.18) \qquad \xi_1/s \to \zeta, \quad \overline{\xi}_2/s \to -\zeta, \quad \xi/s \to \zeta, \quad as\ s \to \infty.$$

Using Proposition 2.2 and 2.3 we can construct

$$(2.19) \qquad u_j(x, \xi_j) = e^{\xi_j \cdot x + \phi_j(x, \xi_j/|\xi_j|)} (1 + \omega_j(x, \xi_j))$$

solution of $H_{\vec{A}_j, q_j} u_j = 0$, $j = 1, 2$, where ϕ_j solves

$$(2.20) \qquad \xi_j \cdot \nabla \phi_j = -i\xi_j \cdot \vec{A}_j, \quad j = 1, 2$$

and ω_j, $j = 1, 2$, satisfies

$$(2.21) \qquad \|\, \omega_j \,\|_{L^2(\Omega)} \leq C |\xi_j|^{-1} \ and \ \|\, \nabla \omega_j \,\|_{L^2(\Omega)} \leq C,$$

where C depends only on Ω, $\|\, \vec{A}_j \,\|_{W^{2,\infty}(\Omega)}$ and $\|\, q_j \,\|_{L^\infty(\Omega)}$, $j = 1, 2$.

Substituting (2.19) into (2.14) and using (2.21) we get

$$(2.22) \qquad \int_\Omega e^{ikx + \phi_1 + \overline{\phi}_2} \xi \cdot (\vec{A}_1 - \vec{A}_2) dx + O(1) = 0$$

Dividing (2.22) by s and using Proposition 2.2, we get

$$(2.23) \qquad \int_\Omega e^{ikx + \phi_1^* + \overline{\phi_2^*}} \, \zeta \cdot (\vec{A}_1 - \vec{A}_2) dx = 0,$$

where $\phi_j^* = \phi_j^*(x, \zeta)$ are solutions (given by the formula (2.6)) to

$$(2.24) \qquad \zeta \cdot \nabla \phi_1^* = -i\zeta \cdot \vec{A}_1, \quad \zeta \cdot \nabla \overline{\phi_2^*} = i\zeta \cdot \vec{A}_2.$$

(2.23) and (2.24) are all we need to prove $rot \, (\vec{A}_1) = rot \, (\vec{A}_2)$. Adding two equations in (2.24) together gives

$$(2.25) \qquad \zeta \cdot (\vec{A}_1 - \vec{A}_2) = i\zeta \cdot \nabla(\phi_1^* + \overline{\phi_2^*}).$$

Substituting (2.25) into (2.23) and using an integration by parts argument (notice that $k \perp \zeta$), we get

$$(2.26) \qquad \int_{\partial\Omega} e^{ikx}(\zeta \cdot N)e^{\Psi(x,\zeta)} ds = 0,$$

where

$$(2.27) \qquad \Psi(x, \zeta) = \phi_1^*(x, \zeta) + \overline{\phi_2^*}(x, \zeta).$$

We shall use (2.26) to prove our result. The remaining proof can be divided into three steps.

Sept 1.

$$(2.28) \qquad \int_{\partial\Omega} e^{ikx}(\zeta \cdot N)(\zeta \cdot x)^m e^{\Psi(x,\zeta)} ds = 0.$$

holds for any integer $m \geq 0$

The main idea used in this step is to choose in (2.26) two special families of frames $\{\zeta^{(1)}(\theta), k^{(1)}(\theta)\}$ and $\{\zeta^{(2)}(\theta), k^{(2)}(\theta)\}$, depending smoothly on a parameter $\theta \in [0, \frac{\pi}{4}]$. More precisely, we define $\zeta^{(j)}(\theta) = \gamma_1^{(j)}(\theta) + i\gamma_2^{(j)}(\theta)$, $j = 1, 2$ and define $\gamma_1^{(1)}(\theta) = \gamma_2$ and $\gamma_1^{(2)}(\theta) = \gamma_1$ for all θ. We then construct $\gamma_1^{(1)}(\theta)$ and $k^{(1)}(\theta)$ (resp. $\gamma_2^{(2)}(\theta)$ and $k^{(2)}(\theta)$) by rotating the (right-handed) two dimensional frame $\{\gamma_1, k\}$ (resp. $\{\gamma_2, k\}$) clockwise with a angle θ in the plane spanned by γ_1 and k (resp. by γ_2 and k). Then, letting $k = k^{(j)}(\theta)$ and $\zeta = \zeta^{(j)}(\theta)$, $j = 1, 2$, in (2.26) and differentiating with respect to θ we get

$$0 = \frac{d}{d\theta} \Big(\int_{\partial\Omega} e^{ik^{(1)}(\theta)x}(\zeta^{(1)}(\theta) \cdot N)e^{\Psi(x,\zeta^{(1)}(\theta))} ds \Big) \Big|_{\theta=0}$$

$$+i\frac{d}{d\theta}\Big(\int_{\partial\Omega}e^{ik^{(2)}(\theta)x}(\zeta^{(2)}(\theta)\cdot N)e^{\Psi(x,\zeta^{(2)}(\theta))}ds\Big)\Big|_{\theta=0}$$

$$=|k|i\int_{\partial\Omega}e^{ikx}(\zeta\cdot N)(\zeta\cdot x)e^{\Psi(x,\zeta)}ds$$

$$+\int_{\partial\Omega}e^{ikx}(\zeta\cdot N)e^{\Psi(x,\zeta)}\frac{d}{d\theta}\big(\Psi(x,\zeta^{(1)}(\theta))+i\Psi(x,\zeta^{(2)}(\theta))\big)\Big|_{\theta=0}ds.$$

Now, based on the way of our construction of $\zeta^{(1)}(\theta)$ and $\zeta^{(2)}(\theta)$ as well as the structure of Ψ, we are able to show that the last integral in the above formula equals to zero. This proves (2.28) with $m=1$. Repeating this procedure gives (2.28) for a arbitary m.

Step 2. *Let T be any two dimensional plane that is parallel to γ_1 and γ_2. Then*

$$(2.29)\qquad\qquad \int_{\partial\Omega\cap T}(\zeta\cdot N_T)^m\Psi(x,\zeta)ds_T=0$$

for any integer $m\geq 1$, where N_T is the outer normal of $\partial\Omega\cap T$ in T.

Notice that $k\perp\gamma_1$ and γ_2. It is easy to see that (2.28) implies

$$\int_{\partial\Omega\cap T}(\zeta\cdot N)(\zeta\cdot x)^m e^{\Psi(x,\zeta)}ds_T=0$$

for any integer $m\geq 0$. without loss of generality we assume that $\Omega=\{x\in\mathbb{R}^n,\ |x|<R\}$ is a ball and thus $\partial\Omega\cap T$ is a circle with origin as its center in the plane T. Therefore

$$(2.30)\qquad\qquad \int_{\partial\Omega\cap T}(\zeta\cdot N_T)^m e^{\Psi(x,\zeta)}ds_T=0$$

for any integer $m\geq 1$. If we denote by θ, $0\leq\theta<2\pi$, the angle between γ_1 and N_T, then $\zeta\cdot N_T=e^{i\theta}$ and (2.30) reads $\int_0^{2\pi}e^{im\theta}e^{f(\theta)}d\theta=0$, $\forall$ integer $m\geq 1$, where $f=\Psi|_{\partial\Omega\cap T}$. This implies that there exists a holomorphic function u defined on $D=\{x\in\mathbb{R}^2;\ |x|<1\}$ such that $u|_{\partial D}=e^l$. By the mapping property of the exponential function one sees that u maps D into the right half plane $\mathbb{R}^2_+=\{x\in\mathbb{R}^2;\ x_1>0\}$. Hence, $log\ u$ is well defined on D. Since $log\ u$ is also holomorphic in D and $log\ u=l$ in ∂D, we must have $\int_0^{2\pi}e^{im\theta}f(\theta)d\theta=0$, $\forall$ integer $m\geq 1$, which leads to (2.29).

Step 3. $rot\,(\vec{A}_1)=rot\,(\vec{A}_1).$

Multiplying e^{ikx} to both sides of (2.25) and integrating by parts we find that

(2.31) $$\gamma_1 \cdot \int_{\Omega} e^{ikx} \, (\vec{A}_1 - \vec{A}_2) dx = - \int_{\partial\Omega} e^{ikx} \, Im((\zeta \cdot N)\Psi(x,\zeta)) dx.$$

Using (2.29) with $m = 1$ and an integration by parts argument we see that the right side of (2.31) equals to zero. Thus

$$\gamma_1 \cdot \int_{\Omega} e^{ikx} \, (\vec{A}_1 - \vec{A}_2) dx = 0$$

for any k and γ_1 with $k \perp \gamma_1$. This is sufficient to conclude $rot\,(\vec{A}_1) = rot\,(\vec{A}_1)$.

To prove $q_1 = q_2$. We first notice that $rot\,(\vec{A}_1) = rot\,(\vec{A}_1)$ implies that there exists $p \in W_0^{1,\infty}(\Omega)$ so that $\vec{A}_1 - \vec{A}_2 = \nabla p$ in Ω. Since $\Lambda_{\vec{A},q}$ is invariant under gauge transformations $\vec{A} \to \vec{A} + \nabla p$, we deduce that $\Lambda_{\vec{A}_1,q_2} = \Lambda_{\vec{A}_2,q_2}$. Then by the hypothesis we must have $\Lambda_{\vec{A}_1,q_1} = \Lambda_{\vec{A}_1,q_2}$. Thus, we may assume without loss of generality that $\vec{A}_1 = \vec{A}_2 = \vec{A}$ in the rest of our proof. Under this assumption the proposition (2.1) reads

(2.32) $$\int_{\Omega} (q_1 - q_2) u_1 \bar{u}_2 dx = 0.$$

Substituting the solution (2.19) into (2.32), letting $s \to \infty$ and using the fact $\phi_1 + \bar{\phi}_2 \to \phi_1^* + \bar{\phi}_2^* = 0$ in $L^\infty(\Omega)$ as $s \to \infty$, one gets $(q_1 - q_2)^\wedge(k) = 0$ for any $k \in \mathbb{R}^n$. Therefore $q_1 = q_2$ in Ω.

3. Proof of Theorem 2

The proof of Theorem 2, which relies on a different idea, can be sketched as follows. We first construct a special class of exponentially growing solutions in the null space of $H_{\vec{A},q}$. We then subtitute these solutions into the orthogonality identity of $H_{\vec{A},q}$ to build two new identities. One of them carries information about the unknown function $curl\,(\vec{A}) + q$ and the other carries information about $- curl\,(\vec{A}) + q$. From these identities we then construct for each q a bounded linear operator $\mathcal{A}_q$, mapping $L^2(\Omega) \times L^2(\Omega)$ into $L^2(\Omega) \times L^2(\Omega)$, and show that if the (nonlinear) map Λ fails to be injective modulo gauge transformations near $curl\,(\vec{A}) = 0$ and q, then $\mathcal{A}_q$ must have nontrivial kernel. Finally, we show that $\mathcal{A}_q$ is a Fredholm operator with zero index and $\mathcal{A}_q$ depends analytically on q. We then prove Theorem 2 using the analytic Fredholm theorem.

The computations needed to carry out the above arguments are technical and lengthy. We shall discuss only the major steps.

We assume in this section that $\vec{A} \in W_\Omega^{3,\infty}$, $q \in W_\Omega^{1,\infty}$. We begin with an orthogonality identity. Let $\vec{A}_j$, q_j, $j = 1,2$, be potential functions and let u_j be the $H^1(\Omega)$ solution to $H_{\vec{A}_j, q_j} u_j = 0$, $j = 1,2$. Then, as in the case when dimension $n \geq 3$, we have

$$(3.1) \qquad i \int_\Omega (\vec{A}_1 - \vec{A}_2) \cdot (u_1 \nabla \overline{u}_2 - \overline{u}_2 \nabla u_1) dx + \int_\Omega (\vec{A}_1^2 - \vec{A}_2^2 + q_1 - q_2) u_1 \overline{u}_2 dx = 0$$

provided $\Lambda_{\vec{A}_1, q_1} = \Lambda_{\vec{A}_2, q_2}$.

Given potential functions $\vec{A}$, q, The two-dimensional analogue of the exponentially growing solution takes the form

$$(3.2) \qquad u(x, \xi) = e^{\xi \cdot x + \phi(x)}(1 + \omega(x, \xi)),$$

where $\xi \in \mathbb{C}^2$ is a complex vector satisfying $\xi \cdot \xi = 0$ and $\omega(x, \xi)$ behaves like $|\xi|^{-1}$ as $|\xi| \to \infty$. In this case the phase factor $\phi(x)$ depends only on x. In dimension three or greater similar solutions were constructed in the Sobolev space $H^2(\Omega)$ in the previous section. However, in the case of two dimensions we need to construct global solutions. More preccisely, we need to construct solutions of the form (3.2) so that $\phi \in H_\delta^3$ and $\omega \in H_\delta^2$ with $-1 < \delta < 0$, where H_δ^2 and H_δ^3 are Sobolev spaces based on the weighted L^2 space on $\mathbb{R}^2$:

$$(3.3) \qquad L_\delta^2 = \{f; \; \int_{\mathbb{R}^2} (1 + |x|^2)^\delta |f|^2 dx < \infty\}.$$

One of the advantages of using H_δ^2 space is that in this case the solution to (3.2) is unique. The uniqueness of solutions plays an important role in some steps of the proof of Theorem 2.

As in Section 2 we can show that there exist constants $\epsilon = \epsilon(\Omega) > 0$ and $K = K(\Omega, \delta, \|\vec{A}\|_{W^{2,\infty}(\Omega)}, \|q\|_{W^{1,\infty}(\Omega)}) > 0$ so that if $\|curl\,(\vec{A})\|_{W^{1,\infty}(\Omega)} < \epsilon$ and $|k| > K$, then there exists a uniqueness solution u of the form (3.2). Moreover, ϕ and ω satisfy the following estimates.

$$(3.4) \qquad \|\phi\|_{L_\delta^2} + \|\nabla \phi\|_{H_\delta^2 + 1} \leq C \|\vec{A}\|_{H^2(\Omega)},$$

$$(3.5) \qquad \|\phi\|_{H_\delta^3} \leq C \|curl\,(\vec{A})\|_{H^1(\Omega)},$$

$$(3.6) \qquad \|\omega\|_{H_\delta^1} \leq \frac{C}{|\xi|} \quad and \quad \|\omega\|_{H_\delta^2} \leq C,$$

for some constant which is independent of ξ.

Let $\vec{A}_j$, q_j, $j = 1, 2$, be potential functions so that $\Lambda_{\vec{A}_1, q_1} = \Lambda_{\vec{A}_2, q_2}$. We assume that the $W_\Omega^{1,\infty}$ norm of $curl\ (\vec{A}_j)$ is less than ϵ, $j = 1, 2$. We set $\xi_1^\pm = \xi^\pm$ and $\xi_2^\pm = -\xi^\pm$, with $\xi^\pm = 2^{-1}(\pm Jk + ik)$, where

$$k \in \mathbb{R}^2 \ with \ |k| > K \quad and \ J = \begin{bmatrix} 0 & 1 \\ -1 & 0 \end{bmatrix}.$$

We then construct expoentially growing solutions

$$(3.7) \qquad u_j^\pm(x, k) = e^{\xi_j^\pm \cdot x + \phi_j^\pm(x)}(1 + \omega_j^\pm(x, \xi)), \qquad j = 1, 2,$$

for the equation $H_{\vec{A}_j, q_j} u_j^\pm = 0$, $j = 1, 2$. We now substitute the solutions (3.1) into the identity (3.1). After a long computation we get two new identities, one of them carries explicit information of $curl\ (\vec{A}_1 - \vec{A}_2) + (q_1 - q_2)$ and the other carries explicit information of $-curl\ (\vec{A}_1 - \vec{A}_2) + (q_1 - q_2)$. Namely, we have

Proposition 3.1. *Assume* $\Lambda_{\vec{A}_1, q_1} = \Lambda_{\vec{A}_2, q_2}$. *Then*

$$(3.8) \qquad \int_\Omega e^{ikx + \phi_1^\pm + \overline{\phi}_2^\pm}[\pm curl\ (\vec{A}_1 - \vec{A}_2) + (q_1 - q_2)](1 + \omega_1^\pm + \overline{\omega}_2^\pm + \omega_1^\pm \overline{\omega}_2^\pm)dx$$

$$= \int_\Omega e^{ikx + \phi_1^\pm + \overline{\phi}_2^\pm}(\vec{A}_1 - \vec{A}_2)F^\pm dx,$$

where,

$$F^\pm = \mp [\nabla^\tau \omega_1^\pm + \nabla^\tau \overline{\omega}_2^\pm + \nabla^\tau(\omega_1^\pm \overline{\omega}_2^\pm)] - i(\nabla \overline{\omega}_2^\pm - \nabla \omega_1^\pm + \omega_1^\pm \nabla \overline{\omega}_2^\pm - \overline{\omega}_2^\pm \nabla \omega_1^\pm).$$

To construct the operator $\mathcal{A}_q$ mentioned above, one first needs to establish a compactness result for potentials functions which satisfy $\Lambda_{\vec{A}_1, q_1} = \Lambda_{\vec{A}_2, q_2}$. The following type of estimate would be sufficient:

$$\| \pm curl\ (\vec{A}_1 - \vec{A}_2) + q_1 - q_2 \|_{H^s(\Omega)} \le C\big(\| curl(\vec{A}_1 - \vec{A}_2) \|_{L^2(\Omega)} + \| q_1 - q_2 \|_{L^2(\Omega)} \big)$$

for $s, 0 < s < 1$.

We achive this goal in an indirect way. Using estimates (3.4)-(3.6) and some more detailed information about the structure of the solution (3.2), we can deduce from (3.8) the following compactness result, which is weaker than the above one.

Proposition 3.2. *Assume* $\Lambda_{\vec{A}_1, q_1} = \Lambda_{\vec{A}_2, q_2}$. *Then*

$$\| \pm curl\,(\vec{A}_1 - \vec{A}_2) + q_1 - q_2 \|_{H^s(\Omega)} \leq C\big(\| \vec{A}_1 - \vec{A}_2 \|_{H^1(\Omega)} + \| q_1 - q_2 \|_{L^2(\Omega)} \big)$$

for s $0 < s < 1$, where C is a constant depending only on the $W_{\Omega}^{2,\infty}$ norms of $\vec{A}_j$, the $W_{\Omega}^{1,\infty}$ norms of q_j, $j = 1,2$, and Ω.

The term $\| \vec{A}_1 - \vec{A}_2 \|_{H^1(\Omega)}$ in the right side of (3.9) is the main source of troubles, since this term cannot be majorized by $\| curl(\vec{A}_1 - \vec{A}_2) \|_{L^2(\Omega)}$ in general. To overcome this difficult, we need the following adjustment result, whose proof is based on uniqueness of the exponentially growing solutions in the weighted Sobolev's spaces..

Proposition 3.3. *Assume that* $\Lambda_{\vec{A}_1, q_1} = \Lambda_{\vec{A}_2, q_2}$. *Then there exist potential functions* $\vec{A}_1'$, $\vec{A}_2'$ *such that*

$$\Lambda_{\vec{A}_1', q_1} = \Lambda_{\vec{A}_2', q_2}, \quad \nabla \cdot \vec{A}_1' = \nabla \cdot \vec{A}_2', \quad curl\,(\vec{A}_j) = curl\,(\vec{A}_j'), \quad j = 1,2$$

and

$$\| \vec{A}_j' \|_{W^{2,\infty}(\Omega)} \leq C(\| \vec{A}_1 \|_{W^{3,\infty}(\Omega)} + \| \vec{A}_2 \|_{W^{3,\infty}(\Omega)}).$$

The construction of $\mathcal{A}_q$ involves a limit process. We indicate briefly the procedure. Assuming that the map Λ: $(\vec{A}, q) \rightarrow \Lambda_{\vec{A}, q}$ is not locally injective modulo gauge transformations near q for small $curl\,(\vec{A})$. Then there exist four sequences of potential functions $\vec{A}_{1,l}, \vec{A}_{2,l} \in W_{\Omega}^{3,\infty}$ and $q_{1,l}, q_{2,l} \in W_{\Omega}^{1,\infty}$, $l \in \mathbf{Z}^+$, such that

$$curl\,(\vec{A}_{j,l}) \rightarrow 0 \;\; in \; W_{\Omega}^{3,\infty} \;\; and \;\; q_{j,l} \rightarrow 0 \;\; in \; W_{\Omega}^{1,\infty} \;\; as \; l \rightarrow 0, \;\; j = 1,2,$$

$$either \quad curl\,(\vec{A}_{1,l}) \neq curl\,(\vec{A}_{2,l}) \quad or \quad q_{1,l} \neq q_{2,l} \;\; \forall l \in \mathbf{Z}^+,$$

$$\Lambda_{\vec{A}_{1,l}, q_{1,l}} = \Lambda_{\vec{A}_{2,l}, q_{2,l}} \qquad \forall l \in \mathbf{Z}^+.$$

We now adjust the above sequences $\{\vec{A}_{j,l}\}$ to $\{\vec{A}_{j,l}'\}$, $j = 1,2$, using Proposition 3.3. For simplicity, we still denote $\{\vec{A}_{j,l}'\}$ by $\{\vec{A}_{j,l}\}$. The result of the adjustment is that the new sequences $\{\vec{A}_{j,l}\}$, $j = 1,2$, have one more property:

$$\nabla \cdot \vec{A}_{1,l} = \nabla \cdot \vec{A}_{2,l}.$$

This new property , together with Proposition 3.2, gives us desired control:

(3.9)
$$\| \pm curl\,(\vec{A}_{1,l} - \vec{A}_{2,l}) + q_{1,l} - q_{2,l} \|_{H^s(\Omega)}$$

$$\leq C\big(\| curl(\vec{A}_{1,l} - \vec{A}_{2,l}) \|_{L^2(\Omega)} + \| q_{1,l} - q_{2,l} \|_{L^2(\Omega)} \big).$$

We now substitute corresponding exponentially growing solutions

$$u_{j,l}^{\pm}(x,k) = e^{\xi_j^{\pm}\cdot x + \phi_{j,l}^{\pm}(x)}(1 + \omega_{j,l}^{\pm}(x,k)), \quad j = 1,2, \quad l \in \mathbb{Z}^+$$

into (3.8). Dividing (3.8) by $\| curl\,(\vec{A}_1 - \vec{A}_2) \|_{L^2(\Omega)} + \| q_1 - q_2 \|_{L^2(\Omega)}$ and then letting $l \to \infty$, we seek to study the limiting expression of (3.8). (3.9) guarantees that the limit exists for a subsequence. This limit can be expressed as the following equation for $V^{\pm} \in L^2(\Omega)$ with $\| V^+ \|_{L^2(\Omega)} + \| V^- \|_{L^2(\Omega)} = 1$.

(3.10)
$$\int_{\Omega} e^{ikx} V^{\pm}(1 + \omega_q^+ + \omega_q^- + \omega_q^+ \omega_q^-)dx \pm H = 0,$$

(3.11)
$$H = i\int_{\Omega} e^{ikx}[(-k_2 + ik_1)\overline{W}\omega_q^+ - (k_2 + ik_1)W\omega_q^-]dx$$

$$+ i\int_{\Omega} e^{ikx}[(\overline{\partial}\overline{W})\omega_q^+ - (\partial W)\omega_q^- - i\overline{W}\omega_q^- \overline{\partial}\omega_q^+ + iW\omega_q^+ \partial\omega_q^-]dx.$$

with $\omega_q^{\pm}$ solutions of

$$\triangle\omega_q^{\pm} + 2\xi^{\pm}\cdot\nabla\omega_q^{\pm} - q\omega_q^{\pm} = q$$

and

$$2W = \nabla^{\tau}\triangle_{\Omega}^{-1}(V^+ - V^-),$$

where $\triangle_{\Omega}^{-1}$ denote the inverse of $\triangle$ with zero Dirichlet boundary data.

The above analysis shows that if the map $\Lambda: (\vec{A}, q) \to \Lambda_{\vec{A},q}$ is not locally injective modulo gauge transformations near q for small $curl\,(\vec{A})$, then the equation (3.10) must have nonzero solution $V^{\pm} \in L^2(\Omega)$. We now rewrite (3.10) in terms of operators. Let's denote by $\mathcal{F}$ and $\mathcal{F}^{-1}$ the Fourier and its inverse, respectively. Define linear operator $\mathcal{Z} = \chi_{\Omega}\mathcal{F}^{-1}\Phi_K\mathcal{F}$, which can be shown to be an isomorphism between $L^2(\Omega)$ and $L^2(\Omega)$, where χ_{Ω} is the indicator function of Ω and Φ_K is a function of k (the dual of x) satisfying $\Phi(k) = 0$ for $|k| \leq K$, $\Phi(k) = 1$ for $|k| > K$. Given $q \in W^{1,\infty}(\Omega)$, define linear operators $\mathcal{P}_q$ and $\mathcal{K}_q$ as follows: for $f \in L^2(\Omega)$,

$$\mathcal{P}_q(f) = \chi_{\Omega}\mathcal{F}^{-1}\Big(\Phi_K\int_{\Omega} e^{ikx}f(\omega_q^+ + \omega_q^- + \omega_q^+ \omega_q^-)dx\Big)$$

$$\mathcal{H}_q(f) = i\chi_\Omega \mathcal{F}^{-1}\left(\Phi_K \int_\Omega e^{ikx}[(-k_2 + ik_1)\overline{f}\omega_q^+ - (k_2 + ik_1)f\omega_q^-]dx\right)$$

$$+ i\chi_\Omega \mathcal{F}^{-1}\left(\Phi_K \int_\Omega e^{ikx}[(\overline{\partial}f)\omega_q^+ - (\partial f)\omega_q^-]dx\right)$$

$$- i\chi_\Omega \mathcal{F}^{-1}\left(\Phi_K \int_\Omega e^{ikx}[\overline{f}\omega_q^- \overline{\partial}\omega_q^+ - f\omega_q^+ \partial\omega_q^-]dx\right).$$

Finally, we define linear operator $\mathcal{A}_q$ on $L^2(\Omega) \times L^2(\Omega)$ by

$$(3.12) \qquad \mathcal{A}_q = \begin{bmatrix} \mathcal{Z} + \mathcal{P}_q + \mathcal{H}_{q_0}\mathcal{W} & -\mathcal{H}_{q_0}\mathcal{W} \\ -\mathcal{H}_{q_0}\mathcal{W} & \mathcal{Z} + \mathcal{P}_q + \mathcal{H}_{q_0}\mathcal{W} \end{bmatrix}$$

Then one can check that (3.10) can be rewritten as $\mathcal{A}_q(V^+, V^-) = 0$. This complete the construction of $\mathcal{A}_q$. Our conclusion at this stage is

Proposition 3.4. *Let $q \in W^{1,\infty}(\Omega)$. If $\mathcal{A}_q$ is injective in $L^2(\Omega) \times L^2(\Omega)$, then Λ is locally injective modulo gauge transformations near $\mathrm{curl}\ (\vec{A}) = 0$ and near q in the $W_\Omega^{3,\infty} \times W_\Omega^{1,\infty}$ topology.*

The significance of this result is that it reduces the local injectivity problem of the nonlinear operator Λ to the injectivity problem of the linear operator $\mathcal{A}_q$. To prove Theorem 2, we need only to show that there exists an open and dense set $\mathcal{O}$ in $W_\Omega^{1,\infty}$ such that $\mathcal{A}_q$ is injective for $q \in \mathcal{O}$. If $q = 0$, we have $\omega^\pm \equiv 0$ and hence $\mathcal{A}_q = \mathcal{Z}$. This means that $V^+ = V^- = 0$. This show that Λ is injective modulo gauge transformations near $q = 0$ and near $\mathrm{curl}\ (\vec{A}) = 0$. In general, however, it is totally nontrivial to see whether V^+ and V^- must equal to zeroes.

So far we have considered only real-valued potential function q. If we define $\mathcal{A}: q \to \mathcal{A}_q$, then $\mathcal{A}$ is map defined only in the space of real potential functions. In order to continue our proof we need to extend the map $\mathcal{A}$ to the space of complex-valued potential functions. Once this is done, we then show that the operator $\mathcal{A}_q$ has the following three properties.

(3.13)   *For any q, $\mathcal{A}_q$ is a Fredholm operator with zero index, mapping $L^2(\Omega) \times L^2(\Omega)$ into $L^2(\Omega) \times L^2(\Omega)$.*

(3.14)   *$\mathcal{A}_q$ depends analytically on q,*

(3.15)   *$\mathcal{A}_0$ is an isomorphism.*

Properties (3.13) − (3.15) enable us to apply the analytic Fredholm theorem [R-S] to the map $\mathcal{A}$ so that we get desired result. See [Su,II] for details.

REFERENCES

[C]   A.P. Calderon, *On an inverse boundary value problem*, Seminar on Numerical Analysis and its Applications to Continuum Physics, Soc. Brasileira de Matematica. Rio de Jfaneiro, (1980), 65-73.

[L-N]   R.B. Lavine and A. Nachman, *Global uniqueness in inverse problems with singular potentials*, in preparation.

[N-S-U]   A. Nachman, J. Sylvester and G. Uhlmann, *An n-dimensional Borg-Levinson theorem*, Comm. Math. Physics, 115 (1988), 595-605.

[No-H]   R. Novikov and G. Henkin, $\bar{\partial}$-*equation in the multidimensional inverse scattering problem*, Uspekhi Mat. Nauk., 42 (3), (1987), 93-152.

[R-S]   M. Reed and B. Simon, *Methods of mathematical physics, Vol. I, functional analysis*, Academic Press, New York, 1980.

[S-U,I]   J. Sylvester and G. Uhlmann, *A global uniqueness theorem for an inverse boundary value problems*, Ann. of Math., 125 (1987), 153-169.

[S-U,II] _____, *A uniqueness theorem for an inverse boundary value problem in electrical prospection*, Comm. Pure Appl. Math, 39 (1986), 91-112.

[Su,I]   Z. Sun, *An inverse boundary value problem for the Schrödinger operator with vector potentials*, Trans. of AMS. (to appear)

[Su,II]   _____, *An inverse boundary value problem for the Schrödinger operator with vector potentials in two dimensions*. Comm. in PDE (to appear)

[Su-U]   Z. Sun and G. Uhlmann, *Generic uniqueness for an inverse boundary value problem*, Duke Math. J, 62 (1), (1991), 131-155.

Linearizations of Anisotropic Inverse Problems

*John Sylvester**

University of Washington, Seattle, WA 98195, USA

In this lecture we will discuss inverse boundary value problems. The problem is to determine the coefficients of an equation or system of equations from measurements of solutions at the boundary.

The first general formulation of this problem was given by Calderon [C] for the impedance tomography problem. In this problem, Ω is a bounded smooth domain in $\mathbb{R}^n$; $\gamma(x)$ is a smooth strictly positive function; $u(x)$ is a voltage potential, a solution to

$$L_\gamma u := \nabla \cdot \gamma \nabla u = 0 \quad \text{in } \Omega. \tag{1}$$

The inverse problem is to deduce the function γ from knowledge of the Cauchy data on $\partial\Omega$ of all solutions to (1). More specifically, define

$$\mathcal{M}_\gamma = \{(f,\alpha)|fu|_{\partial\Omega}, \alpha = \gamma\frac{\partial u}{\partial \nu}|_{\partial\Omega}, L_\gamma u = 0\} \tag{2}$$

We call $\mathcal{M}_\gamma$ the linear space of boundary measurements. In the impedance tomography problem, $\mathcal{M}_\gamma$ is the graph of the linear operator Λ_γ, the so-called Dirichlet to Neumann, or voltage to current, operator.

What Calderon did in [C] was to compute the derivative at $\gamma \equiv 1$ of the map Φ:

$$\gamma \overset{\Phi}{\mapsto} \mathcal{M}_\gamma. \tag{3}$$

He calculated $D\Phi|_{\gamma=1}$ and exhibited its inverse. Although, this calculation did not permit any direct conclusions about the full nonlinear problem (the range of $D\Phi$ is not closed, so that the implicit function theorem does not apply), this calculation was the first convincing evidence that the impedance tomography problem was solvable. Subsequent work by Kohn and Vogelius, Sylvester and Uhlmann, and others (see the review paper [S-U 1]) have borne out these conclusions in various cases.

The purpose of this talk is to compute the kernel of the derivative of the map Φ in (3) for certain anisotropic inverse problems. We will begin with the anisotropic impedance tomography problem, then the anisotropic fixed energy wave equation, and lastly anisotropic Maxwell's equations. We begin with

* Partially supported by NSF grant DMS–9123757.

1 Impedance tomography

$$L_\gamma u = d\gamma du = \frac{\partial}{\partial x^i}\gamma^{ij}\frac{\partial}{\partial x^j}u = 0 \quad \text{in } \Omega$$
$$u|_{\partial\Omega} = f \tag{4}$$
$$\gamma du|_{\partial\Omega} = \nu_j\gamma^{ij}\frac{\partial u}{\partial p^j}dS|_{\partial\Omega} = \alpha$$

In the description above, we have used the differential geometric notation; d is the exterior derivative, γ is a map from Λ^1, exterior one forms, to Λ^{n-1} exterior $n-1$ forms, ν_i are the components of the unit euclidean exterior conormal, and dS is the euclidean surface measure on $\partial\Omega$.

$$\mathcal{M}_\gamma = \{(f,\alpha)|f = u|_{\partial\Omega}, \alpha = \gamma du|_{\partial\Omega}, L_\gamma u = 0\}. \tag{5}$$

$\mathcal{M}_\gamma$ is a subspace of $\mathcal{H} = C^\infty(\partial\Omega) \oplus C^\infty(\partial\Omega, \Lambda^{n-1}(\partial\Omega))$. If we define the symplectic form on $\mathcal{H}$

$$A(p,q) := \int_{\partial\Omega} p_1q_2 - q_1p_2 \tag{6}$$

Suppose that

$$
\begin{array}{ll}
L_{\gamma_0} u = 0 & L_{\gamma_1} v = 0 \\
u|_{\partial\Omega} = p_1 & v|_{\partial\Omega} = q_1 \\
\gamma_0 du|_{\partial\Omega} = p_2 & \gamma_1 dv|_{\partial\Omega} = q_2
\end{array}
$$

Then we have the basic relation

$$A(p,q) = \int_{\partial\Omega} u\gamma_1 dv - v\gamma_0 du = \int_\Omega du_\wedge(\gamma_1 - \gamma_0)dv. \tag{7}$$

If we take $\gamma_0 = \gamma_1$ in (7) we see that

$$A|_{\mathcal{M}_\gamma} = 0 \tag{8}$$

and a little more checking will show that if $p_* \in \mathcal{H}$ and $A(p,q) = 0$ for every $q \in \mathcal{M}_\gamma$, then $q \in \mathcal{M}_\gamma$; that is $\mathcal{M}_\gamma$ are maximal subspaces on which A vanishes, Lagrangian subspaces. In particular, we note that

$$A(p,q) = 0 \text{ for all } p \in \mathcal{M}_{\gamma_0}, \text{ and } q \in \mathcal{M}_{\gamma_1} \iff \mathcal{M}_{\gamma_0} = \mathcal{M}_{\gamma_1} \tag{9}$$

This is slightly more general than the formulation (10) below which uses the Dirichilet to Neumann map

$$\int_{\partial\Omega} f(\Lambda_0 - \Lambda_1)g = 0 \; \forall \; f,g \in C^\infty(\partial\Omega), \tag{10}$$

as it applies directly to cases where the Dirichlet problem does not have a unique solution.

Suppose now that we have a family of conductivities, $\gamma(s)$, depending on the single parameter s, and two family of solutions, $u(s)$ and $v(s)$, with boundary data $p(s)$ and $q(s)$, i.e.,

$$
\begin{aligned}
L_\gamma u &= 0 & L_\gamma v &= 0 \\
u|_{\partial\Omega} &= p_1 & v|_{\partial\Omega} &= q_1 \\
\gamma du|_{\partial\Omega} &= p_2 & \gamma dv|_{\partial\Omega} &= q_2.
\end{aligned}
\tag{11}
$$

Then, via (7)

$$
A(p(s), q(0)) = \int_\Omega du(s)(\gamma(s) - \gamma(0)dv(0).
\tag{12}
$$

If we divide (12) by s and let $s \to 0$, we obtain

$$
\frac{d}{ds}|_{s=0} A(p(s), q(0)) = \int_\Omega du(0)\dot{\gamma}(0)dv(0).
\tag{13}
$$

If the curve $\gamma(s)$ of conductivities all give rise to the same boundary measurements, then the left hand side of (13) is zero, even if p depends on s. Thus, we have shown that if

$$
D\Phi\dot{\gamma} = 0,
\tag{14}
$$

then

$$
0 = \int_\Omega du \wedge \dot{\gamma} dv
\tag{15}
$$

for all u and v satisfying $L_\gamma u = L_\gamma v = 0$. We shall compute the derivative in (14) at $\gamma \equiv 1$, so that u and v may be arbitrary harmonic functions. It is an old theorem of Runge that all solutions to constant coefficient PDE's can be approximated by the complex exponential and polynomial solutions. We shall restrict ourselves here to considering only exponential solutions. We choose

$$
u = e^{x \cdot \zeta_1} \qquad v = e^{x \cdot \zeta_2},
$$

and insist that[2] $\zeta \in \mathbb{C}^n$ satisfies

$$
\zeta \cdot \zeta = 0
\tag{16}
$$

to guarantee that u and v are harmonic. With this choice, (15) becomes

$$
\begin{aligned}
0 &= \int_\Omega e^{x \cdot (\zeta_1 + \zeta_2)} \zeta_1^T \dot{\gamma} \zeta_2, \\
0 &= \frac{1}{4} \int_\Omega e^{x \cdot (\zeta_1 + \zeta_2)} \{(\zeta_1 + \zeta_2)^T \dot{\gamma}(\zeta_1 + \zeta_2) - (\zeta_1 - \zeta_2)^T \dot{\gamma}(\zeta_1 - \zeta_2)\}.
\end{aligned}
\tag{17}
$$

In coordinates, a conductivity is represented by a positive definite symmetric matrix, hence the notation $\zeta_1^T \dot{\gamma} \zeta_2$ above just means the matrix product.

If we constrain

$$
\zeta_1 + \zeta_2 = ik; \qquad k \in \mathbb{R}^n
$$

---

[2] We use the notation ζ without subscript to refer to both ζ_1 and ζ_2.

$$0 = \int_\Omega e^{x \cdot ik} \{ -k^T \dot{\gamma} k - (\zeta_1 - \zeta_2)^T \dot{\gamma} (\zeta_1 - \zeta_2) \} \tag{18}$$

Now, in the isotropic case (γ is a multiple of the identity matrix) which Calderon considered, a consequence of (16) is that

$$(\zeta_1 - \zeta_2)^T \dot{\gamma} (\zeta_1 - \zeta_2) = \dot{\gamma}(\zeta_1 - \zeta_2)^T (\zeta_1 - \zeta_2) = -\dot{\gamma}(\zeta_1 + \zeta_2)^T (\zeta_1 + \zeta_2) = \dot{\gamma} k^T k \tag{19}$$

so that (18) becomes

$$0 = \int_\Omega \dot{\gamma} \cdot e^{ix \cdot k} k \cdot k,$$

$$0 = (X_\Omega \dot{\gamma})^\wedge (k) \cdot |k|^2,$$

where X_Ω is the characteristic function of Ω and $^\wedge$ denotes the Fourier Transform. Hence we conclude that $(X_\Omega \dot{\gamma})^\wedge \equiv 0$ and therefore that $X_\Omega \dot{\gamma} = 0$[3]. Thus we have proved that

Theorem (Calderon). *Suppose that $\gamma(s)$ is a smooth family of isotropic conductivities, $\gamma(0) = 1$ and, for all s, the boundary measurements are equal, then $\dot{\gamma}(0) = 0$.*

For anisotropic conductivities, the kernel of $D\Phi$ is not empty. We may return to (18), but (19) does not hold. Instead, we note that,

$$\{l \in \mathbb{C}^n | l \cdot k = 0, l \cdot l = k \cdot k\} \tag{20}$$

$$= \{l \in \mathbb{C}^n | l = \zeta_1 - \zeta_2; \zeta_1 + \zeta_2 = ik, \zeta \cdot \zeta = 0\},$$

that is, (18) reads

$$0 = -k^T \widehat{\gamma}(k) k - l^T \widehat{\gamma}(k) l \tag{21}$$

for all l satisfying $l \cdot l = k \cdot k, l \cdot k = 0$. We rewrite (21) as

$$\widehat{\gamma}(k) \Big|_{\{l | l \cdot k = 0\}} = -\frac{k^\wedge \widehat{\gamma}(k) k}{k^T k} I \tag{22}$$

that is, the quadratic form $\widehat{\gamma}(k)$, restricted to the subspace $\{l \cdot k = 0\}$ is a scalar multiple of the identity matrix. Since (21) is rotationally equivariant and homogeneous we may choose a basis where k has coordinates $(1, 0, \ldots 0)$ then (22) reads

$$\widehat{\gamma} \Bigg|_{\begin{pmatrix} 1 \\ 0 \\ \vdots \\ 0 \end{pmatrix}^\perp} = -\widehat{\gamma}_{11} I$$

----

[3] We cheat here slightly and will cheat in this way throughout this talk, we should conclude that $(X_\Omega \dot{\gamma})^\wedge$ is a distribution supported at the origin and then use (15) with harmonic polynomials in place of u and v to conclude that $(X_\Omega \dot{\gamma})^\wedge(0) = 0$

or

$$\widehat{\gamma} = \begin{pmatrix} \begin{array}{c|ccc} v_1 & v_2 & \cdots & v_n \\ \hline v_2 & & & \\ \vdots & & -v_1 I & \\ v_n & & & \end{array} \end{pmatrix}$$

where $v = (v_1, \ldots, v_n)$ is an arbitrary vector. This can be rewritten

$$\widehat{\gamma} = \begin{pmatrix} v_1 \\ \vdots \\ v_n \end{pmatrix} (1, 0, \ldots, 0) + \begin{pmatrix} 1 \\ 0 \\ \vdots \\ 0 \end{pmatrix} (v_1 \cdots v_n) - tr \left[\begin{pmatrix} 1 \\ 0 \\ \vdots \\ 0 \end{pmatrix} (v_1, \ldots, v_n) \right] I.$$

For general nonzero k this can be expressed as

$$\widehat{\gamma}(k) = \widehat{v}(k)k^T + k\widehat{v}(k)^T - tr(k\widehat{v}(k)^T)I \tag{23}$$

where $v = v(k)$ is an arbitrary vector valued function of k and tr denotes the trace.

What we have shown is that any curve of conductivities with the same boundary measurements has derivative, $\dot{\gamma}$, satisfying (23).

It is possible to recognize these linear deformations as coming from the natural action of diffeomorphisms on conductivities for the full nonlinear problem. Namely, if Φ is a diffeomorphism from Ω to Ω and $\Phi|_{\partial\Omega}$ is the identity map, then Φ pushes forward γ to another conductivity

$$\Phi_* \gamma := \frac{D\Phi^T \gamma D\Phi}{\det D\Phi}, \tag{24}$$

where $D\Phi$ is the Jacobian matrix.

It is easy to check that

$$M_\gamma = M_{\Phi_* \gamma}.$$

If we choose, in euclidean coordinates

$$\gamma = I \qquad \Phi = I + sV$$

where V is a vector field on $\mathbb{R}^n$, then

$$D\Phi = I + sDV$$

and

$$\frac{d}{ds}\Big|_{s=0} \Phi_* \gamma = \frac{d}{ds}\Big|_{s=0} \frac{(I + sDV^T)(I + sDV)}{\det(I + sDv)} = DV^T + DV - trDV \tag{25}$$

Taking the Fourier Transform yields,

$$\left(\frac{d}{ds}\Big|_{s=0} \Phi_* \gamma \right)^{\wedge} (k) = \widehat{V}k^T + k\widehat{V}^T - tr(k\widehat{V}^T)I. \tag{26}$$

Notice that (26) is exactly (23) with $v = \widehat{V}$. Thus we have shown

Theorem. *Let $\gamma(s)$ be a curve of conductivities with $\gamma(0) = I$ and with constant boundary measurements. Then the tangent $\dot\gamma(0)$ has the form (25), i.e. $\dot\gamma(0)$ is tangent to the orbit of the identity under the diffeomorphism action (24).*

This leads to the following

Conjecture. *Let γ_0 and γ_1 be conductivities with the same boundary measurements. Then there exists $\Phi : \Omega \to \Omega$ with $\Phi|_{\partial\Omega} = I$ and $\Phi_*\gamma_0 = \gamma_1$.*

This conjecture is known to be true if $n = 2$ and $\gamma \in C^3$ and is close to the identity [S], and if γ is real analytic [L-U].

2 Anisotropic wave equation at fixed energy

We will next consider the equation

$$(26) \qquad \rho V_{tt} - d\gamma dV = 0$$

where $\gamma(x)$ a conductivity, mapping 1-forms to $n-1$ forms, and ρ is an n-form. We shall consider the equation at fixed frequency, i.e., we set

$$V = e^{i\omega t} v(x)$$

where $v(x)$ satisfies

$$L_{\gamma,\rho} v := -\omega^2 \rho v - d\gamma dv = 0 \qquad (27)$$

$\mathcal{M}_{\gamma,\rho}$ is defined exactly as in (5), namely

$$\mathcal{M}_{\gamma,\rho} = \{(f,\alpha)| f = u|_{\partial\Omega}, \alpha = \gamma du|_{\partial\Omega}; L_{\gamma,\rho} v = 0\}$$

The basic relation which replaces (7) is

$$A(p,q) = \int_{\partial\Omega} u\gamma_0 dv - v\gamma_1 du = \int_\Omega du_\wedge (\gamma_0 - \gamma_1) dv - \omega^2(\rho_0 - \rho_1) uv \qquad (28)$$

If we now let $\rho = \rho(s)$, $\gamma = \gamma(s)$ depend on a parameter s with $\rho(0) = \gamma(0) = 1$, and let $u(s)$ and $v(s)$ be solutions with Cauchy data p and q, just as we did in (11), then $(\dot\rho, \dot\gamma) \in \text{kernel } (D\Phi)$ at $(\rho,\gamma) = (1,1)$ if and only if

$$0 = \int_\Omega \omega^2 \dot\rho uv - du^T \dot\gamma dv \qquad (29)$$

We take u and v to be exponential solutions

$$u = e^{ik\cdot\zeta_1} \qquad\qquad v = e^{ik\cdot\zeta_2}$$

where

$$\zeta \cdot \zeta = -\omega^2$$

and (29) becomes

$$0 = \int_\Omega e^{x\cdot(\zeta_1+\zeta_2)}\{\dot\rho\omega^2 - \zeta_1^T \dot\gamma \zeta_2\}$$

or equivalently,

$$0 = 2\omega^2\widehat{\rho} - k^T\widehat{\gamma}k - l^T\widehat{\gamma}l \tag{30}$$

where $\widehat{\rho}$ and $\widehat{\gamma}$ are computed at the point k and

$$l \cdot k = 0 \qquad \text{and} \qquad |l|^2 = |k|^2 - (2\omega)^2.$$

Just as in the impedance tomography problem (22), (30) implies

$$\dot{\gamma}\Big|_{\{l|l\cdot k=0\}} = -\alpha I$$

where

$$\alpha = \frac{(2\omega)^2\widehat{\rho} - k^T\widehat{\gamma}k}{(2\omega)^2 - k^Tk}$$

If we choose coordinates so that k has coordinates $(1, 0, \ldots, 0)$, then

$$\widehat{\gamma} = \left(\begin{array}{c|ccc} v_1 & v_2 & \cdots & v_n \\ \hline v_2 & & & \\ \vdots & & -\alpha I & \\ v_n & & & \end{array} \right)$$

In case $\alpha = v_1$, this kernel corresponds to the same diffeomorphism action as in (24) for impedance tomography.

However, the kernel here is larger, the case where we choose $v_2 = v_3 = \cdots = v_n = 0$ and $\alpha = -v_1$ corresponds to an isotropic perturbation i.e.,

$$\widehat{\gamma} = v_1 I, \tag{31}$$

and

$$\widehat{\rho} = v_1 \left(\frac{1 + k^2 - (2\omega)^2}{2\omega^2} \right). \tag{32}$$

We notice that $\dot{\rho}$ depends on ω, i.e. this perturbation is only possible if we observe at a single frequency, since ρ is independent of ω in our model.

Just as we can recognize the case $\alpha = v_1$ as coming from the action of the diffeomorphism group, we can recognize the perturbation in (31) and (32) as resulting from a change of dependent variables. Suppose that

$$\rho\omega^2 u + d\gamma du = 0$$

and that γ is isotropic. Let

$$\omega = \gamma^{\frac{1}{2}}u.$$

Then

$$\Delta w + \left(\frac{\Delta\gamma^{\frac{1}{2}}}{\gamma^{\frac{1}{2}}} + \frac{\rho}{\gamma}\omega^2 \right) w = 0, \tag{33}$$

and furthermore (q_1, q_2) is Cauchy data for (33) on $\partial\Omega$ if and only if

$$\left(\gamma^{-\frac{1}{2}}q_1, \gamma^{\frac{1}{2}}\left(q_2 - \frac{1}{2}\frac{\partial \log \gamma}{\partial \nu}\right)\right) \in \mathcal{M}_{\gamma, \rho}. \tag{34}$$

Therefore, if we choose $\gamma(s)$ so that

$$\gamma(s)\Big|_{\partial\Omega} = \gamma(0)\Big|_{\partial\Omega}$$
$$\frac{\partial\gamma(s)}{\gamma\nu}\Big|_{\partial\Omega} = \frac{\partial\gamma(0)}{\partial\nu}\Big|_{\partial\Omega}$$

and choose

$$\rho(s) = \frac{\gamma^{\frac{1}{2}}\Delta\gamma^{\frac{1}{2}}}{\omega}.$$

then (33) remains unchanged and therefore, by (34), $\mathcal{M}_{\gamma,\rho}$ is independent of s.

3 Anisotropic Maxwell's equations

In this section, we let $\mathcal{E}$ and $\mathcal{H}$ represent the electric and magnetic 1-forms $(\mathcal{E}, \mathcal{H} \in \Lambda^1(\mathbb{R}^3))$, and ε, μ, and σ represent maps from 1-forms to 2-forms with the symmetry property

$$\text{(34)} \qquad\qquad\qquad \alpha \wedge \varepsilon\beta = \varepsilon\alpha \wedge \beta$$

for any 1-forms α and β. We assume that (34) holds with ε replaced by μ or σ as well. Maxwells equations take the form

$$d\mathcal{E} = \mu\mathcal{H}_t, \qquad d\mathcal{H} = -\varepsilon\mathcal{E}_t - \sigma\mathcal{E}. \tag{35}$$

In the following we shall take $\sigma = 0$ in order to simplify the analysis and fix the frequency equal to ω. We let

$$\mathcal{E} = e^{iwt}E(x), \qquad \mathcal{H} = e^{iwt}H(x),$$

so that E and H satisfy the time independent equations:

$$dE = iw\mu H, \qquad dH = -iw\varepsilon E. \tag{36}$$

For the system (36), the set of boundary measurements consists of pairs of 1-forms

$$\mathcal{M}_{\mu,\varepsilon} = \{(\alpha, \beta)|\alpha = E|_{\partial\Omega}, \beta = H|_{\partial\Omega}, E, H \text{ satisfy (36)}\}$$

The inverse problem is to identify μ, and ε from $\mathcal{M}_{\mu,\varepsilon}$. If Φ denotes the map

$$\begin{pmatrix} \varepsilon \\ \mu \end{pmatrix} \overset{\Phi}{\mapsto} \mathcal{M}_{\mu,\varepsilon}$$

then we shall attempt to compute the kernel of $D\Phi$ at $\varepsilon = \mu =$ the identity.

For the isotropic case the kernel of this map is empty ([S-I-C]) and it is shown in [O-P-S] that Φ is in fact $1 - 1$ in this case. We begin our analysis with the anisotropic version of the [S-I-C] calculations. Let $(E_0(s), H_0(s))$ and $(E(s), H(s))$ solve Maxwell's equations with coefficients $(\varepsilon(s), \mu(s))$ depending on the parameter s

$$\frac{1}{i\omega} \int_{\partial\Omega} E_0(s) \wedge H_1(0) = \frac{1}{i\omega} \int_\Omega dE_0(s) \wedge H_1(0) - E_0(s) \wedge dH_1(0)$$

$$= \int_\Omega \mu(s) H_0(s) \wedge H_1(0) + E_0(s) \wedge \varepsilon(0) E(0).$$

Hence, we have the basic relation

$$\frac{1}{i\omega} \int_{\partial\Omega} E_0(s) \wedge H_1(0) - E_1(0) \wedge H_0(s)$$
$$= \int_\Omega H_0(s) \wedge (\mu(s) - \mu(0)) H_1(0) - E_0(s) \wedge (\varepsilon(s) - \varepsilon(0)) E_1(s),$$
(37)

The left hand side of (37) is zero iff $\mathcal{M}_{\varepsilon,\rho}$ is independent of s; in this case dividing by s and letting s tend to zero gives

$$0 = \int_\Omega H_0 \wedge \dot{\mu} H_1 - E_0 \wedge \dot{\varepsilon} E_1,$$
(38)

where $\dot{\mu} = \frac{d}{ds}\big|_{s=0} \mu(s)$ and (E_0, H_0) and (E_1, H_1) are solutions to Maxwell's equations with coefficients $(\varepsilon(0), \mu(0))$. We will take $\varepsilon(0) = \mu(0) =$ identity and seek exponential solutions in the form

$$E = e^{i\omega\eta\cdot x} e, \qquad H = e^{i\omega\eta\cdot x} h,$$

with η, e and h constant covectors

$$dE = i\omega H, \qquad dH = -i\omega E,$$
(39)

iff

$$\eta_\wedge e = h, \qquad \eta_\wedge h = -e.$$
(40)

Let N denote the linear transformation

$$v \xmapsto{N} \eta_\wedge v;$$

We note that

$$\ker(N) = \{\eta\},$$
$$\text{Range}(N) = \{v | v \cdot \eta = 0\},$$
(41)

and (40) has a solution if and only if

$$\eta \cdot \eta = 1, \qquad e \cdot \eta = 0, \qquad h = Ne.$$
(42)

Now let (η_0, e_0, h_0) and (η_1, e_1, h_1) be any two solutions to (42), then (39) becomes

$$0 = \int_\Omega e^{x \cdot i\omega(\eta_0 + \eta_1)} \{e_0 N_0^T \dot{\mu} N_1 e_1 - e_0 \dot{\varepsilon} e_1\}.$$
(43)

If we decide to fix

$$k = (\eta_0 + \eta_1),$$

then (43) reads

$$0 = e_0 N_0^T \widehat{\mu}(\omega k) N_1 e_1 - e_0 \widehat{\varepsilon} e_1 \tag{44}$$

for every (e, η) satisfying (42). If we use (41) we may write $e = Nf$ so (44) becomes

$$0 = f_0 N_0^T (N_0^T \widehat{\mu}(\omega k) N_1 - \widehat{\varepsilon}(\omega k)) N_1 f_1$$

for every f_0 and $f_1 \in \mathbb{C}^3$ so that the matrix equation below holds

$$0 = N_0^T (N_0^T \widehat{\mu} N_1 - \widehat{\varepsilon}) N_1. \tag{45}$$

Because (45) depends holomorphically on $(\eta_0 - \eta_1)$ $(\eta_0 + \eta_1$ is fixed), it suffices to check (45) when $\text{Im}(\eta_0 - \eta_1) = 0$. That is, we may take $\eta_0 = -\overline{\eta}_1$ to obtain

$$0 = N^*)1(N_1^* \widehat{\mu}(wk) - \widehat{\varepsilon}(wk)) N_1, \tag{46}$$

where $*$ denotes the conjugate transpose.

Since (46) is rotationally invariant we may choose coordinates so that

$$\text{Re } \eta_1 = \begin{pmatrix} \frac{|k|}{2} \\ 0 \\ 0 \end{pmatrix}, \quad \text{Im } \eta_1 = \left(\frac{1 - |k|^2}{4} \right)^{\frac{1}{2}} \begin{pmatrix} 0 \\ \cos\theta \\ \sin\theta \end{pmatrix}. \tag{47}$$

Inserting (47) into (46) yields a 4th order trigonometric polynomial in θ; equating coefficients in that polynomial gives

$$\widehat{\varepsilon} + \widehat{\mu} = \widehat{v} k^T - tr(\widehat{v} k^T) I,$$
$$\widehat{\varepsilon} - \widehat{\mu} = (k^2 - (2\omega)^2) \widehat{\phi} I - 2k^T k \widehat{\phi},$$

or

$$\dot{\varepsilon} + \dot{\mu} = Dv^T + Dv - tr(Dv) I,$$
$$\dot{\varepsilon} - \dot{\mu} = (\Delta - (2\omega)^2) \phi I - 2H(\phi),$$

where ϕ is a scalar valued function, v is a vector field, and $H(\phi)$ is the Hessian of ϕ.

In case we take $\dot{\varepsilon} = \dot{\mu}$, this is a deformation which arises from the (by now familiar) diffeomorphism action, but the case $\dot{\varepsilon} - \dot{\mu}$ yields a different sort of perturbation. It is always anisotropic because of the Hessian term. At present, we cannot give an example of a deformation of the full inverse problem which has this linearization.

References

[C]   Calderón, A. P.: On an inverse boundary value problem. In: *Seminar on Nu-merical Analysis and its Applications to Continuum Physics*. Soc. Brasileira de Matemática, Río de Janeiro, (1980), 65–73.

[L–U]   Lee, J. M., Uhlmann, G.: Determining anisotropic real-analytic conductivity by boundary measurements. comm. Pure appl. Math. **42** (1989) 1097–1112.

[S]   Sylvester, J.: An anisotropic inverse boundary value problem. Comm. Pure Appl. Math. **43** (1990) 201–233.

[S-U 1]   Sylvester, J., Uhlmann, G.: The Dirichlet to Neumann Map and Applications. In: Colton,D., Ewing, R., Rundell, W. (eds.): *Partial Differential Equations*. SIAM Proceedings Series, Philadelphia 1990, 101–139,

[S-U 2]   Sylvester, J., Uhlmann, G.: Inverse problems in anisotropic media. Contemp. Math. **122** (1991) 105–117.

[S-I-C]   Somersalo, E., Isaacson, D., Cheney, M.: A linearized inverse boundary value problem for Maxwell's equations. In : *Seventh Annual Review of Progress in Applied Computational Electromagnetics*. Naval Postgraduate School, Mon-terey 1991.

[O-P-S]   Ola, P., Päivärinta, L., Somersalo, E.: An inverse boundary value problem in electrodynamics. To appear in Duke J. Math. 1993.

Optimal Parameter Choice for Tikhonov Regularization in Hilbert Scales

Ulrich Tautenhahn

Fachbereich Mathematik, Universität Chemnitz, O–9010 Chemnitz , PSF 964, Germany

Abstract *In this paper we consider the method of Tikhonov regularization in Hilbert scales for finding solutions q^* of ill–posed linear and nonlinear operator equations $A(q) = z$, which consists in minimizing the functional $J_\alpha(q) = \|A(q) - z^\delta\|^2 + \alpha \|q - \bar{q}\|_s^2$. Here z^δ are the available perturbed data with $\|z - z^\delta\| \leq \delta$, $\bar{q}$ is a suitable approximation of q^* and $\| \cdot \|_s$ is the norm in a Hilbert scale Q_s. Assuming $\|A(q) - A(q^*)\| \sim \|q - q^*\|_{-a}$ and $q^* - \bar{q} \in Q_{2\gamma}$ for some $a \geq 0$ and $\gamma \geq 0$ we discuss questions of the appropriate choice of the norm $\| \cdot \|_s$ and the regularization parameter $\alpha > 0$. We prove that the choice of the regularization parameter by Morozov's discrepancy principle leads to optimal convergence rates if $2\gamma \in [s, 2s + a]$ and discuss convergence rate results for the case $2\gamma \notin [s, 2s + a]$.*

1 Introduction

In this paper the method of Tikhonov regularization in Hilbert scales for solving explicit and implicit ill–posed inverse problems is investigated. A large class of explicit ill-posed inverse problems can be described by linear or nonlinear integral equations of the first kind $A(q) = z$, where $A : Q \rightarrow Z$ denotes an integral operator between Hilbert spaces Q and Z. Implicit ill–posed inverse problems arise e.g. in problems connected with the identification of unknown coefficients q (which are in general functions) in distributed systems from certain observations z. Distributed systems are governed by differential equations, in general, which may be described by an operator equation of the form

$$F(q, u) = b \tag{1.1}$$

where F maps the couple (q, u) from the product space $Q \times U$ into the space B of the right hand side of equation (1.1). This is of course formal and has to be made precise in each particular case.

In the direct problem associated with (1.1), for given $b \in B$ and given coefficient $q \in Q_{ad} \subseteq Q$ of the set of (physically) admissible parameters the solution $u \in U$ is to be determined. We suppose that for each $q \in Q_{ad}$ of the convex closed subset of Q there exists a unique solution $u \in U$ to (1.1) and denote this solution by $u = G(q)$. Here, G denotes the solution operator of the direct problem.

In (1.1), $F(q, \cdot)$ can be linear or nonlinear and can be stationary or an evolution operator where corresponding initial- and boundary conditions should be also incorporated in the equation (1.1).

In the inverse problem, equation (1.1) and some information on the state u of the form

$$C(u) = z \tag{1.2}$$

is given where the observation operator C of the observation equation (1.2) maps the state $u \in U$ into the observation space Z and z is the noise–free observation. Hence, inverse problems of this type consist in finding $q \in Q_{ad}$ from the equation

$$A(q) = z \quad , \quad A(q) := C(G(q)) , \tag{1.3}$$

where generally, z is unknown and z^δ are the (given) noisy data with $\| z - z^\delta \| \leq \delta$.

A number of applications lead to problems (1.3) which are ill–posed in sense of Hadamard. Let us mention one linear and one nonlinear problem of this kind.

Example 1. Consider the heat equation

$$\begin{aligned} u_t - \Delta u &= f(x,t) \text{ in } \Omega \times (0,T] \\ u(x,0) &= q(x) \quad \text{in } \Omega \\ u(x,t) &= 0 \qquad \text{on } \Gamma \times (0,T] \end{aligned}$$

where $\Omega \subset R^2$ denotes some bounded and open domain with a smooth boundary Γ. If we ask for the initial condition $q(x)$ from additional observations of $z(x) = u(x,T)$ we are led to a linear problem of type (1.3).

Example 2. The identification of the function $q(x)$ in an elliptic problem

$$\begin{aligned} - \operatorname{div}(q \nabla u) + c\, u &= f \text{ in } \Omega \subset R^2 \\ u &= g \text{ on } \Gamma \end{aligned}$$

from observations of $z(x) = u(x)$, $x \in \Omega$, of the state u can be described by a nonlinear equation of the form (1.3).

2 Tikhonov regularization for linear problems

Let A be a bounded linear injective operator between two infinite dimensional Hilbert spaces Q and Z and Q_s be a Hilbert scale (cf. Krein and Petunin (1966)), i.e. Q_s is a Hilbert space with norm

$$\|q\|_s = \left\| B^{\frac{s}{2}} q \right\|_Q \quad , \ s \in R \tag{2.1}$$

where B is an unbounded self–adjoint strictly positive definite operator in Q.

We assume in this Chapter

(A1) there exist constants $M \geq m > 0$ and $a \geq 0$ such that

$$m \|q\|_{-a} \leq \|Aq\| \leq M \|q\|_{-a} \quad \text{for all } q \in Q_s,$$

(A2) there holds $Aq^* = z$ and $\|q^* - \bar{q}\|_{2\gamma} \leq E$ for some $\gamma \geq 0$ and $\bar{q} \in Q$,

(A3) $\|z - z^\delta\| \leq \delta$ with known noise level δ.

Obviously there holds $m = M = 1$ and $a = 1$ for the special case $B = (A^*A)^{-1}$; in this case we denote the Hilbert scale Q_s by $Q_s = Q_s(A)$, the so called Hilbert scale generated by $A : Q \to Z$. Further examples for Q_s are Sobolev spaces of various kinds, in which case a is the "degree of ill–posedness" of the operator equation (1.3) (cf. e.g. Wahba (1980), Hofmann (1986)).

The numerical treatment of ill–posed inverse problems in which A^{-1} is unbounded requires special regularization methods; surveys of regularization methods are given for example in the books of Baumeister (1987) and Louis (1989).

A standard regularization method to deal with ill–posed inverse problems is the method of Tikhonov regularization in Hilbert scales (cf. Natterer (1984)) which consists in searching for approximate solutions q_α^δ by solving the minimization problem

$$\inf_{q \in Q_s} J_\alpha(q); \; J_\alpha(q) = \|Aq - z^\delta\|^2 + \alpha\|q - \bar{q}\|_s^2. \tag{2.2}$$

Here $\alpha > 0$ denotes the regularization parameter to be chosen appropriately, $\|\cdot\|$ denotes the norm in Z and $\bar{q}$ denotes a suitable approximation of the unknown solution q^* of the operator equation (1.3). It is well known that for $\alpha > 0$ problem (2.2) is uniquely solvable. The minimizer q_α^δ is obviously the solution of the Euler equation

$$(A^*A + \alpha B^s)q = A^*z^\delta + \alpha B^s\bar{q}.$$

Objective a priori information on the solution q_α^δ obtained from physical considerations may be taken into consideration by using a programming algorithm to minimize (2.2) subject to $q \in Q_{ad} \cap Q_s$ where Q_{ad} describes the set of physically admissible solutions. Physically meaningful minimizers can be positive solutions or monoton increasing solutions, for example. Assuming that (A1) holds for all $q \in Q_{ad} \cap Q_s$ one can see that the error estimates of that Chapter remain true for the constrained case; the only difference is that we obtain estimates with "better" constants m and M. For the sake of simplicity we treat the unconstrained case in this Chapter.

The following Theorem shows that the accuracy of Tikhonov regularization in Q_s is asymptotically best possible provided that α and s are chosen properly.

Theorem 1. *If (A1)–(A3) are satisfied, then, for* $2\gamma \leq 2s + a$ *and* $\alpha = \left(\frac{\delta}{E}\right)^{\frac{(2s+2a)}{(2\gamma+a)}}$ *there holds*

$$\|q_\alpha^\delta - q^*\|_r \leq c E^{\frac{a+r}{2\gamma+a}} \delta^{\frac{2\gamma-r}{2\gamma+a}} \qquad \forall r \in [-a, 2\gamma]. \tag{2.3}$$

In the special case $r = 0$ and $\bar{q} = 0$ Theorem 1 has been obtained by Natterer (1984). Following the ideas of Natterer it is possible to find an explicit expression of the constant $c = c(m, M, r, s, a, \gamma)$ of the estimate (2.3).

Theorem 1 shows that there is nothing wrong with high order regularization if the regularization parameter α is chosen properly. Unfortunately, the a priori parameter choice given in Theorem 1 is only possible if δ, E, a and γ are known. In the following considerations we discuss Morozov's discrepancy principle for choosing the regularization parameter α (cf. Morozov (1966)). This a posteriori parameter choice strategy does not require to know E, a and γ, here one determines α from the equation

$$. \, h(\alpha) = \|Aq_\alpha^\delta - z^\delta\|^2 - \delta^2 = 0 \, . \tag{2.4}$$

It is well known that the nonlinear equation (2.4) has a unique solution $\alpha = \alpha(\delta)$ provided $\|A\bar{q} - z^\delta\| > \delta$ holds. We prove in the next Theorem that the choice of the regularization parameter α from (2.4) leads to optimal convergence rates provided the order s in the regularization functional (2.2) is chosen properly.

Theorem 2. *Assume (A1)–(A3) and* $\|A\bar{q} - z^\delta\| > \delta$. *If s is chosen such that $2\gamma \in [s, 2s + a]$ holds and α is chosen from Morozov's discrepancy principle (2.4), then there holds the error estimate*

$$\|q_\alpha^\delta - q^*\|_r \leq 2 \, E^{\frac{a+r}{2\gamma+a}} \left(\frac{\delta}{m}\right)^{\frac{2\gamma-r}{2\gamma+a}} \qquad \forall \, r \in [-a, s] \, . \tag{2.5}$$

Proof. From (A1), the triangle inequality and (2.4) we obtain

$$\|q_\alpha^\delta - q^*\|_{-a} \leq \frac{1}{m} \|Aq_\alpha^\delta - Aq^*\|$$

$$\leq \frac{1}{m} \{\|Aq_\alpha^\delta - z^\delta\| + \|z^\delta - Aq^*\|\}$$

$$\leq \frac{2\delta}{m} \, . \tag{2.6}$$

Since q_α^δ solves (2.2) and since $2\gamma \geq s$ we have $J_\alpha(q_\alpha^\delta) \leq J_\alpha(q^*)$, hence

$$\|Aq_\alpha^\delta - z^\delta\|^2 + \alpha \|q_\alpha^\delta - \bar{q}\|_s^2 \leq \|Aq^* - z^\delta\|^2 + \alpha \|q^* - \bar{q}\|_s^2 \, .$$

Since α satisfies (2.4) we obtain

$$\delta^2 \quad + \alpha \|q_\alpha^\delta - \bar{q}\|_s^2 \leq \quad \delta^2 \quad + \alpha \|q^* - \bar{q}\|_s^2 \, ,$$

hence

$$\|q_\alpha^\delta - \bar{q}\|_s^2 \leq \|q^* - \bar{q}\|_s^2 \, .$$

From this inequality, the assumption $2\gamma \geq s$ and (A2) we obtain

$$\|q_\alpha^\delta - q^*\|_s^2 \leq \|q_\alpha^\delta - q^*\|_s^2 - \|q_\alpha^\delta - \bar{q}\|_s^2 + \|q^* - \bar{q}\|_s^2$$

$$= 2 \, (q_\alpha^\delta - q^*, \bar{q} - q^*)_s$$

$$\leq 2 \, E \, \|q_\alpha^\delta - q^*\|_{2s-2\gamma} \, . \tag{2.7}$$

For $2\gamma \in [s, 2s + a]$ the interpolation inequality for Hilbert scales yields

$$\|q_\alpha^\delta - q^*\|_{2s-2\gamma} \le \|q_\alpha^\delta - q^*\|_{-a}^{\frac{2\gamma-s}{s+a}} \|q_\alpha^\delta - q^*\|_s^{\frac{2s+a-2\gamma}{s+a}} . \tag{2.8}$$

We combine (2.6), (2.7) and (2.8) to obtain the inequality

$$\|q_\alpha^\delta - q^*\|_s^2 \le 2 E \|q_\alpha^\delta - q^*\|_s^{\frac{2s+a-2\gamma}{s+a}} \left(\frac{2\delta}{m}\right)^{\frac{2\gamma-s}{s+a}} .$$

Rearranging terms yields

$$\|q_\alpha^\delta - q^*\|_s^{\frac{a+r}{a+s}} \le (2E)^{\frac{a+r}{a+2\gamma}} \left(\frac{2\delta}{m}\right)^{\frac{(2\gamma-s)(a+r)}{(s+a)(a+2\gamma)}} . \tag{2.9}$$

Finally we apply for $r \in [-a, s]$ the interpolation inequality and obtain together with (2.6) and (2.9) the estimate

$$\|q_\alpha^\delta - q^*\|_r \le \|q_\alpha^\delta - q^*\|_{-a}^{\frac{s-r}{s+a}} \|q_\alpha^\delta - q^*\|_s^{\frac{a+r}{a+s}}$$

$$\le (2E)^{\frac{a+r}{2\gamma+a}} \left(\frac{2\delta}{m}\right)^{\frac{s-r}{s+a} + \frac{(2\gamma-s)(a+r)}{(a+s)(2\gamma+a)}}$$

which gives the desired result (2.5).   □

If the parameter s in the functional (2.2) is not chosen properly, i.e. one has $2\gamma \notin [s, 2s + a]$, then one can obtain some nonoptimal convergence rate results which are collected in Figures 1 and 2 (cf. Tautenhahn (1992)).

Remark 1. We do not know if Theorem 1 holds for the region A of Figure 1. If $A^* A$ and B commute, then (2.3) is also valid for the region A of Figure 1.

Remark 2. We do not know if Theorem 2 holds for the regions B and C of Figure 2. If one replaces (2.4) by a more general discrepancy principle $h_c(\alpha) = \|Aq_\alpha^\delta - z^\delta\|^2 - c\delta^2 = 0$ with $c > 1$, then optimal convergence rate results are also valid for the regions B and C of Figure 2.

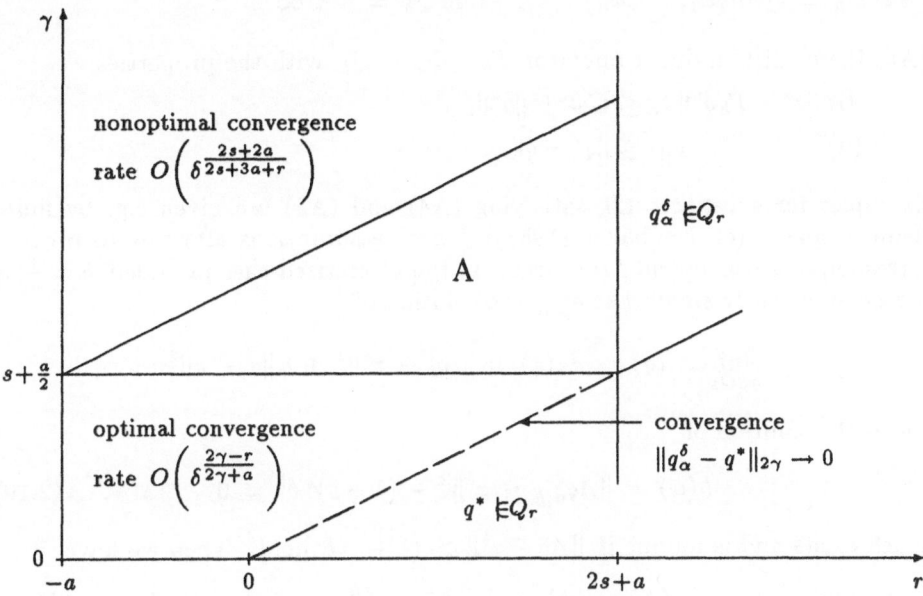

Fig. 1. Convergence rates in case of a priori parameter choice

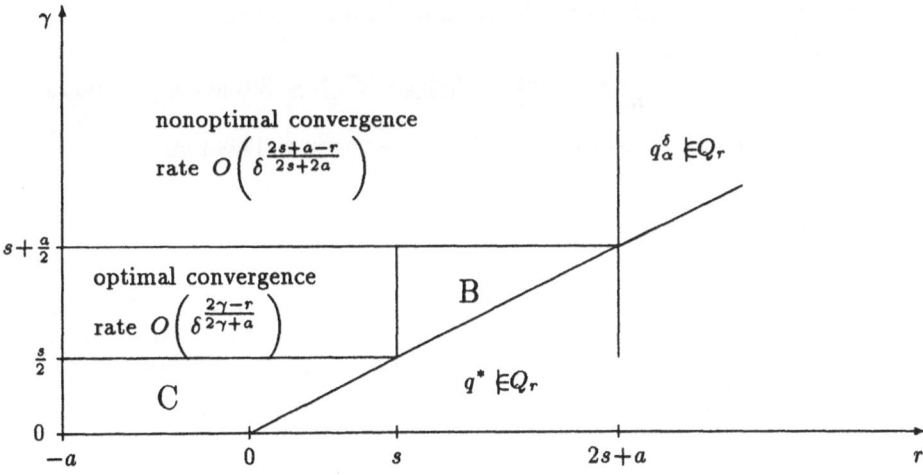

Fig. 2. Convergence rates in case of a posteriori parameter choice

The computation of regularized solutions q_α^δ requires the numerical realization of (2.2) and (2.4) in finite dimensional spaces. We suppose

(A4) $Q_h = span\{\phi_1, \ldots, \phi_n\} \subset Q_s$, $\dim Q_h = n < \infty$

(A5) there exists a linear operator $P_h : Q_s \to Q_h$ with the properties

(i) $\|q^* - P_h q^*\|_{-a} \leq C h^{a+s} \|q^*\|_s$,

(ii) $\|P_h q^* - \bar{q}\|_s \leq \|q^* - \bar{q}\|_s$.

Examples for subspaces Q_h satisfying (A4) and (A5) are given e.g. by finite element spaces (cf. Neubauer (1988)). These assumptions allow us to prove a corresponding convergence rate result in the discretized case provided $h = \frac{1}{n}$ is chosen sufficiently small. Let $q_{\alpha,h}^\delta$ the solution of

$$\inf_{q \in Q_h} J_\alpha(q) ; \quad J_\alpha(q) = \|Aq - z^\delta\|^2 + \alpha \|q - \bar{q}\|_s^2$$

and α the solution of

$$h(\alpha) = \|Aq_{\alpha,h}^\delta - z^\delta\|^2 - (1 + \varepsilon)^2 \delta^2 = 0 \tag{2.10}$$

which exists and is unique if $\|A\bar{q} - z^\delta\| > (1 + \varepsilon) \delta$ holds. Then we have

Theorem 3. *Assume (A1)–(A5) and* $\|A\bar{q} - z^\delta\| > (1 + \varepsilon)\delta$. *If* s *is chosen such that* $2\gamma \in [s, 2s + a]$ *holds and* α *is chosen from (2.10) with* $\varepsilon\delta \geq M C h^{a+s} \|q^*\|_s$, *then there holds the error estimate*

$$\|q_{\alpha,h}^\delta - q^*\|_r \leq (2E)^{\frac{a+r}{2\gamma+a}} \left(\frac{(2+\varepsilon)\delta}{m} \right)^{\frac{2\gamma-r}{2\gamma+a}} \qquad \forall r \in [-a, s] . \tag{2.11}$$

Proof. Instead of (2.6), (2.7) one shows that the inequalities

$$\|q_{\alpha,h}^\delta - q^*\|_{-a} \leq \frac{(2+\varepsilon)\delta}{m} \quad \text{and} \quad \|q_{\alpha,h}^\delta - q^*\|_s^2 \leq 2E \|q_{\alpha,h}^\delta - q^*\|_{2s-2\gamma}$$

are valid. Now one proceeds analogously to the proof of Theorem 2.   □

3 Tikhonov regularization for nonlinear problems

In this Chapter we consider nonlinear ill–posed problems $A(q) = z$ where q and z are elements of infinite dimensional Hilbert spaces Q and Z , respectively, and $A : D(A) \subset Q \to Z$ is a nonlinear operator. Let z^δ are the available perturbed data with $\|z - z^\delta\| \leq \delta$, then we search for approximate solutions q_α^δ by solving the minimization problem

$$\inf_{q \in D(A)} J_\alpha(q) ; J_\alpha(q) = \|A(q) - z^\delta\|^2 + \alpha \|q - \bar{q}\|_s^2 \tag{3.1}$$

where $\| \cdot \|_s$ denotes the norm (2.1).

We suppose the conditions

(B1) there exist constants $M \geq m > 0$ and $a \geq 0$ such that
$$m \|q - q^*\|_{-a} \leq \|A(q) - A(q^*)\| \leq M \|q - q^*\|_{-a} \quad \text{for all } q \in D(A) \cap Q_s ,$$
$$(3.2)$$

(B2) the equation $A(q) = z$ possesses a unique solution $q^* \in D(A)$
with $\|q^* - \bar{q}\|_{2\gamma} \leq E$ for some $\gamma \geq 0$ and $\bar{q} \in Q$,

(B3) there holds $\|z - z^\delta\| \leq \delta$ with known noise level δ,

(B4) the regularized problem (3.1) is solvable for $\alpha > 0$;
the solutions q_α^δ satisfy the inequality $J_\alpha(q_\alpha^\delta) \leq J_\alpha(q)$ for all
$q \in D(A) \cap Q_s$,

(B5) for given $\varepsilon \geq 0$ there exists at least one $\alpha > 0$ satisfying
$$\delta \leq \|A(q_\alpha^\delta) - z^\delta\| \leq (1 + \varepsilon)\delta . \qquad (3.3)$$

Theorem 4. *Let (B1)–(B5) be satisfied and $2\gamma \in [s, 2s + a]$. If $\alpha = \alpha(\delta)$ is an arbitrary solution of (3.3), then there holds the error estimate*
$$\|q_\alpha^\delta - q^*\|_r \leq (2E)^{\frac{a+r}{2\gamma+a}} \left(\frac{(2+\varepsilon)\delta}{m} \right)^{\frac{2\gamma-r}{2\gamma+a}} \qquad \forall r \in [-a, s] .$$

Proof. The proof can be done along the lines of the proof of Theorem 2, where instead of (2.6) we have
$$\|q_\alpha^\delta - q^*\|_{-a} \leq \frac{1}{m} \|A(q_\alpha^\delta) - A(q^*)\|$$
$$\leq \frac{1}{m} \left\{ \|A(q_\alpha^\delta) - z^\delta\| + \|z^\delta - A(q^*)\| \right\} \leq \frac{(2+\varepsilon)\delta}{m} .$$
□

Remark. Let assumption (B1) be satisfied where Q_s denotes a Sobolev scale. In this case the number a in (3.2) can be interpreted as a "degree of ill–posedness" of the nonlinear operator equation (1.3). In the case that A is a linear mapping the assumption (A1) and (B1) are equivalent.

Remark. In order to satisfy (B4) we assume that A is continuous and weakly closed. In this case it can be shown that problem (3.1) has at least one solution $q_\alpha^\delta \in D(A) \cap Q_s$ (cf. Engl, Kunisch and Neubauer (1989)).

Remark. For nonlinear operators A it is well known that the functional $\|A(q_\alpha^\delta) - z^\delta\|$ is not strong monoton increasing and not continuous with respect to α in general (cf. Tikhonov and Arsenin (1977)). In this case Morozov's discrepancy principle (2.4) has to be generalized, e.g. by the a posteriori parameter choice criterion (3.3) which coincides with (2.4) in case $\varepsilon = 0$.

Remark. Under the additional assumptions (A4), (A5) it is possible to give error estimates for discretized versions of (3.1) which coincide with the results of Theorem 3 in the case of linear ill–posed inverse problems (1.3).

References

1. Baumeister, J.:*Stable Solution of Inverse Problems*. Vieweg, Braunschweig, 1987.
2. Engl, H. W. , Kunisch, K. , Neubauer, A.: Convergence rates for Tikhonov regularisation of nonlinear ill–posed problems. Inverse Problems **5** (1989)523–540.
3. Groetsch, C. W.:*The Theory of Tikhonov Regularization for Fredholm Integral Equations of the First Kind*. Pitman, Boston, 1984.
4. Hofmann, B. (1986) *Regularization for Applied Inverse and Ill–Posed Problems*. Teubner Bd. 85, Leipzig, 1986.
5. Krein, S. G. , Petunin, J. I.:Scales of Banach spaces. Russian Math. Surveys **21** (1966) 85–160 .
6. Louis, A. K.:*Inverse und schlecht gestellte Probleme*. Teubner, Stuttgart, 1989.
7. Morozov, V. A.: On the solution of functional equations by the method of regularization. Soviet Math. Doklady **7** (1966) 414–417 .
8. Natterer, F.:Error bounds for Tikhonov regularization in Hilbert scales. Applic. Analysis **18** (1984) 29–37 .
9. Neubauer, A.: An a posteriori parameter choice for Tikhonov regularization in Hilbert scales leading to optimal convergence rates. Siam J. Numer. Anal. **25** (1988) 1313–1326 .
10. Neubauer, A. , Scherzer, O.: Finite–dimensional approximation of Tikhonov regularized solutions of non–linear ill–posed problems. Numer. Funct. Anal. and Optimiz. , **11** (1990) 85–99 .
11. Plato, R. , Vainikko, G.: On the regularization of projection methods for solving ill–posed problems. Numer. Math. **57** (1990) 63–79 .
12. Tautenhahn, U.: Optimal convergence rates for Tikhonov regularization in Hilbert scales. Numer. Math. (1992) (submitted).
13. Tikhonov, A. N. , Arsenin, V. Y.: *Solution of Ill–Posed Problems*. Wiley, New York, 1977
14. Wahba, G.: Ill–posed problems : Numerical and statistical methods for mildy, moderately and severely ill–posed problems with noisy data. Technical Report 595, Univ. Wisconsin, 1980.
15. Wahba, G.: *Spline Models for Observational Data*, Siam, Philadelphia, 1990.

Identification of the Filtration Coefficient

Gennadi Vainikko

University of Tartu, Estonia

1 Inverse problem

Let $\Omega \subset \mathbb{R}^n$ $(n \geq 2)$ be an open bounded region with a piecewise smooth boundary $\partial\Omega$: we denote by ν the outer unit normal to $\partial\Omega$. Let $\Gamma \subseteq \partial\Omega$ be a relatively open set. Consider the following inverse problem: find the coefficient $a = a(x)$ such that

$$
\begin{aligned}
-\mathrm{div}(a(x)\nabla u(x)) &= f(x), & x \in \Omega, \\
[a(x)\nabla u(x)] \cdot \nu(x) &= g(x), & x \in \Gamma,
\end{aligned}
\tag{1}
$$

where $u \in W^{1,\infty}(\Omega)$, $f \in L^2(\Omega)$, $g \in L^2(\Gamma)$ are given functions. We do not exclude the cases $\Gamma = \partial\Omega$ and $\Gamma = \emptyset$. The boundary condition is omitted in (1) in the case $\Gamma = \emptyset$. Physically, u can be interpreted as the piezometrical head of the ground water in Ω; the function f characterizes the sources and sinks in Ω and the function g characterizes the inflow and outflow trough $\Gamma \subseteq \partial\Omega$. The filtration (transmissivity) coefficient a is, physically, positive and piecewise smooth with possible discontinuities of the first kind on some surfaces in Ω.

Conditions (1) can be understood in the sense of distributions. We prefer to deal with the weak formulation of the problem. Introduce the subspace

$$
H^1(\Omega, \Gamma) = \{w \in H^1(\Omega) : w(x) = 0 \text{ for } x \in \partial\Omega \setminus \Gamma\} \subseteq H^1(\Omega).
$$

We obtain the following weak formulation of inverse problem (1): given $u \in W^{1,\infty}(\Omega)$, $f \in L^2(\Omega)$, $g \in L^2(\Gamma)$ find $a \in L^2(\Omega)$ such that

$$
\int_\Omega a\nabla u \cdot \nabla w\, dx = \int_\Omega fw\, dx + \int_\Gamma gw\, dS, \ \forall w \in H^1(\Omega, \Gamma).
\tag{2}
$$

Problem (2) is ill–posed: small perturbations of $u \in W^{1,\infty}(\Omega)$, $f \in L^2(\Omega)$ and $g \in L^2(\Gamma)$ may cause a large perturbation of $a \in L^2(\Omega)$.

2 Discretization

A natural way to discretize the inverse problem (1) is to apply finite element approximations to the weak formulation (2) of the problem. Introduce finite-dimensional subspaces $S_h \subset H^1(\Omega, \Gamma)$ depending on a discretization parameter $h > 0$: we assume that S_h is complete in $H^1(\Omega, \Gamma)$ as $h \to 0$, i.e. for every $w \in H^1(\Omega, \Gamma)$, there exist $w_h \in S_h$ such that $w_h \to w$ in $H^1(\Omega)$ as $h \to 0$. We introduce the following discrete version of problem (2):

$$\left.\begin{array}{l} \text{find } a_h \in L^2(\Omega) \text{ of minimal } L^2(\Omega) \text{ norm such that} \\[2mm] \displaystyle\int_\Omega a_h \nabla u \cdot \nabla w_h dx = \int_\Omega f w_h dx + \int_\Gamma g w_h dS, \quad \forall w_h \in S_h. \end{array}\right\} \tag{3}$$

Problem (3) has never more than one solution. If problem (2) is solvable then (3) is solvable, too, and the solutions satisfy the relation $a_h = P_{h,u}a$ where $P_{h,u}$ is the orthoprojector in $L^2(\Omega)$ corresponding to the subspace $\{a_h \in L^2(\Omega) : a_h = \nabla u \cdot \nabla v_h, \; v_h \in S_h\}$; a consequence is that $a_h \to a_0$ in $L^2(\Omega)$ as $h \to 0$ where $a_0 \in L^2(\Omega)$ is the normal solution (the solution of minimal $L^2(\Omega)$ norm) of problem (2). Conversely if (2) is non–solvable in $L^2(\Omega)$, but (3) is solvable, then $\|a_h\|_{L^2(\Omega)} \to \infty$ as $h \to 0$.

Choosing a basis $w_j = w_{j,h}$ $(j = 1, \ldots, l = l_h)$ of S_h, problem (3) can be reformulated as follows: find

$$a_h = \sum_{j=1}^l c_j \nabla u \cdot \nabla w_j$$

solving the system of linear equations

$$Ac = d$$

where c is l–vector with components c_j, d is l–vector with components

$$d_i = \int_\Omega f w_i dx + \int_\Gamma g w_i dS, \quad i = 1, \ldots, l,$$

and $A = (a_{ij})$ is an $l \times l$–matrix with elements

$$a_{ij} = \int_\Omega (\nabla u \cdot \nabla w_j)(\nabla u \cdot \nabla w_i) dx, \quad i, j = 1, \ldots, l.$$

Note that method (3) as well its regularized version introduced in the next Section need to restore only the first derivatives of u given usually on a sparse grid. The method of characteristics (see e.g. [2]) needs the second derivatives.

3 Regularization

Consider a case where, instead of exact data denoted here by u_0, f_0 and g_0, we have noisy data $u = u_\eta \in W^{1,\infty}(\Omega)$, $f = f_\delta \in L^2(\Omega)$ and $g = g_\delta \in L^2(\Gamma)$ at our disposal. Then numerical difficulties should be expected especially for fine grids, and a precedent regularization of problem (3) is needed. The Tikhonov regularization yields the following numerical scheme:

$$a_{\alpha,h} = \sum_{j=1}^{l} c_{j,\alpha} \nabla u \cdot \nabla w_j, \quad (\alpha B + A) c_\alpha = d.$$

Here d is l–vector with components d_i defined above, c_α is l–vector with components $c_{j,\alpha}$ $(j = 1, \ldots, l)$, $A = (a_{ij})$ and $B = (b_{ij})$ are $l \times l$–matrices with elements a_{ij} defined above and

$$b_{ij} = \int_\Omega \nabla w_j \cdot \nabla w_i dx \quad (i, j = 1, \ldots, l).$$

A suitable value of regularization parameter $\alpha > 0$ depends on the error level of the data. Assume that

$$\sup_{x \in \Omega} |\nabla u_\eta(x) - \nabla u_0(x)| \le \eta \tag{4}$$

$$\|f_\delta - f_0\|_{L^2(\Omega)} \le \delta_1, \quad \|g_\delta - g_0\|_{L^2(\Omega)} \le \delta_2,$$

where $\delta = (\delta_1^2 + \delta_2^2)^{1/2}$ and η are small positive numbers. Then an a priori choice of $\alpha = \alpha(\delta, \eta)$ such that

$$\alpha(\delta, \eta) \to 0, \quad (\delta^2 + \eta^2)/\alpha(\delta, \eta) \to 0 \text{ as } \delta \to 0, \ \eta \to 0$$

guarantees the convergence $a_{\alpha(\delta,\eta),h} \to a_0$ in $L^2(\Omega)$ norm as $h \to 0$, $\delta \to 0$, $\eta \to 0$ where a_0 is the normal solution to (2) corresponding to exact data u_0, f_0, g_0 (we assume that (2) with the exact data is solvable in $L^2(\Omega)$). The same result holds if $\alpha = \alpha(h, \delta, \eta)$ is chosen, according to the residual principle, so that

$$\tilde{\delta} + \langle Ac_\alpha, c_\alpha \rangle^{1/2} \eta \le \langle Ac_\alpha - d, B^{-1}(Ac_\alpha - d) \rangle^{1/2} \le \beta(\tilde{\delta} + \langle Ac_\alpha, c_\alpha \rangle^{1/2} \eta)$$

where $\langle \cdot, \cdot \rangle$ denotes the scalar product in $\mathbb{R}^l$, $\beta \ge 1$ is a constant not depending on h, δ, η and

$$\tilde{\delta} = \sup\{\|\nabla \psi\|_{(L^2(\Omega))^n} : -\Delta \psi(x) = \tilde{f}(x) \text{ for } x \in \Omega,$$
$$\nabla \psi(x) \cdot \nu(x) = \tilde{g}(x) \text{ for } x \in \Gamma,$$
$$\psi(x) = 0 \text{ for } x \in \partial\Omega \setminus \Gamma,$$
$$\|\tilde{f}\|_{L^2(\Omega)} \le \delta_1, \|\tilde{g}\|_{L^2(\Gamma)} \le \delta_2\}.$$

The convergence result concerning the a priori parameter choice remains valid if (4) is replaced by the conditions

$$a_0 \in L^\infty(\Omega), \quad \sup_{x \in \Omega} |u_\eta(x)| \le c, \quad \|\nabla u_\eta - \nabla u_0\|_{(L^2(\Omega))^n} \le \eta$$

where the constant c does not depend on η.

For more details see [3,4].

4 Identifiability

The methods considered in Sections 2 and 3 converge to the normal solution of problem (2). A question about the uniqueness (identifiability) of the filtration coefficient among the functions of the class $L^2(\Omega)$ acutely arises. For sufficiently smooth data, the identifiability of the filtration coefficient among smooth functions is sufficiently fully analysed by G.R. Richter [2], C. Chicone and J. Gerlach [1]. Here, imposing only physically realistic assumptions, we concentrate on identifiability within the class of $L^2(\Omega)$ or more generally $L^1(\Omega)$ functions.

We slightly generalize our inverse problem (cf. (2)): having the data $u \in W^{1,\infty}(\Omega)$, $f \in L^1(\Omega)$ and $g \in L^1(\Omega)$ at our disposal, we look for $a \in L^1(\Omega)$ such that

$$\int_\Omega a\nabla u \cdot \nabla w dx = \int_\Omega fw dx + \int_\Gamma gw dS, \quad \forall w \in W^{1,\infty}(\Omega, \Gamma), \qquad (5)$$

where

$$W^{1,\infty}(\Omega, \Gamma) = \{w \in W^{1,\infty}(\Omega) : w(x) = 0 \text{ for } \partial\Omega \setminus \Gamma\}.$$

We say that the filtration coefficient a is L^1–identifiable from problem (5) on a subregion $\Omega' \subseteq \Omega$ if, for any solutions $a_1 \in L^1(\Omega)$ and $a_2 \in L^1(\Omega)$ to (5), $a_1(x) = a_2(x)$ for a.e. $x \in \Omega'$. It is clear that L^1–identifiability of a from (5) on Ω' implies L^2–identifiability of a from (2) on the same set Ω'.

Now we assume that $\Gamma \subset \partial\Omega$ is a relatively open subset of $\partial\Omega$ with a piecewise smooth (relative) boundary on $\partial\Omega$. We also assume that

$$u \in W^{1,\infty}(\Omega) \cap W^{2,\infty}(\Omega_\varepsilon), \quad \forall \varepsilon > 0, \qquad (6)$$

where Ω_ε consists of all points $x \in \Omega$ such that the distance from x to a nearest non–smoothness point of $\partial\Omega$ exceeds ε; if $\partial\Omega$ is smooth, then (6) means that $u \in W^{2,\infty}(\Omega)$. Introduce the flow curve $x = \varphi(t, y)$ through a point $y \in \Omega$ as the maximal solution (the solution on the maximal time interval) to the Cauchy problem

$$\frac{dx}{dt} = -\nabla u(x), \quad x(0) = y. \qquad (7)$$

Physically, the ground water flows along those curves but the speed depends on ∇u and the filtration coefficient (Darcy's Law).

Due to (6), ∇u is bounded and locally Lipschitz continuous on Ω, with possible singularities of Lipschitz coefficient only as x tends to a non–smoothness point of $\partial\Omega$. Therefore, problem (7) is uniquely solvable and $\varphi(t, y)$ is defined on a finite or infinite time interval (t_y^-, t_y^+); if t_y^- or t_y^+ is finite, then $\varphi(t, y)$ tends to a point on $\partial\Omega$ as $t \downarrow t_y^-$, respectively, $t \uparrow t_y^+$. If $\nabla u(y) \neq 0$, then $\nabla u(\varphi(t, y)) \neq 0$ for all $t \in (t_y^-, t_y^+)$ and $u(\varphi(t, y))$ is strictly decreasing:

$$du(\varphi(t, y))/dt = \nabla u(\varphi(t, y)) \cdot d\varphi(t, y)/dt = -|\nabla u(\varphi(t, y))|^2 < 0, \quad t \in (t_y^-, t_y^+).$$

A corollary is that problem (7) allows no periodic solutions.

Introduce further the following subsets of Ω:

$\Omega_C = \{y \in \Omega : \nabla u(y) = 0\}$,

$\Omega^+ = \{y \in \Omega : \nabla u(y) \neq 0, \quad t_y^+ = \infty\}$,

$\Omega^- = \{y \in \Omega : \nabla u(y) \neq 0, \quad t_y^- = -\infty\}$,

$\Omega_\Gamma^+ = \{y \in \Omega : \nabla u(y) \neq 0, \quad t_y^+ < \infty, \ \varphi(t,y)$ transversely (non − tangentially)

$\qquad$ reaches a smoothness point of $\Gamma \subseteq \partial\Omega$ as $t \uparrow t_y^+\}$,

$\Omega_\Gamma^- = \{y \in \Omega : \nabla u(y) \neq 0, \quad t_y^- > -\infty, \ \varphi(t,y)$ transversely reaches

$\qquad$ a smoothness point of $\Gamma \subseteq \partial\Omega$ as $t \downarrow t_y^-\}$.

Since $\Gamma \subset \partial\Omega$ is relatively open, Ω_Γ^+ and Ω_Γ^- are open subsets of Ω. The interior of $\Omega^{\pm}$ will be denoted by $int\ \Omega^{\pm}$.

The function

$$a(x) = \begin{cases} 1 & \text{if } \nabla u(x) = 0 \\ 0 & \text{elsewhere in } \Omega \end{cases}$$

satisfies the homogeneous problem corresponding to (5). Thus, a is non−L^1−identifiable from (5) on Ω_C if $meas\ \Omega_C > 0$.

Theorem 1. *Under condition (6), the transmissivity coefficient a is L^1–identifiable from problem (5) on the sets $int\ \Omega^+$, $int\ \Omega^-$, Ω_Γ^+ and Ω_Γ^-; on $int\ \Omega^+$ and $int\ \Omega^-$, L^1–identifiability holds even if $\Gamma = \emptyset$.*

Figure 1 illustrates a case where Ω^+ and Ω^- cover Ω except isolated critical points of u. In this case we can identify a putting $\Gamma = \emptyset$. Figure 2 illustrates a case where a boundary condition on a part Γ of $\partial\Omega$ is necessary to identify a on whole Ω.

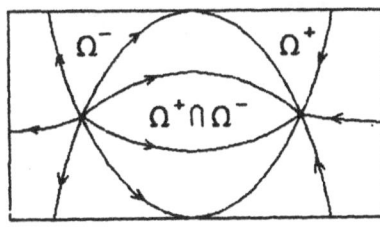

Fig.1

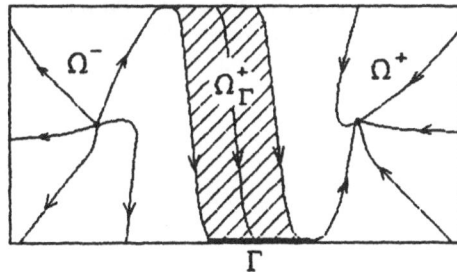

Fig.2

A result of C. Chicone and J. Gerlach [1] says the following: if $u \in C^2(\tilde{\Omega})$ where $\tilde{\Omega}$ is open and contains $\bar{\Omega}$ (the closure of Ω), then a is C^1–identifiable

(identifiable among functions a of the class $C^1(\bar{\Omega})$) from problem (1) with $\acute{\Gamma} = \emptyset$ on the closure of the set $int\ \Omega^+ \cup int\ \Omega^-$. A result of G.R. Richter [2] can be interpreted as C–identifiability on the closures of Ω_Γ^+ and Ω_Γ^- (the smoothness conditions and a priori assumptions on a are not explicitly formulated but a must be differentiable at least in the direction of ∇u). Theorem 1 extends these results to the case where no a priori smoothness of a is assumed.

Remark. Theorem 1 fails if assumption (6) is replaced by $u \in W^{1,\infty}(\Omega) \cap W^{2,p}(\Omega)$, $p < \infty$. Thus, for L^1–identifiability, the conditions of Theorem 1 are rather close to the necessary ones. For the L^2-identifiability of a on $int\overset{+}{\Omega^-}$ and $\overset{+}{\Omega_\Gamma^-}$ the conditions of Theorem 1 are only sufficient and seem to be far from the necessity – examining the examples one can conjecture that here (6) may be replaced by $u \in C^1(\bar{\Omega}) \cap H^2(\Omega)$.

For the proof of Theorem 1 and comments, see [5]. Also the case of piecewise Lipschitz continuous ∇u is examined in [5]. Physically, this case corresponds to possible jumps of the "physical" solution a on some surfaces in Ω. Theorem 1 remains true in its essence assuming that a strictly positive piecewise smooth ("physical") solution of inverse problem exists.

References

1. Chicone, C., Gerlach, J.: Identifiability of distributed parameters. In: Inverse and Ill–Posed Problems, Engl H.W. and Groetsch (Ed.). Boston etc.: Academic Press, 1987, 513–521.
2. Richter, G.R.: An inverse problem for the steady state diffusion equation. SIAM J. Appl. Math.**41** (1981) 210–221.
3. Vainikko, E., Vainikko, G.: Some numerical scemes for the identification of the filtration coefficients. Acta Comm. Univ. Tartuensis **937** (1992) 3–14.
4. Vainikko, G.: On the discretization and regularization of ill–posed problems with non–compact operators. Numer. Funct. Anal. Optim. **13** (1992) 381–396.
5. Vainikko, G. Kunisch, K.: L^1–identifiability of the transmissivity coefficient. To appear in Z. Anal. Anw.